Collins

Student Book

EDEXCEL INTERNATIONAL GCSE (9-1) PHYSICS

Steve Bibby, Malcolm Bradley and Susan Gardner

William Collins' dream of knowledge for all began with the publication of his first book in 1819. A self-educated mill worker, he not only enriched millions of lives, but also founded a flourishing publishing house. Today, staying true to this spirit, Collins books are packed with inspiration, innovation and practical expertise. They place you at the centre of a world of possibility and give you exactly what you need to explore it.

Collins. Freedom to teach.

Published by Collins
An imprint of HarperCollins*Publishers*
The News Building
1 London Bridge Street
London
SE1 9GF

10 9 8 7 6 5 4 3 2 1

ISBN 978-0-00-823620-5

Steve Bibby, Malcolm Bradley and Susan Gardner assert their moral right to be identified as the authors of this work.

A special thanks is extended to Katherine Richard for her helpful feedback.

British Library Cataloguing in Publication Data
A catalogue record for this publication is available from the British Library.

Authors: Steve Bibby, Malcolm Bradley and Susan Gardner
Original material by Malcolm Bradley and Susan Gardner
New material by Steve Bibby Commissioning Editor: Joanna Ramsay
Development Editor: Gillian Lindsey
Project manager: Maheswari PonSaravanan
Project editor: Vicki Litherland
Copy editor: Gwynneth Drabble
Proofreader: Mitch Fitton
Indexer: Jane Henley
Answer checker: Peter Batty
Typesetting: Jouve India
Artwork: Jouve India
Cover design: ink-tank
Production: Rachel Weaver
Printed by: Grafica Veneta

In order to ensure that this resource offers high-quality support for the associated Pearson qualification, it has been through a review process by the awarding body. This process confirms that this resource fully covers the teaching and learning content of the specification or part of a specification at which it is aimed. It also confirms that it demonstrates an appropriate balance between the development of subject skills, knowledge and understanding, in addition to preparation for assessment.

Endorsement does not cover any guidance on assessment activities or processes (e.g. practice questions or advice on how to answer assessment questions), included in the resource nor does it prescribe any particular approach to the teaching or delivery of a related course.

While the publishers have made every attempt to ensure that advice on the qualification and its assessment is accurate, the official specification and associated assessment guidance materials are the only authoritative source of information and should always be referred to for definitive guidance.

Pearson examiners have not contributed to any sections in this resource relevant to examination papers for which they have responsibility.

Examiners will not use endorsed resources as a source of material for any assessment set by Pearson.

Endorsement of a resource does not mean that the resource is required to achieve this Pearson qualification, nor does it mean that it is the only suitable material available to support the qualification, and any resource lists produced by the awarding body shall include this and other appropriate resources.

Contents

Getting the best from the book

Welcome to *Edexcel International GCSE Physics.*

This textbook has been designed to help you understand all of the requirements needed to succeed in the Edexcel International GCSE Physics course. Just as there are eight sections in the Edexcel specification, there are eight sections in the textbook: Forces and motion, Electricity, Waves, Energy resources and energy transfers, Solids, liquids and gases, Magnetism and electromagnetism, Radioactivity, fission and fusion and Astrophysics.

Each section is split into topics. Each topic in the textbook covers the essential knowledge and skills you need. The textbook also has some very useful features which have been designed to really help you understand all the aspects of Physics that you will need to know for this specification.

SAFETY IN THE SCIENCE LESSON

This book is a textbook, not a laboratory or practical manual. As such, you should not interpret any information in this book that relates to practical work as including comprehensive safety instructions. Your teachers will provide full guidance for practical work and cover rules that are specific to your school.

A brief introduction to the section to give context to the science covered in the section.

Starting points will help you to revise previous learning and see what you already know about the ideas to be covered in the section.

The section contents shows the separate topics to be studied matching the specification order.

Knowledge check shows the ideas you should have already encountered in previous work before starting the topic.

Learning objectives cover what you need to learn in this topic.

The blue side panels and background shading indicate content for Physics International GCSE students only.

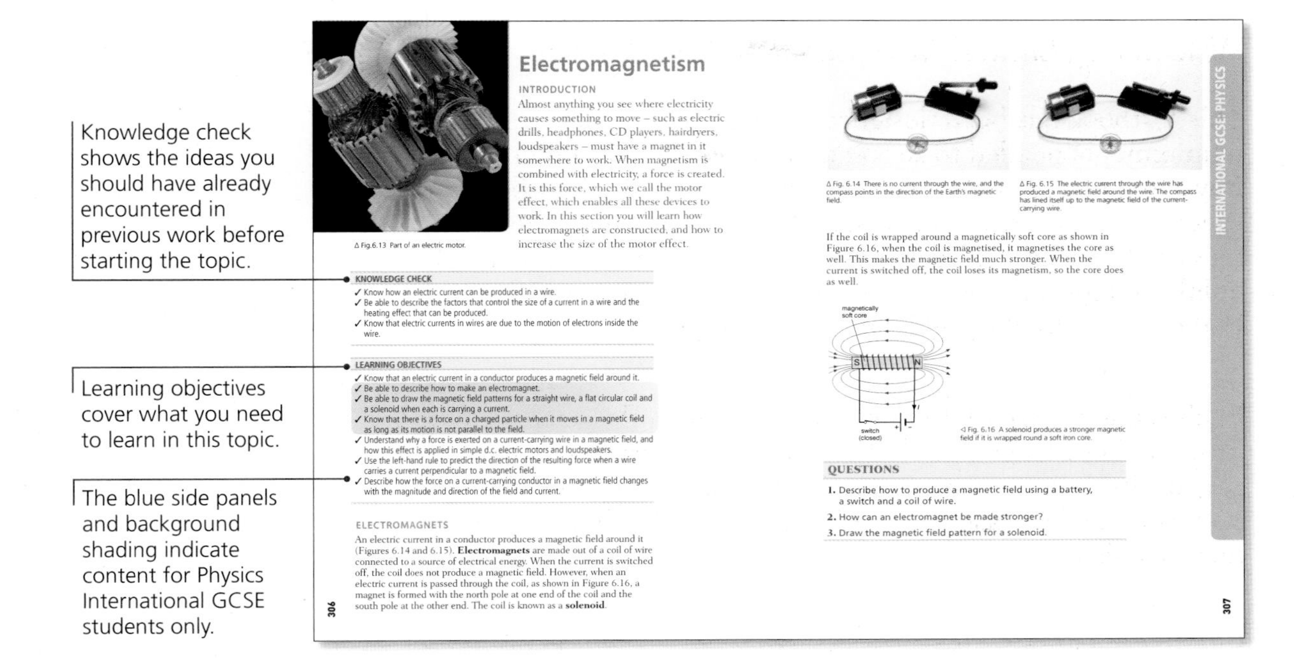

Electromagnetism

INTRODUCTION

Almost anything you see where electricity causes something to move – such as electric drills, headphones, CD players, hairdryers, loudspeakers – must have a magnet in it somewhere to work. When magnetism is combined with electricity, a force is created. It is this force, which we call the motor effect, which enables all these devices to work. In this section you will learn how electromagnets are constructed, and how to increase the size of the motor effect.

Δ Fig.6.13 Part of an electric motor.

KNOWLEDGE CHECK

- ✓ Know how an electric current can be produced in a wire.
- ✓ Be able to describe the factors that control the size of a current in a wire and the heating effect that can be produced.
- ✓ Know that electric currents in wires are due to the motion of electrons inside the wire.

LEARNING OBJECTIVES

- ✓ Know that an electric current in a conductor produces a magnetic field around it.
- ✓ Be able to describe how to make an electromagnet.
- ✓ Be able to draw the magnetic field patterns for a straight wire, a flat circular coil and a solenoid when each is carrying a current.
- ✓ Know that there is a force on a charged particle when it moves in a magnetic field as long as its motion is not parallel to the field.
- ✓ Understand why a force is exerted on a current-carrying wire in a magnetic field, and how this effect is applied in simple d.c. electric motors and loudspeakers.
- ✓ Use the left-hand rule to predict the direction of the resulting force when a wire carries a current perpendicular to a magnetic field.
- ✓ Describe how the force on a current-carrying conductor in a magnetic field changes with the magnitude and direction of the field and current.

ELECTROMAGNETS

An electric current in a conductor produces a magnetic field around it (Figures 6.14 and 6.15). **Electromagnets** are made out of a coil of wire connected to a source of electrical energy. When the current is switched off, the coil does not produce a magnetic field. However, when an electric current is passed through the coil, as shown in Figure 6.16, a magnet is formed with the north pole at one end of the coil and the south pole at the other end. The coil is known as a **solenoid**.

306

Δ Fig. 6.14 There is no current through the wire, and the compass points in the direction of the Earth's magnetic field.

Δ Fig. 6.15 The electric current through the wire has produced a magnetic field around the wire. The compass has lined itself up to the magnetic field of the current-carrying wire.

If the coil is wrapped around a magnetically soft core as shown in Figure 6.16, when the coil is magnetised, it magnetises the core as well. This makes the magnetic field much stronger. When the current is switched off, the coil loses its magnetism, so the core does as well.

◁ Fig. 6.16 A solenoid produces a stronger magnetic field if it is wrapped round a soft iron core.

QUESTIONS

1. Describe how to produce a magnetic field using a battery, a switch and a coil of wire.
2. How can an electromagnet be made stronger?
3. Draw the magnetic field pattern for a solenoid.

INTERNATIONAL GCSE: PHYSICS

307

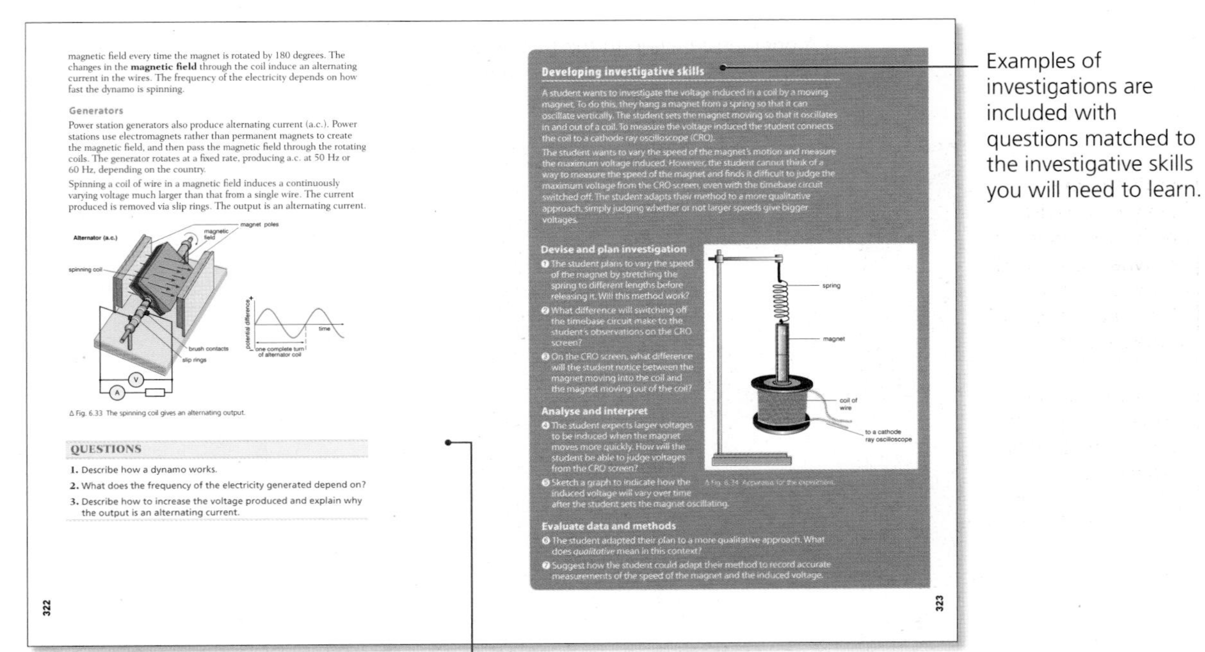

magnetic field every time the magnet is rotated by 180 degrees. The changes in the **magnetic field** through the coil induce an alternating current in the wires. The frequency of the electricity depends on how fast the dynamo is spinning.

Generators

Power station generators also produce alternating current (a.c.). Power stations use electromagnets rather than permanent magnets to create the magnetic field, and then pass the magnetic field through the rotating coils. The generator rotates at a fixed rate, producing a.c. at 50 Hz or 60 Hz, depending on the country.

Spinning a coil of wire in a magnetic field induces a continuously varying voltage much larger than that from a single wire. The current produced is removed via slip rings. The output is an alternating current.

Δ Fig. 6.33 The spinning coil gives an alternating output.

QUESTIONS

1. Describe how a dynamo works.
2. What does the frequency of the electricity generated depend on?
3. Describe how to increase the voltage produced and explain why the output is an alternating current.

322

Developing investigative skills

A student wants to investigate the voltage induced in a coil by a moving magnet. To do this, they hang a magnet from a spring so that it can oscillate vertically. The student sets the magnet moving so that it oscillates in and out of a coil. To measure the voltage induced the student connects the coil to a cathode ray oscilloscope (CRO).

The student wants to vary the speed of the magnet's motion and measure the maximum voltage induced. However, the student cannot think of a way to measure the speed of the magnet and finds it difficult to judge the maximum voltage from the CRO screen, even with the timebase circuit switched off. The student adapts their method to a more qualitative approach, simply judging whether or not larger speeds give bigger voltages.

Devise and plan investigation

1. The student plans to vary the speed of the magnet by stretching the spring to different lengths before releasing it. Will this method work?
2. What difference will switching off the timebase circuit make to the student's observations on the CRO screen?
3. On the CRO screen, what difference will the student notice between the magnet moving into the coil and the magnet moving out of the coil?

Analyse and interpret

4. The student expects larger voltages to be induced when the magnet moves more quickly. How will the student be able to judge voltages from the CRO screen?
5. Sketch a graph to indicate how the induced voltage will vary over time after the student sets the magnet oscillating.

Δ Fig. 6.34 Apparatus for the experiment.

Evaluate data and methods

6. The student adapted their plan to a more qualitative approach. What does *qualitative* mean in this context?
7. Suggest how the student could adapt their method to record accurate measurements of the speed of the magnet and the induced voltage.

323

Examples of investigations are included with questions matched to the investigative skills you will need to learn.

Extension boxes take your learning further.

Getting the best from the book *continued*

Science in context boxes put the ideas you are learning into a historical or modern context.

QUESTIONS

1. What causes the pressure in a fluid in a container?
2. Why do our lungs not collapse under atmospheric pressure?
3. Why does a plastic bottle collapse if air is removed from inside it?

Pressure difference, height and density

If you dive below the water, the height of the water above you also puts pressure on you. At a depth of 10 m of water, the pressure has increased by 100 kPa, and for each further 10 m of depth the pressure increases by another 100 kPa. The rapid increase in pressure explains why scuba divers cannot go down more than 20 m without taking extra safety precautions.

PRESSURE AND SUBMARINES

Δ Fig. 5.20 Submarines have two hulls to withstand the water pressure at depths down to about 300 m.

Early military submarines had propulsion systems that could not operate well when submerged, so these submarines spent most of their time on the surface, with hull designs that balanced the need for a relatively streamlined structure with the ability to move on the surface. Late in World War II, technological advances meant that longer and faster submerged operations were possible.

Submarines actually have two hulls. The external hull, which forms the shape of the submarine, is sometimes called the light hull. (This term is particularly appropriate for Russian submarines, whose external hull is usually made of steel that is only 2 to 4 millimetres thick.) The pressure hull, which is inside the external hull, is designed to withstand the pressure outside it from the water around the submarine. It has normal atmospheric pressure inside, which allows the submarine crew to breathe normally. The dive depth (the maximum depth at which the submarine can operate) depends on the strength of the hull. Submarines used in World War I had hulls made of carbon steel and could not dive below 100 m. In World War II, high-strength alloyed steel was used, and the dive depth increased to 200 m. This is still the main material used today, with a current limit of 250–300 m dive depth. A few submarines have been built with titanium hulls and the deepest diving submarine was the Soviet *Komsomolets*, which dived to about 1000 m.

258

The increase in pressure below the surface of a liquid depends on the depth below the surface and the density of the liquid. The pressure is much higher at a certain depth below the surface of mercury than it is below the same depth of water. It does not depend on anything else, and note in particular that the pressure does not depend on the width of the water.

If a diver goes to inspect a well, the pressure 10 m below the surface is the same as the pressure 10 m below the surface of a large lake. This explains why an engineer who is designing a dam needs to make it the same thickness whether the lake is going to be 100 m long or 100 km long.

Scuba divers breathe compressed air at high pressure to prevent their lungs collapsing due to the high pressure from the water above them. This is a safe sport, but only because new divers are trained to a very high standard before they are allowed to dive.

Δ Fig. 5.21 Scuba divers.

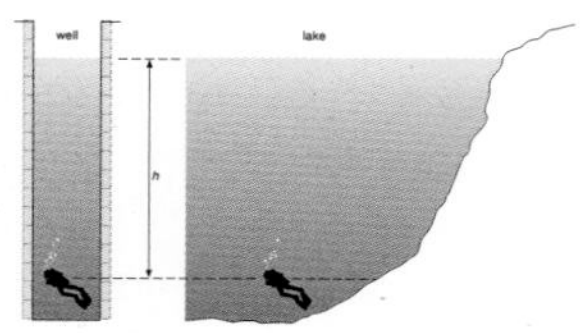

Δ Fig. 5.22 The pressure on the diver is the same in the well and in the lake. In both cases it depends only on the density of the liquid and his depth, *h*.

The pressure below the surface of a fluid – and between any two points in the fluid – can be calculated by the following equation:

pressure difference = height × density × gravitational field strength

$$p = h \times \rho \times g$$

where p = pressure difference in pascals (Pa), ρ = density in kilograms per cubic metre (kg/m^3) and g = gravitational field strength (N/kg).

259

Write down the formula with F as the subject:	$F = \frac{W}{d}$
Substitute the values for W and d:	$F = \frac{3000\ \text{J}}{50\ \text{m}}$
Work out the answer and write down the unit:	$F = 60$ N

QUESTIONS

1. Calculate the work done when a 50 N force moves an object 5 m.
2. Calculate the force required to move an object 8 m by transferring 4000 J of energy.
3. Calculate the work done when a force of 40 N moves a block 2 m.
4. How far does an object move if the force on it is 6 N and the work done is 300 J?
5. What force is needed to move a piano a distance of 2 m when the work done is 800 J?

Questions to check understanding.

EXTENSION

When something slows down because of friction, work is done. The kinetic energy of the motion is transferred to the thermal energy store of the object and its surroundings as the frictional forces slow the object down. For example, when a cyclist brakes, the work done by friction produces an energy transfer. The kinetic energy store of the cyclist decreases and the thermal energy store of the brake pads increases.

Space shuttles were operated by NASA between 1981 and 2011. A space shuttle was a manned, reusable spacecraft that was launched vertically by rocket but landed horizontally on a runway. The space shuttle used friction to do work on its motion upon re-entry into the Earth's atmosphere.

1. A space shuttle has 8.45×10^{12} J of energy to transfer over an 8000 km flight path. What force is applied by the atmosphere?

Use the internet or books to research the landing of a space shuttle and answer the following questions.

2. What happens to the transferred energy on landing?
3. What temperatures are generated by the work being done? How does this relate to the material used for the underside of the shuttle surface – for example, why was it not made from aluminium or iron?

Δ Fig. 4.31 A space shuttle in orbit.

214

GRAVITATIONAL POTENTIAL ENERGY AND KINETIC ENERGY

If a load is raised above the ground, it increases its store of gravitational potential energy (GPE). If the load moves back to the ground, the stored potential energy decreases and is transferred to the store of kinetic energy (KE).

Gravitational potential energy can be calculated using the formula:

gravitational potential energy = mass × gravitational field strength × height

$$\text{GPE} = mgh$$

where GPE = gravitational potential energy in joules (J), m = mass in kilograms (kg), g = gravitational field strength of 10 N/kg and h = height in metres (m).

The kinetic energy of an object depends on its mass and its speed. The kinetic energy can be calculated using the following formula:

kinetic energy = $\frac{1}{2}$ × mass × (speed)^2 or

$$\text{KE} = \frac{1}{2}mv^2$$

where KE = kinetic energy in joules (J), m = mass in kilograms (kg) and v = speed in m/s.

REMEMBER

An object gains gravitational potential energy as it gains height. Work has to be done to increase the height of the object above the ground. Therefore:

gain in gravitational potential energy of an object = work done on that object against gravity.

Remember boxes provide tips and guidance to help you during the course and in your exam.

WORKED EXAMPLES

Learn to apply formulae through worked examples.

1. A skier has a mass of 70 kg and travels up a ski lift a vertical height of 300 m. Calculate the change in the skier's gravitational potential energy.

Write down the formula:	GPE = mgh
Substitute values for m, g and h:	GPE = 70 × 10 × 300
Work out the answer and write down the unit:	GPE = 210 000 J or 210 kJ

2. An ice skater has a mass of 50 kg and travels at a speed of 5 m/s. Calculate the skater's kinetic energy.

Write down the formula:	KE = $\frac{1}{2}mv^2$
Substitute the values for m and v:	KE = $\frac{1}{2}$ × 50 × 5 × 5
Work out the answer and write down the unit:	KE = 625 J

215

End of topic questions allow you to apply the knowledge and understanding you have learned in the topic to answer the questions.

A full checklist of all the information you need to cover the complete specification requirements for each topic.

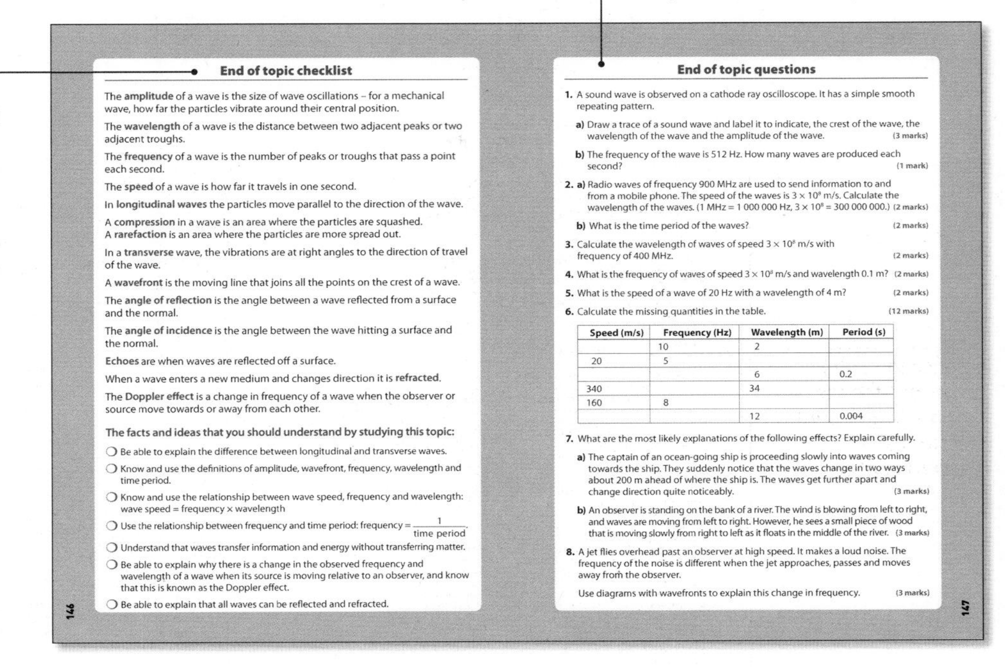

End of topic checklist

The **amplitude** of a wave is the size of wave oscillations – for a mechanical wave, how far the particles vibrate around their central position.

The **wavelength** of a wave is the distance between two adjacent peaks or two adjacent troughs.

The **frequency** of a wave is the number of peaks or troughs that pass a point each second.

The **speed** of a wave is how far it travels in one second.

In **longitudinal waves** the particles move parallel to the direction of the wave.

A **compression** is an area where the particles are squashed.
A **rarefaction** is an area where the particles are more spread out.

In a **transverse** wave, the vibrations are at right angles to the direction of travel of the wave.

A **wavefront** is the moving line that joins all the points on the crest of a wave.

The **angle of reflection** is the angle between a wave reflected from a surface and the normal.

The **angle of incidence** is the angle between the wave hitting a surface and the normal.

Echoes are when waves are reflected off a surface.

When a wave enters a new medium and changes direction it is **refracted**.

The **Doppler effect** is a change in frequency of a wave when the observer or source move towards or away from each other.

The facts and ideas that you should understand by studying this topic:

- ❍ Be able to explain the difference between longitudinal and transverse waves.
- ❍ Know and use the definitions of amplitude, wavefront, frequency, wavelength and time period.
- ❍ Know and use the relationship between wave speed, frequency and wavelength: wave speed = frequency × wavelength
- ❍ Use the relationship between frequency and time period: frequency = $\frac{1}{\text{time period}}$
- ❍ Understand that waves transfer information and energy without transferring matter.
- ❍ Be able to explain why there is a change in the observed frequency and wavelength of a wave when its source is moving relative to an observer, and know that this is known as the Doppler effect.
- ❍ Be able to explain that all waves can be reflected and refracted.

146

End of topic questions

1. A sound wave is observed on a cathode ray oscilloscope. It has a simple smooth repeating pattern.
   a) Draw a trace of a sound wave and label it to indicate, the crest of the wave, the wavelength of the wave and the amplitude of the wave. (3 marks)
   b) The frequency of the wave is 512 Hz. How many waves are produced each second? (1 mark)
2. a) Radio waves of frequency 900 MHz are used to send information to and from a mobile phone. The speed of the waves is 3×10^8 m/s. Calculate the wavelength of the waves. (1 MHz = 1 000 000 Hz, 3×10^8 = 300 000 000.) (2 marks)
   b) What is the time period of the waves? (2 marks)
3. Calculate the wavelength of waves of speed 3×10^8 m/s with frequency of 400 MHz. (2 marks)
4. What is the frequency of waves of speed 3×10^8 m/s and wavelength 0.1 m? (2 marks)
5. What is the speed of a wave of 20 Hz with a wavelength of 4 m? (2 marks)
6. Calculate the missing quantities in the table. (12 marks)

Speed (m/s)	Frequency (Hz)	Wavelength (m)	Period (s)
	10	2	
20	5		
		6	0.2
340		34	
160	8		
		12	0.004

7. What are the most likely explanations of the following effects? Explain carefully.
   a) The captain of an ocean-going ship is proceeding slowly into waves coming towards the ship. They suddenly notice that the waves change in two ways about 200 m ahead of where the ship is. The waves get further apart and change direction quite noticeably. (3 marks)
   b) An observer is standing on the bank of a river. The wind is blowing from left to right, and waves are moving from left to right. However, he sees a small piece of wood that is moving slowly from right to left as it floats in the middle of the river. (3 marks)
8. A jet flies overhead past an observer at high speed. It makes a loud noise. The frequency of the noise is different when the jet approaches, passes and moves away from the observer.

   Use diagrams with wavefronts to explain this change in frequency. (3 marks)

147

The first question is a student sample with examiner's comments to show best practice.

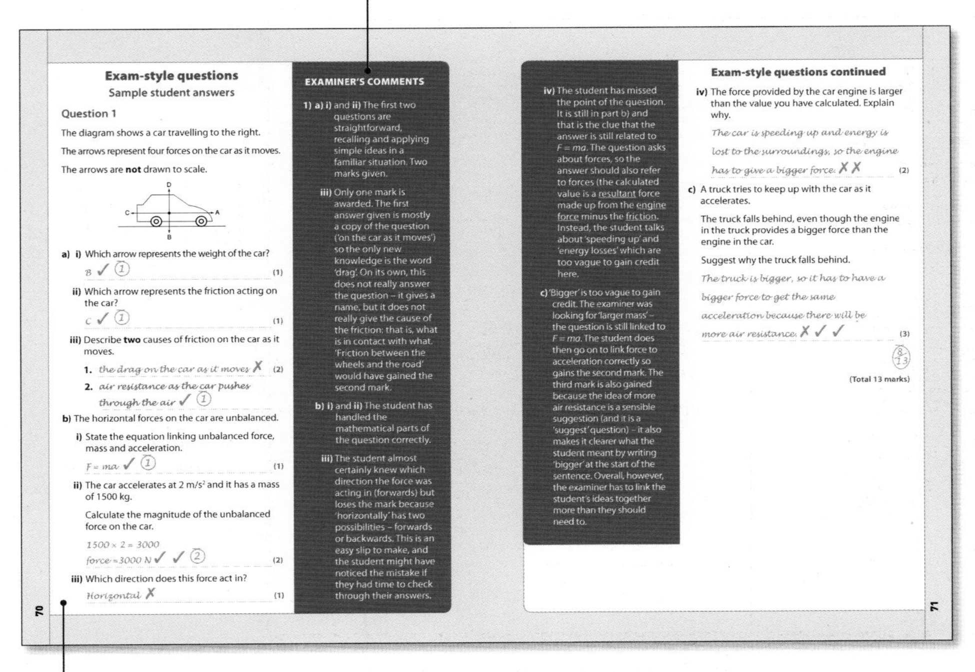

Exam-style questions

Sample student answers

Question 1

The diagram shows a car travelling to the right.

The arrows represent four forces on the car as it moves.

The arrows are **not** drawn to scale.

a) i) Which arrow represents the weight of the car?

B ✓ ① (1)

ii) Which arrow represents the friction acting on the car?

C ✓ ① (1)

iii) Describe **two** causes of friction on the car as it moves.

1. *the drag on the car as it moves* ✗ (2)
2. *air resistance as the car pushes through the air* ✓ ①

b) The horizontal forces on the car are unbalanced.

i) State the equation linking unbalanced force, mass and acceleration.

F = ma ✓ ① (1)

ii) The car accelerates at 2 m/s² and it has a mass of 1500 kg.

Calculate the magnitude of the unbalanced force on the car.

1500 × 2 = 3000

force = 3000 N ✓ ✓ ② (2)

iii) Which direction does this force act in?

Horizontal ✗ (1)

70

EXAMINER'S COMMENTS

1) a) i) and ii) The first two questions are straightforward, recalling and applying simple ideas in a familiar situation. Two marks given.

iii) Only one mark is awarded. The first answer given is mostly a copy of the question ('on the car as it moves') so the only new knowledge is the word 'drag'. On its own, this does not really answer the question – it gives a name, but it does not really give the cause of the friction: that is, what is in contact with what. 'Friction between the wheels and the road' would have gained the second mark.

b) i) and ii) The student has handled the mathematical parts of the question correctly.

iii) The student almost certainly knew which direction the force was acting in (forwards) but loses the mark because 'horizontally' has two possibilities – forwards or backwards. This is an easy slip to make, and the student might have noticed the mistake if they had time to check through their answers.

iv) The student has missed the point of the question. It is still in part b) and that is the clue that the answer is still related to $F = ma$. The question asks about forces, so the answer should also refer to forces (the calculated value is a resultant force made up from the engine force minus the friction. Instead, the student talks about 'speeding up' and 'energy losses' which are too vague to gain credit here.

c) 'Bigger' is too vague to gain credit. The examiner was looking for 'larger mass' – the question is still linked to $F = ma$. The student does then go on to link force to acceleration correctly so gains the second mark. The third mark is also gained because the idea of more air resistance is a sensible suggestion (and it is a 'suggest' question) – it also makes it clearer what the student meant by writing 'bigger' at the start of the sentence. Overall, however, the examiner has to link the student's ideas together more than they should need to.

Exam-style questions continued

iv) The force provided by the car engine is larger than the value you have calculated. Explain why.

The car is speeding up and energy is lost to the surroundings, so the engine has to give a bigger force. ✗ ✗ (2)

c) A truck tries to keep up with the car as it accelerates.

The truck falls behind, even though the engine in the truck provides a bigger force than the engine in the car.

Suggest why the truck falls behind.

The truck is bigger, so it has to have a bigger force to get the same acceleration because there will be more air resistance. ✗ ✓ ✓ (3)

8/13

(Total 13 marks)

71

Each section includes exam-style questions to help you prepare for your exam in a focused way and get the best results.

In this section you will learn about the physics of motion: about how forces affect motion. Every time you go on a journey you are applying the physics of motion, and every time you move something from one place to another you are applying the physics of forces.

You may previously have met forces as interactions between objects that can affect their shape and motion.

STARTING POINTS

1. What happens when the driving force on a car is greater than the friction force on the car? Why?
2. An animal has a mass of 28 kg. If the animal were able to travel to the Moon, would it weigh more or less there than it does on Earth? Give a reason for your answer.
3. If an object is stationary, what must be true about the forces acting on it?
4. In physics, what do we mean when we say an object is accelerating?
5. What is the force that keeps all the bodies in the Universe moving as they do?
6. A person is standing on the floor. What forces are acting on the person? Is the person creating a force on anything?

CONTENTS

1 Forces and motion

Δ In this chapter you will learn the physics behind the launch and orbit of this satellite.

Units

INTRODUCTION

Physics is a measuring science. All theories need to be checked and inspired by experimental work, and the results of experiments compared. Experiments can only be compared if there is an agreed system of units in which to make measurements.

Every physical quantity that can be measured has a standard unit of measurement. As you add to your knowledge of physics, you should always take care to use the correct units. As you will notice, unit names start with a lower case letter. The unit symbol can be either upper case or lower case. It depends on the symbol (newton is always upper case N whereas metre is always lower case m). For the topics of forces and motion, you will need to be familiar with:

Quantity	Unit	Symbol
mass	kilogram	kg
length, distance	metre	m
time	second	s
speed, velocity	metre per second	m/s
acceleration	metre per second^2	m/s^2
force	newton	N
gravitational field strength	newton per kilogram	N/kg
momentum	kilogram metre per second	kg m/s
moment	newton metre	N m

In practice, physicists use a range of units that match the particular area in which they are working. For example, the kilogram is much too large for physicists working with subatomic particles, so they use a smaller unit such as the atomic mass unit.

LEARNING OBJECTIVES

✓ Use the following units: kilogram (kg), metre (m), metre/second (m/s), metre/second^2 (m/s^2), newton (N), second (s) and newton/kilogram (N/kg).

✓ Use the following units: newton metre (N m), kilogram metre/second (kg m/s).

REMEMBER

The unit for acceleration, metres/second^2 causes some confusion. Students often forget that it is metres/second/second and write metres/second (which is the unit of velocity). In the exam, make sure that you check whether you are talking about velocity (units metres/second), or acceleration (units metres/second^2).

UNITS AND MEASURING

Many units of measure, particularly to do with measuring motion, have an origin related to the human body. The cubit is the length from the elbow to the tip of the middle finger; the fathom is the reach across two outstretched arms. Other units, such as measuring weight in stones, have simple links to the world around us.

As scientific ideas have developed, so have the measuring systems used. Physics uses the SI system, which was originally developed in 1960 and has been further developed since. It is a set of agreed standards that allow all measurements to be compared and so scientists can look at the findings of different groups in a coherent way.

The SI system is based upon seven *base units*, with all others being described as *derived units*. The base units describe the key measuring unit for seven *base quantities*. These are features such as length, time and mass to which all other quantities are connected (a *quantity* is a property that can be measured). The seven base quantities, with their base units, form the minimum set of definitions that we need to build the entire system.

It is important that, as far as possible, the base units are defined in terms of physical properties that can be reproduced at any place at any time. This is so that the unit will remain constant and the science we do will remain comparable. For example, the cubit mentioned above depends on the particular size of the human body and will be different for different people. If you decide to use one particular person that will be fine as long as the person is alive and as long as they don't grow at all, but it is not satisfactory. Copying that length onto a standard material, such as a metal bar, would be an improvement, but then expansion in higher temperatures and oxidation in the air will cause problems.

QUESTIONS

1. Name three base quantities.

2. Why is it important that the base units are defined in terms of physical properties that can be reproduced at any place at any time?

3. From this list, which units are:

a) base units?

b) derived units?

kilogram, metre/second, newton, second, metre

Δ Fig. 1.1 You can use a stop watch to measure the time taken to run a certain distance.

Movement and position

INTRODUCTION

To study almost anything about the surrounding world, or out in space, it helps to describe where things are, where they were and where they are expected to go. It is even better to measure these things. When there is an organised system for doing this, it becomes possible to look for patterns in the way things move – these are the laws of motion. Then you can go a step further and suggest *why* things move as they do, using ideas about forces.

KNOWLEDGE CHECK

✓ Know how to describe motion using simple everyday language.
✓ Know how to read graph scales.
✓ Know how to calculate the gradient of a straight-line graph.

LEARNING OBJECTIVES

✓ Be able to plot, interpret and explain graphs of straight line motion.
✓ Be able to calculate average speed from total distance/total time.
✓ Be able to plot and explain a distance–time graph and a speed–time graph.
✓ Be able to calculate the area under a speed–time graph to determine the distance travelled.
✓ Be able to calculate acceleration using change of velocity/time taken.
✓ Use graphs to calculate accelerations and distances.
✓ Be able to investigate the motion of everyday objects.

USING GRAPHS TO STUDY MOTION: DISTANCE–TIME GRAPHS

Think about riding in a car travelling at 90 kilometres per hour. This, of course, means that the car (if it kept travelling at this speed for one hour) would travel 90 km. During one second, the car travels 25 metres, so its speed can also be described as 25 metres per second. Scientists prefer to measure time in seconds and distance in metres. So they prefer to measure speed in metres per second, usually written as m/s.

Journeys can be summarised using graphs. The simplest type is a **distance–time graph** where the distance travelled is plotted against the time of the journey.

At the beginning of an experiment, time is usually given as 0 seconds, and the position of the object as 0 metres. If the object is not moving, then time increases, but distance does not. This gives a horizontal line. If the object is travelling at a constant speed, then both time and distance increase steadily, which gives a straight line. If the speed varies, then the line will not be straight.

Developing investigative skills

Some students investigate the time taken for a tennis ball to roll 1 metre down a ramp. They investigate how the height of the ramp affects the time taken. They use a ruler to measure the height of the ramp and to mark a distance of 1 m along the ramp. They start a stopwatch when the ball is released and stop the stopwatch when the ball has travelled a distance of 1 m.

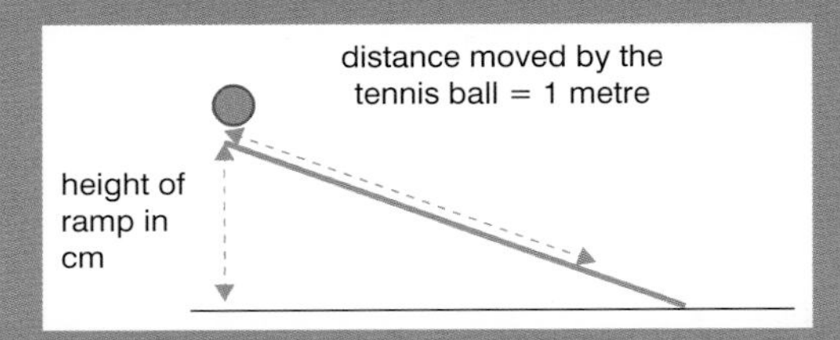

Δ Fig. 1.2 Investigating the motion of a ball rolling down a ramp.

The results are shown below.

Height of ramp (cm)	Time for tennis ball to roll 1 m (s)				
	A	B	C	D	Mean
10					
15					
20					
25					
30					
35					

Analyse and interpret

1. Suggest how the height of the slope may affect the time taken for the tennis ball to roll 1 m.
2. Why is it important to do four measurements of time for each height?
3. Why is it important that the ball is rolled rather than pushed down the slope?
4. Plot a distance–time graph for the motion of the ball.

Analysing distance–time graphs

Look at Fig. 1.3, which shows a distance–time graph for a bicycle journey. The graph slopes when the bicycle is moving. The slope gets steeper when the bicycle goes faster. The slope is straight (has a constant **gradient**) when the bicycle's speed is constant. The cyclist falls off 150 m from the start. After this, the graph is horizontal because the bicycle is not moving.

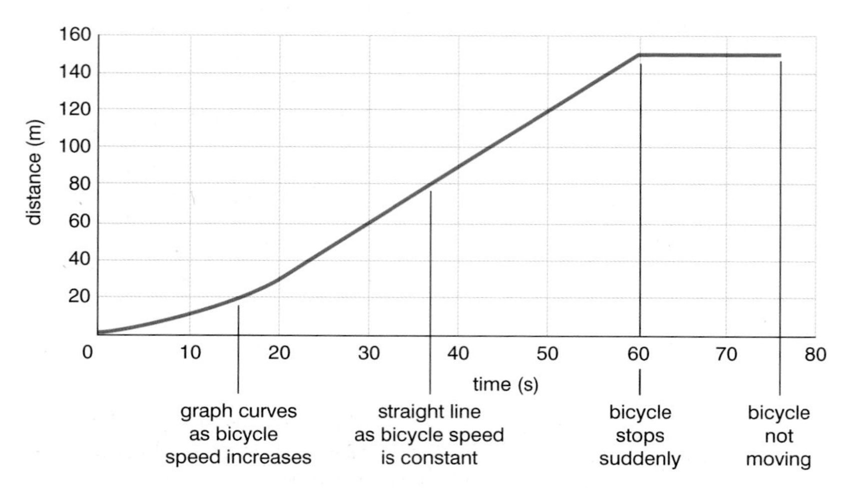

Δ Fig. 1.3 A distance–time graph for a bicycle journey.

In Figure 1.4, the left hand distance–time graph has a constant slope. This means that the bicycle is travelling at a constant speed. The right hand distance–time graph has a slope that gets gradually steeper. This means that the speed of the bicycle is increasing with time. Since the speed of the bicycle is changing it is accelerating.

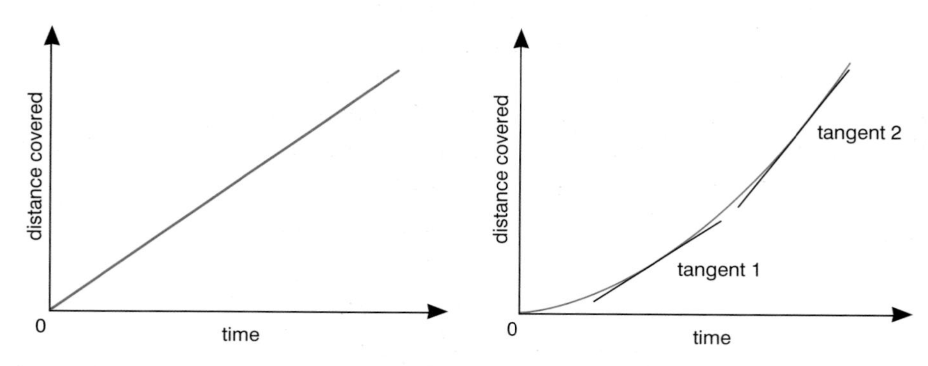

Δ Fig. 1.4 Steady speed is shown by a straight line. Acceleration is shown by a smooth curve of increasing gradient.

The gradient of the curved line on the right hand distance–time graph is not constant, but we can see how the speed is changing by drawing a **tangent** to the curve at any point. The gradient of the tangent gives the **instantaneous speed** at that point. The greater the slope, the greater the speed.

QUESTIONS

1. On a distance–time graph, what does a horizontal line indicate?
2. If a car is travelling at a constant speed, what shape would the corresponding distance–time graph have?
3. Sketch a distance–time graph for the bicycle shown in the photograph.

△ Fig. 1.5 As this bicycle travels downhill, its speed is likely to increase. Its distance–time graph will be a curve.

CALCULATING AVERAGE SPEED

The **speed** of an object can be calculated using the following formula:

$$\text{speed} = \frac{\text{distance}}{\text{time}} \text{ or } v = \frac{s}{t}$$

where v = speed in m/s, s= distance in m and t= time in s

Most objects speed up and slow down as they travel. An object's **average speed** can be calculated by dividing the total distance travelled by the total time taken.

REMEMBER

Make sure you can explain *why* this is an 'average' speed. You need to talk about the speed not being constant throughout, perhaps giving specific examples of where it changed. For example, you might consider a journey from home to school. You know how long the journey takes, and the distance between home and school. From these, you can work out the average speed using the formula. However, you know that, in any journey, you do not travel at the same speed at all times. You may have to stop to cross the road, or at a road junction. You may be able to travel faster on straight sections of the journey or round corners.

◁ Fig. 1.6 Each car's average speed can be calculated by dividing the distance it has travelled by the time it has taken.

WORKED EXAMPLES

1. Calculate the average speed of a car that travels 500 m in 20 seconds.

Write down the formula: $\text{average speed} = \dfrac{\text{distance moved}}{\text{time taken}}$

Substitute the values: $\text{average speed} = \dfrac{500}{20}$

Work out the answer and write down the units: average speed = 25 m/s

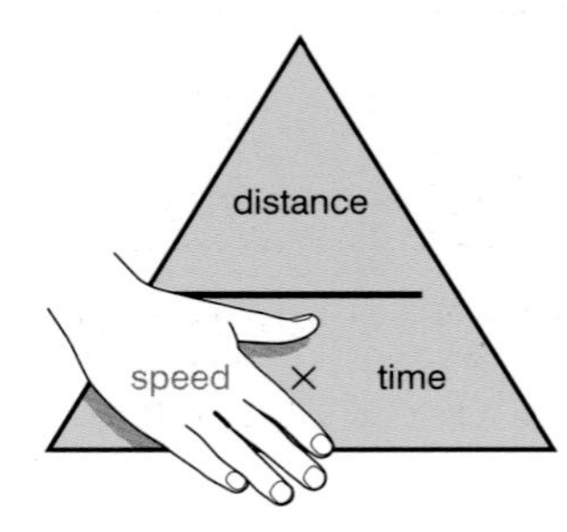

◁ Fig. 1.7 *This triangle is useful for rearranging the equation speed = distance/time to find either distance (if you are given speed and time) or time if you are given speed and distance. Cover speed to find that*

$\text{speed} = \dfrac{\text{distance}}{\text{time}}$

2. A horse canters at an average speed of 5 m/s for 2 minutes. Calculate the distance it travels.

Write down the formula in terms of distance moved:
distance moved = average speed × time taken

Substitute the values for average speed and time taken:
distance moved = 5 × 2 × 60

Work out the answer and write down the units:
distance moved = 600 m

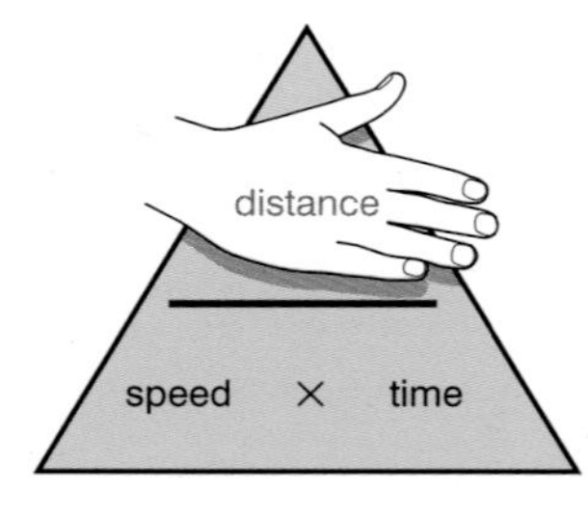

◁ Fig. 1.8 *Cover distance to find that distance = speed × time*

QUESTIONS

1. A journey to school is 10 km. It takes 15 minutes in a car. What is the average speed of the car?
2. How far does a bicycle travelling at 1.5 m/s travel in 15 s?
3. A person walks at 0.5m/s and travels a distance of 1500 m. How long does this take?

SPEED AND VELOCITY

There are many situations in which we want to know the direction that an object is travelling. For example, when a space rocket is launched, it is likely to reach a speed of 1000 km/h after about 30 seconds. However, it is extremely important to know whether this speed is up or down. You want to know the speed *and* the direction of the rocket.

The **velocity** of an object is one piece of information, but it consists of two parts: the speed *and* the direction. In this case, the velocity of the rocket is 278 m/s (its speed) upwards (its direction).

A velocity can have a minus sign. This tells you that the object is travelling in the opposite direction. By convention, we use positive and negative directions to match the coordinate grid. So, direction to the right is positive and direction to the left is negative (as on *x*-axis). Direction upwards is positive and direction downwards is negative (as on *y*-axis) In the diagram, the car at the top is moving with a velocity of +10 m/s as it is moving to the right. The car at the bottom has the same speed (10 m/s) but its velocity is –10 m/s as it is moving to the left.

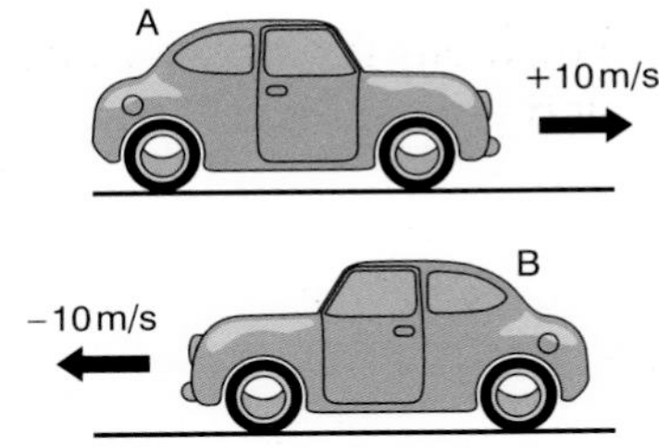

Δ Fig. 1.9 Both cars have the same speed. Car A has a velocity of +10 m/s, car B has a velocity of –10 m/s.

QUESTIONS

1. Describe the difference between speed and velocity.

2. Explain the significance of the positive or negative sign of a velocity.

3. Draw a diagram to show a tennis ball:

a) after it has been thrown upwards with a speed of 5 m/s

b) as it comes down again with the same speed as in part **a)**.

Mark the velocities on your diagrams. Remember to include the sign for each.

Developing investigative skills

Describing motion

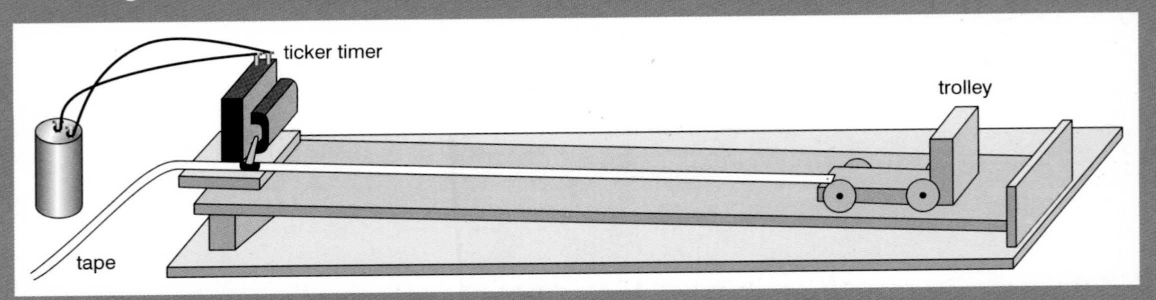

Δ Fig. 1.10 Apparatus for experiment.

A student investigated the motion of a trolley rolling down a ramp. To measure the distance the trolley had travelled at different times, the student used a ticker timer and ticker tape. A ticker timer has a moving arm that bounces up and down 50 times each second. When the arm moves down and hits the tape, it makes a small dot on the tape.

The student attached the tape to the trolley and released the trolley to roll down the ramp. After the test the student divided the tape into strips with 5 dots in each strip – at 50 dots per second this meant that a 5-dot strip had taken 0.1 s.

◁ Fig. 1.11 Five-dot portion of ticker tape.

The student measured the length of each 5-dot strip with a ruler. The results are shown below.

Time (s)	Distance from start (cm)	Distance covered in the last 0.1 s (cm)	Average speed for last 0.1 s (cm/s)
0.0		0.0	
0.1		1.8	
0.2		3.4	
0.3		5.2	
0.4		6.0	
0.5		7.7	
0.6		11.1	
0.7		9.9	
0.8		11.9	
0.9		12.5	
1.0		14.0	

Devise and plan investigations

❶ Suggest how using this method might have changed the motion of the trolley.

❷ How else could the student have measured the position of the trolley every 0.1 seconds?

Analyse and interpret

❸ Copy the table and complete the second column, showing the total distance travelled up to that time.

❹ Draw a distance–time graph using the data in the first two columns. Use your graph to describe the motion of the trolley.

❺ Use the equation average speed = distance/time to complete the final column.

❻ Draw a speed–time graph using the data in the first and fourth columns. Does this graph support the description of the motion you gave in question 4? Explain your answer.

Evaluate data and methods

❼ The student thought they might have made a mistake in measuring the strips. Is there any evidence for an anomaly on either of the graphs?

❽ Would repeating the experiment make the data more reliable? Justify your answer.

CALCULATING ACCELERATION

If the speedometer of a car displays 50 km/h, and then a few seconds later it displays 70 km/h, then the car is accelerating. If the car is slowing down, this is called negative acceleration, or deceleration.

How much an object's speed or velocity changes in a certain time is its **acceleration**. Acceleration can be calculated using the following formula:

$$\text{acceleration} = \frac{\text{change in velocity}}{\text{time taken}} \text{ or } a = \frac{(v - u)}{t}$$

where a = acceleration, v = final velocity in m/s, u = starting velocity in m/s and t = time in s.

REMEMBER

Negative acceleration shows that the object is slowing down.

Imagine that the car is initially travelling at 15 m/s, and that one second later it has reached 17 m/s; and that its speed increases by 2 m/s each second after that. Each second its speed increases by 2 metres per second. We can say that its speed is increasing at 2 metres per second *per second*. It is much more convenient to write this as an acceleration of 2 m/s^2.

We can show this using the formula

$$\text{acceleration} = \frac{\text{change in velocity}}{\text{time taken}} = \frac{17 - 15}{1} = 2 \text{ m/s}^2$$

Earth attracts all objects towards its centre with the force of gravity. Gravity's strength decreases slowly with distance from the Earth's surface, but for objects within a few kilometres of the surface, all free-falling objects have the same constant acceleration of just under 10 m/s^2. If a coconut falls from a tree, then after 1 second (abbreviation 's', so 1 second is written 1 s) it falls at 10 m/s. After 2 s it falls at 20 m/s, if it does not hit the ground first.

We can show this using the formula

$$\text{acceleration} = \frac{\text{change in velocity}}{\text{time taken}} = \frac{20 - 0}{2} = 10 \text{ m/s}^2$$

WORKED EXAMPLE

Calculate the acceleration of a car that travels from 0 m/s to 28 m/s in 10 seconds.

Write down the formula: $a = \frac{(v - u)}{t}$

Substitute the values for v, u and t: $a = \frac{(28 - 0)}{10}$

Work out the answer and write down the units: $a = 2.8 \text{ m/s}^2$

Make sure you remember that acceleration measures how quickly the velocity changes, the *rate of change of velocity*. In physics, acceleration does not mean 'gets faster'. Neither does it measure how much the velocity changes.

QUESTIONS

1. Define acceleration.
2. State an everyday name for negative acceleration.
3. As a stone falls, it accelerates from 0 m/s to 30 m/s in 3 seconds. Calculate the acceleration of the stone and give the units.
4. A racing car slows down from 45 m/s to 0 m/s in 3 s. Calculate the acceleration of the racing car and give the units.

BRINGING THE IDEAS TOGETHER

We have learnt how to calculate speed and acceleration. Here is another equation you will find useful in solving problems:

$$(\text{final speed})^2 = (\text{initial speed})^2 + (2 \times \text{acceleration} \times \text{distance moved})$$

$$v^2 = u^2 + (2 \times a \times s)$$

where a = acceleration in m/s^2, v = final velocity in m/s, u = starting velocity in m/s, t = time in s and s = distance moved in m.

This equation uses the relationship between final speed, initial speed, acceleration and distance moved.

WORKED EXAMPLE

A ball falls from rest over the edge of a building 80 m high. It accelerates at 10 m/s^2.

Calculate the speed of the ball as it hits the ground.

$v^2 = u^2 + (2 \times a \times s)$
$v^2 = u^2 + (2 \times 10 \times 80)$
$v^2 = 0 + 1600$
$v = \sqrt{1600}$

Speed of the ball as it hits the ground, v = 40 m/s

QUESTION

1. An astronaut jumps from a small cliff on the Moon. The acceleration due to gravity on the Moon is 1.6 m/s^2. She falls from rest and falls 1.5 m before she lands on the Moon's surface. Calculate her speed as she hits the ground.

USING VELOCITY–TIME GRAPHS

A **velocity–time graph** provides information on velocity, acceleration and distance travelled.

Finding acceleration from a velocity–time graph

The gradient of a velocity–time graph shows how velocity changes with time – the rate of change of velocity, or acceleration. A steep gradient means the acceleration is large. The object's velocity is changing rapidly.

In the graph, the object is already moving when the graph begins. If the object starts with a velocity of zero, then the line starts from the origin.

Note that the object may not move to begin with. In this case the line will start by going along the time axis, showing that the velocity stays at zero for a while.

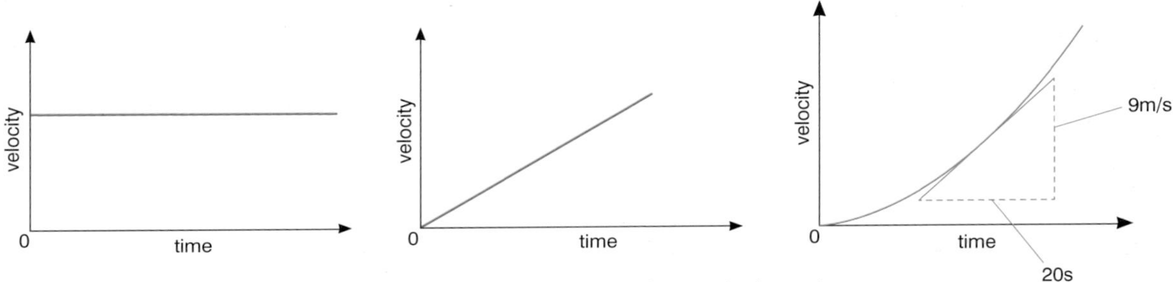

△ Fig. 1.12 Steady speed is shown by a horizontal line. Steady acceleration is shown by a straight line sloping up. Acceleration that is not constant is shown by a curved line.

In Figure 1.12, the right-hand distance–time graph has a slope that gets gradually steeper. The slope of the tangent to the curve can be calculated. The tangent shows that the velocity increased by 9 m/s in 20 s. The slope would be 9/20 = 0.45 m/s^2.

This slope also gives the acceleration.

Velocity–time graphs and speed–time graphs

As long as the motion is in *one direction only,* a speed–time graph and a velocity–time graph will be the same. Both are usually called velocity–time graphs.

However, if the direction of the object changes, the graphs will not look the same. Think about a stone which is thrown upwards at 10 m/s (a velocity of +10 m/s if we set the upwards direction as positive). One second later (ignoring air resistance) the stone will have gone as high as it can and will have a speed of 0 m/s (and a velocity of 0 m/s). When it falls, the speed increases and it will land at a speed of 10 m/s. Because it has changed direction, though, it will have a velocity of –10 m/s when it hits the ground.

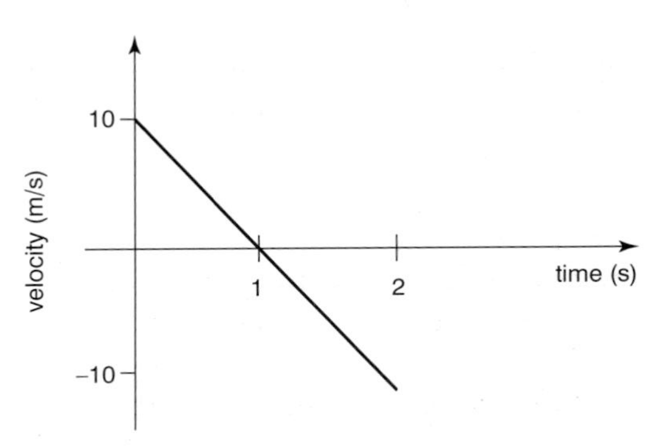

△ Fig. 1.13 Velocity–time graph for a stone.

REMEMBER

For any graph of motion, *always* check carefully to see if it is a distance–time graph or a velocity–time graph. Although the graphs may look similar, you need to interpret them very differently. For instance, the slope of a distance–time graph gives the velocity (since velocity = distance/time) but the slope of a velocity–time graph gives the acceleration (since acceleration = change of velocity/time).

QUESTIONS

1. On a velocity–time graph, what does a horizontal line indicate?
2. Describe the shape of the velocity–time graph for a car travelling with constant acceleration.
3. On a velocity–time graph, what quantity is shown by the slope of the line?

Finding distance from a velocity–time graph

The area under the line on a velocity–time graph gives the distance travelled, because distance = velocity × time. Always make sure the units are consistent: if the speed is in km/h, you must use time in hours too.

Fig. 1.14 shows a velocity–time graph for a car travelling between two sets of traffic lights. It can be divided into three regions.

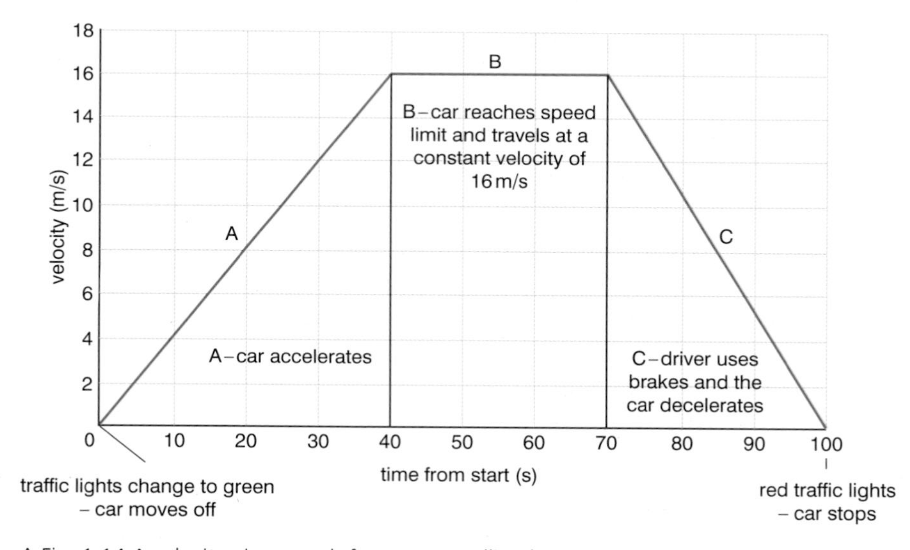

Δ Fig. 1.14 A velocity–time graph for a car travelling between two sets of traffic lights.

In region A, the car is accelerating at a constant rate (the line has a constant positive gradient). The distance travelled by the car can be calculated:

average velocity = $\frac{(16 + 0)}{2}$ = 8 m/s

time = 40 s

so distance moved = average velocity × time taken = 8 × 40 = 320 m

This can also be calculated from the area under the line

($\frac{1}{2}$ base × height = $\frac{1}{2}$ × 40 × 16 = 320 m).

In region B, the car is travelling at a constant velocity (the line has a gradient of zero). The distance travelled by the car can be calculated:

velocity = 16 m/s

time = 30 s

so distance moved = average velocity × time taken = 16 × 30 = 480 m

This can also be calculated from the area under the line (base × height = 30 × 16 = 480 m).

In region C, the car is decelerating at a constant rate (the line has a constant negative gradient). The distance travelled by the car can be calculated:

average velocity = $\frac{(16 + 0)}{2}$ = 8 m/s

time = 30 s

so distance moved = average velocity × time taken = 8 × 30 = 240 m

This can also be calculated from the area under the line

($\frac{1}{2}$ base × height = $\frac{1}{2}$ × 30 × 16 = 240 m).

In the above example, the acceleration and deceleration were constant, and the lines in regions A and C were straight. This is very often not the case. You will probably have noticed that a car can have a larger acceleration when it is travelling at 30 km/h than it can when it is already travelling at 120 km/h.

A people-carrying space rocket does exactly the opposite. If you watch a rocket launch you can see that it has a smaller acceleration to begin with. Because it is burning several tonnes of fuel per second, it quickly becomes lighter and starts to accelerate more quickly.

For simple shapes, the distance travelled can be found by simply adding up the squares underneath the velocity–time graph.

WORKED EXAMPLE

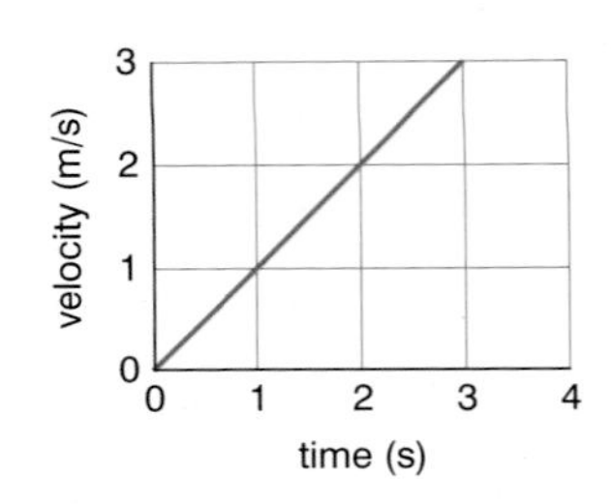

Δ Fig. 1.15

Figure 1.15 shows a velocity–time graph for a toy car steadily increasing its speed.

Calculate the distance travelled by the car from 0 to 3 s.

There are 3 whole squares and 3 half squares. In total, there are $4\frac{1}{2}$ squares.

The distance travelled is 4.5 m.

Another way is to treat it as the area of a triangle. The area under the line is:

$(\frac{1}{2} \times \text{base} \times \text{height} = \frac{1}{2} \times 3 \times 3 = 4.5 \text{ m})$

Similarly using the idea of average velocity:

$$\text{average velocity} = \frac{(0+3)}{2} = 1.5 \text{ m/s}$$

time = 3 s

so distance moved = average velocity × time taken = 1.5 × 3 = 4.5 m.

QUESTIONS

1. Explain how to calculate the acceleration from a velocity–time graph.
2. Explain how to calculate the distance travelled from a velocity–time graph.
3. Two runners complete a 400 m race. Athlete A takes 50 s and athlete B takes 64 s.
   a) Calculate the average speed of each runner.
   b) Sketch a speed–time graph for each runner.
   c) How could you show from the graphs that the runners both cover 400 m?
4. From the graph below, calculate the distance travelled between 2 and 12 seconds.

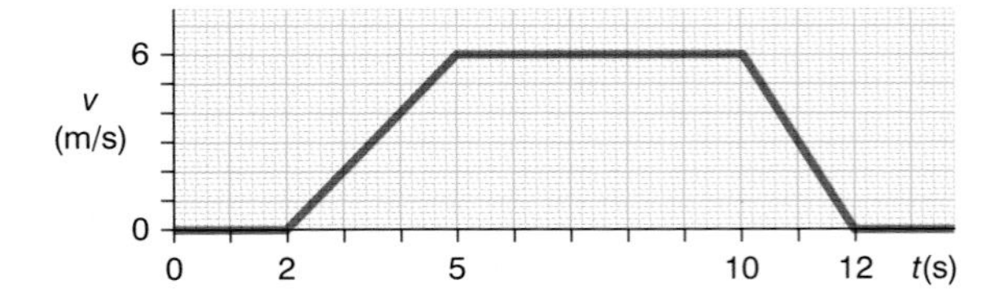

5. Use the velocity–time graph below to answer the questions below:

a) What is the speed of the object at 1 s?

b) What can you say about the speed between points B and C?

c) Between which points is there the greatest acceleration?

d) What is happening to the object between points E and F?

e) What is the distance travelled between points D and E?

f) Calculate the acceleration of the object between points A and B.

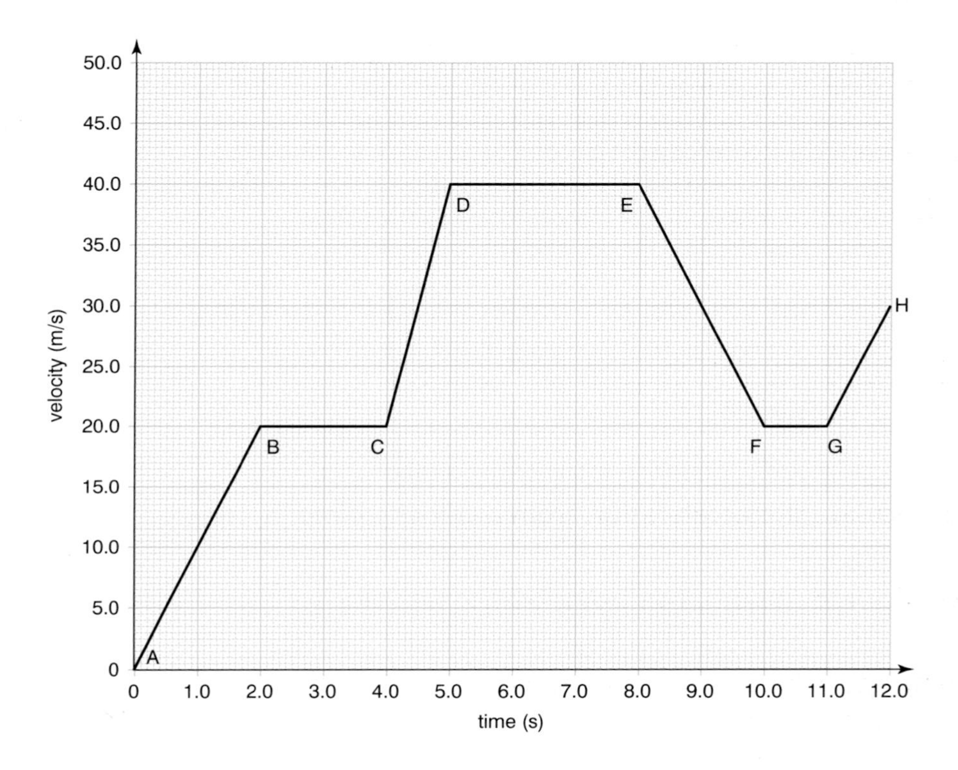

End of topic checklist

A **distance–time graph** is a visual representation of how distance travelled varies with time.

The **speed** of an object is how fast it is moving.

Average speed = distance moved/time taken

The **acceleration** of an object is how much its velocity changes every second. Acceleration = change in velocity/time taken.

A **velocity–time graph** is a visual representation of how velocity varies with time.

The area under a velocity–time graph gives the distance travelled.

The gradient of a velocity–time graph gives the acceleration.

The facts and ideas that you should understand by studying this topic:

- ❍ Know how to plot a distance–time graph.
- ❍ Know that average speed = distance moved/time taken.
- ❍ Know and be able to use the relationship between acceleration, change in velocity and time taken: $a = \frac{(v - u)}{t}$
- ❍ Know and be able to use the relationship $v^2 = u^2 + (2 \times a \times s)$.
- ❍ Know how to plot a velocity–time graph.
- ❍ Know how to find distance travelled by calculating the area under a velocity–time graph.
- ❍ Know how to find the acceleration of an object by finding the gradient of a velocity–time graph.
- ❍ Understand how an object moves from interpreting lines drawn on distance–time and velocity–time graphs.
- ❍ Describe how to investigate the motion for everyday objects in simple linear motion.

End of topic questions

1. John's journey to school takes 10 minutes and is 3.6 km. What is his average speed in km/minute? **(1 mark)**

2. Alice runs 400 m in 1 minute 20 seconds. What is her average speed in m/s? **(1 mark)**

3. At one point Alice is running due west at 6 m/s. Later she is running due east at 4 m/s. How could we write her velocities to show that they are in opposite directions? **(2 marks)**

4. A train moves away from a station along a straight track, increasing its velocity from 0 to 20 m/s in 16 s. What is its acceleration in m/s^2? **(1 mark)**

5. A rally car accelerates from 100 km/h to 150 km/h in 5s. What is its acceleration in

   a) km/h/s **(1 mark)**

   b) m/s^2? **(1 mark)**

6. Imagine two cars travelling along a narrow road where it is not possible for the cars to pass each other.

   Describe what will happen if:

   a) both cars have a velocity of +15 m/s **(1 mark)**

   b) one car has a velocity of +15 m/s and the other –15 m/s **(1 mark)**

   c) both cars have a velocity of –15 m/s. **(1 mark)**

7. You walk to school and then walk home again. What is your average velocity for the whole journey? Explain your answer. **(2 marks)**

8. John cycles to his friend's house. In the first part of his journey, he rides 200 m from his house to a road junction in 20 s. After waiting for 10 s to cross the road, John cycles for 20 s at 8 m/s to reach his friend's house.

   a) What is John's average speed for the first part of the journey?

   b) How far is it from the road junction to his friend's house?

   c) What is John's average speed for the whole journey? **(3 marks)**

9. We often use sketch graphs to illustrate motion. Describe the main differences between a sketch graph and a graph. **(2 marks)**

End of topic questions continued

10. How could you tell from a distance–time graph whether the object is moving away from you or towards you? **(2 marks)**

11. The graph shows a distance–time graph for a journey.

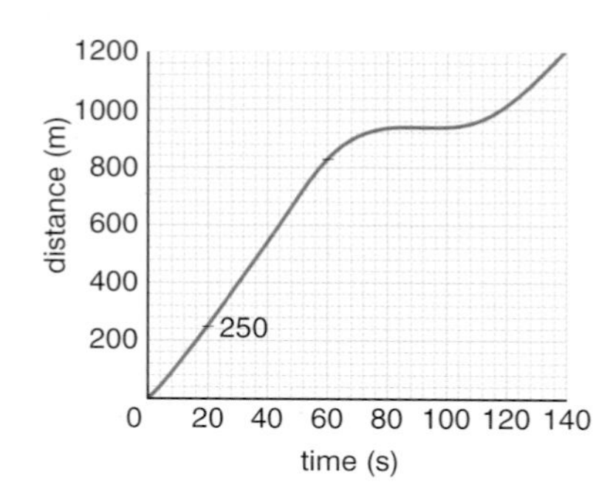

a) What does the graph tell us about the speed of the car between 20 and 60 seconds? **(1 mark)**

b) How far did the car travel between 20 and 60 seconds? **(1 mark)**

c) Calculate the speed of the car between 20 and 60 seconds. **(1 mark)**

d) What happened to the car between 80 and 100 seconds? **(1 mark)**

12. Look at the velocity–time graph for a toy tractor.

a) Calculate the acceleration of the tractor from A to B. **(1 mark)**

b) Calculate the total distance travelled by the tractor from A to C. **(2 marks)**

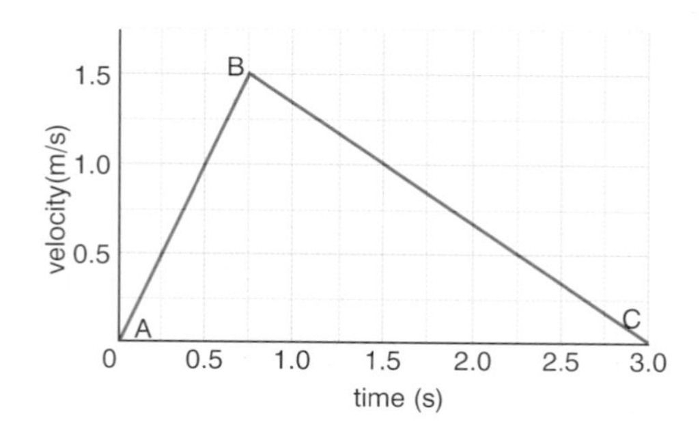

Forces, movement, shape and momentum

Δ Fig. 1.16 These people are applying forces to drag the net from the sea.

INTRODUCTION

We live in a dynamic universe. There is constant motion around us all the time, from the vibrations of the atoms that make up our bodies to the movement of giant galaxies through space, and the motion is constantly changing. Objects do not remain constant: some change size, others change shape. Atoms arrange and rearrange themselves into many different chemicals. Energy moves about through the motion of objects and through the transfer of waves. Behind this constant change lie the forces that determine how all the interactions will play out.

KNOWLEDGE CHECK

✓ Be able to recognise situations with balanced and unbalanced forces.
✓ Describe forces in different situations, such as friction and drag.
✓ Be able to calculate speed and acceleration, including gathering and using experimental data.

LEARNING OBJECTIVES

✓ Identify different types of force such as gravitational or electrostatic.
✓ Understand how vector quantities differ from scalar quantities.
✓ Know how to combine forces that act along a line, including ideas about direction, to calculate resultant forces.
✓ Be able to link the idea of resultant force to the effect of forces, such as changes in speed, shape or direction.
✓ Know that friction is a force that opposes motion.
✓ Know and use the relationship between resultant force, mass and acceleration: $F = ma$.
✓ Know and use the relationship between weight, mass and gravitational field strength: $W = mg$.
✓ Know that the stopping distance of a vehicle is made up of the sum of the thinking distance and the braking distance.
✓ Use ideas about balanced and unbalanced forces to explain the idea of terminal velocity.
✓ Be able to investigate how extension varies with applied force for helical springs, metal wires and rubber bands.
✓ Know and use the relationship between momentum, mass and velocity: $p = mv$.
✓ Use ideas about momentum and conservation of momentum to explain some safety features.

✓ Use the relationship between force, change in momentum and time taken: force = change in momentum/time taken.
✓ Be able to use Newton's third law.
✓ Be able to calculate the moment of a force using the product force × perpendicular distance from the pivot.
✓ Be able to apply the principle of moments to situations where objects are balanced.
✓ Know that the weight of a body acts through its centre of gravity.

WHAT ARE FORCES?

A **force** is a push or a pull. The way that an object behaves depends on all of the forces acting on it. A force may come from the pull of a chain or rope, the push of a jet engine, the push of a pillar holding up a ceiling, or the pull of the gravitational field around the Earth.

Effects of forces

It is unusual for a single force to act on an object. Usually there are two or more. The size and direction of these forces determine whether the object will move and the direction it will move in.

Forces are measured in **newtons** (N). They take many forms and have many effects including pushing, pulling, bending, stretching, squeezing and tearing. Forces can:

- change the speed of an object
- change the direction of movement of an object
- change the shape of an object.

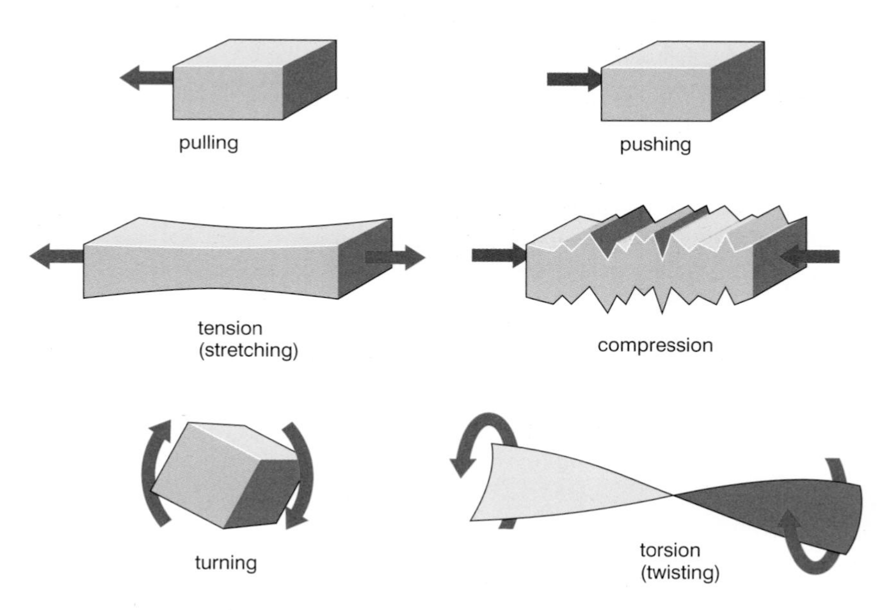

Δ Fig. 1.17 Types of force.

Types of force

There are several different types of force. All objects in the universe attract each other with the extremely feeble force of **gravity**. The strength of the gravitational attraction depends on the mass of the two objects and the distance between their centres. You may think that gravity is strong, but you are, after all, close to the Earth, which is a very massive object compared with you!

Electricity and magnetism both generate forces that are far stronger than gravity. As we shall discover in section 7, you see **magnetic** forces and electric forces combining as an electromagnetic force used every day when an electric motor turns.

Electrostatic forces are the most important in our everyday lives. Electrostatic forces are those between charges such as electrons. Like charges repel (so an electron will repel another electron) and unlike charges attract (so a negatively charged electron will attract a positively charged proton). The reason that you are not sinking into the floor at the moment is that the electrons on the outside of the atoms of your shoes are being repelled by the electrons on the outside of the atoms of the floor.

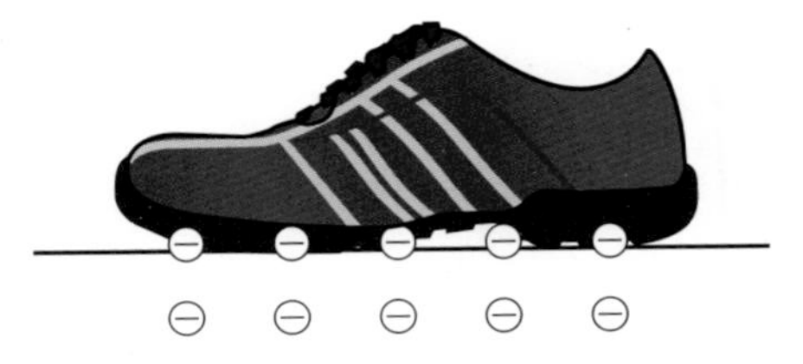

Δ Fig. 1.18 Electrons on the surface of the floor repel electrons on the surface of the sole of the shoe.

The same force is used when your hand lifts something up, or when friction slows down a car. In fact, all of the forces in this section are either gravitational or – ultimately – electrostatic.

And when you consider that it is an electrostatic force that allows a bullet-proof coat to stop a speeding bullet, you'll probably agree that electrostatic forces are much stronger than gravity.

There are a few other types of force apart from these three – for example, the 'strong' force that holds the nucleus of the atom together. However, most of the forces that we feel or notice around us are one of the above three: gravitational, electrostatic or magnetic.

QUESTIONS

1. Describe three effects of a force and three types of force.
2. What is the force that connects all the objects in the universe?
3. What two factors does the strength of this force depend on?
4. What force is seen in a motor?

LINKING THE FUNDAMENTAL FORCES

Scientists are working to find a theory that links all the fundamental forces: gravity, electromagnetism, strong nuclear force and the weak nuclear force. Particle accelerators, in which high energy collisions take place, are useful tools in this search. In 1963, Glashow, Salam and Weinberg predicted that the electromagnetic force and the weak nuclear force might combine (in what would be called the electroweak force) at particle energies higher than about 100 GeV. (This is equivalent to 1.6×10^{-8} J. 1 eV is one electronvolt, which is the energy gained or lost by the charge of a single electron moving across a pd of 1 V.) These particle energies occur at temperatures of about 10^{15} K, which would have occurred shortly after the Big Bang. This prediction was confirmed 20 years later in a particle accelerator.

Δ Fig. 1.19 Inside the Large Hadron Collider.

There are theories that predict that the electroweak and strong forces would combine at energies greater than 10^{15} GeV and that all the forces may combine at energies greater than 10^{19} GeV. At present, the largest particle accelerator is the Large Hadron Collider at CERN in Switzerland. It is able to accelerate protons to 99.99% of the speed of light, and they can reach energies of 1.4×10^{4} GeV, so it is still some way short of the energies needed to test the theory about combining the electroweak and strong forces, and all four forces.

However, science never stands still and it may be that these energies are reached in your lifetime.

SCALARS AND VECTORS

Force, velocity and acceleration are examples of **vector** quantities. A vector has a specific direction as well as a size, with a unit.

Speed and mass are examples of **scalar** quantities. A scalar quantity has size only, with a unit. There are many more scalar quantities: temperature, work, power and electrical resistance are all scalars.

REMEMBER

Displacement is the vector quantity linked to distance, which is a scalar quantity. Displacement is the distance travelled in a particular direction.

The vector nature of force

To describe a force fully, you must state the size of the force and also the direction in which it is trying to move the object. The direction can be described in many different ways such as 'left to right', 'upwards' or 'north'. Sometimes it is useful to describe all of the forces in one direction as positive, and all of the forces in the other direction as negative. For two forces to be equal they must have the same size and the same direction.

Adding forces

If two or more forces are pulling or pushing an object in the same direction, then the effect of the forces will add up; if they are pulling it in opposite directions, then the backwards forces can be subtracted. These forces combine to produce a **resultant force**. When a car is being driven, it has a forward driving force from the engine. It also has a backward force due to friction and air resistance (Fig. 1.20).

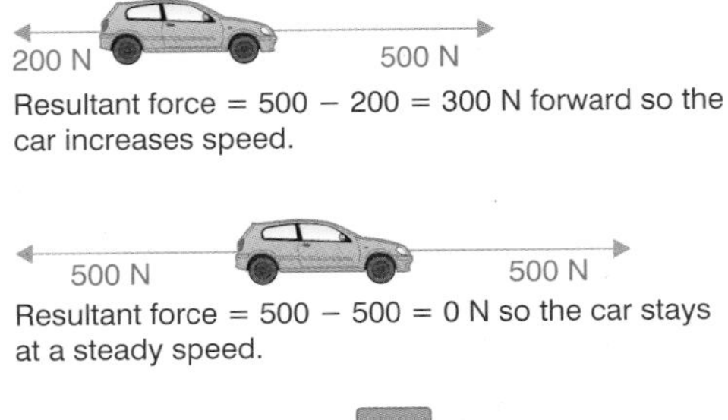

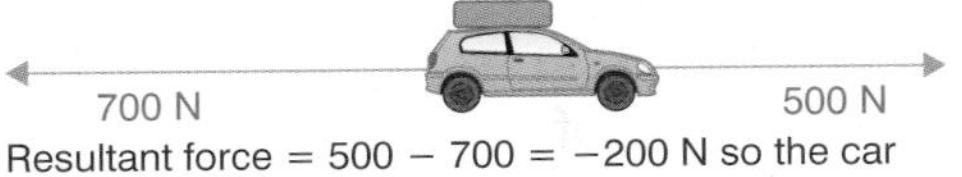

◁ Fig. 1.20 Calculating the resultant of two forces that act along a line.

REMEMBER

Think very carefully about this. Having zero resultant force does not mean the object is stationary. What it does mean is that the object is not accelerating – if it is already moving, it will continue to so at constant speed in a straight line.

QUESTIONS

1. Explain the difference between a scalar quantity and a vector quantity.
2. A car has a forward driving force of 1000 N. Its backward frictional force changes with speed.

   Calculate the resultant force for each of these frictional forces.

   a) 100 N

   b) 500 N

   c) 1000 N

   d) 1200 N
3. What is the resultant force on the hand of a karate expert if the weight of her hand is 32 N, the thrust downwards from her muscles is 2728 N and the air friction against the movement downwards is 200 N?

WHAT IS FRICTION?

A force that opposes motion may not be a bad thing. When you walk, your feet try to slide backwards on the ground. It is only because there is **friction** (working against this sliding) that you can move forwards. Just think how much harder it is to walk on a slippery (that is, low friction) surface such as ice. A force which opposes motion is also very useful in applications such as between the brake pads and a bicycle wheel.

However, in many situations friction can be a disadvantage. For example, there is some friction in the bearings of a bicycle wheel, which causes some energy to be transferred to unwanted thermal energy.

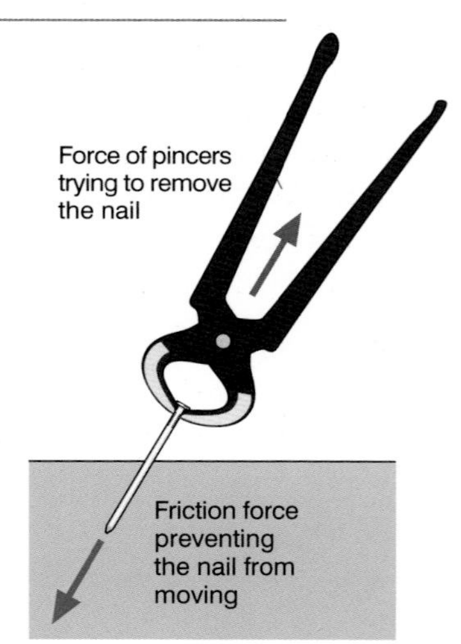

Δ Fig. 1.21 Friction can stop any movement occurring at all, and it is friction that stops a nail coming out of a piece of wood.

QUESTIONS

1. Give an example of where friction may be useful.
2. Give an example of where friction may be a disadvantage.

BALANCED FORCES

Usually there are at least two forces acting on an object. If these two forces are **balanced** then there is no resultant force and the object will either be stationary or moving at a constant speed.

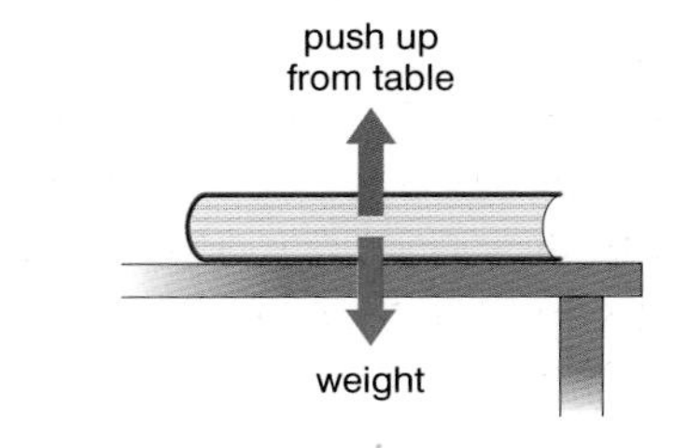

Δ Fig. 1.22 The forces on this book are balanced.

The book is stationary because the push upwards from the table is equal to the weight downwards. If the table stopped pushing upwards, the book would fall.

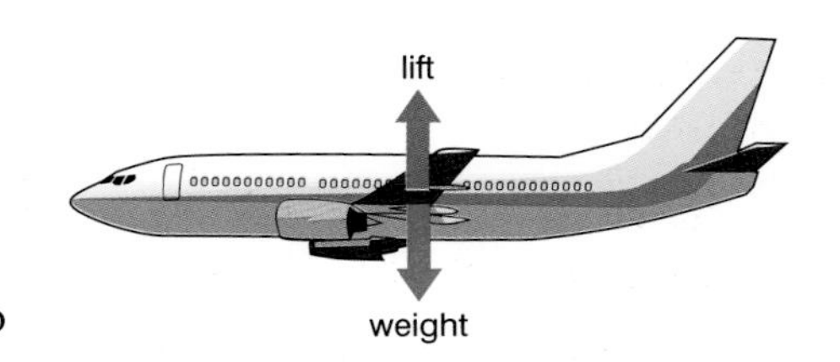

Δ Fig. 1.23 The forces in the vertical direction on this aircraft are balanced so it will not move in the vertical direction.

This aircraft is flying 'straight and level' because the lift generated by the air flowing over the wings is equal and opposite to the weight of the aircraft. This diagram shows that the plane will neither climb nor dive, as it would if the forces were not equal.

This is **Newton's first law**, which simply says that you need a resultant force to change the way something is moving.

UNBALANCED FORCES

For an object's speed or direction of movement to change, the forces acting on it must be **unbalanced**.

As a gymnast first steps on to a trampoline, his or her weight is much greater than the opposing supporting force of the trampoline, so he moves downwards, stretching the trampoline. (Note that we are not talking about the gymnast jumping on to the trampoline – if that were the case the physics would be different!) As the trampoline stretches, its supporting force increases until the supporting force is equal to the gymnast's weight. When the two forces are balanced, the trampoline stops stretching. If an elephant stood on the trampoline, the trampoline would break because it could never produce an equal supporting force.

You see the same effect if you stand on snow or soft ground. If you stand on quicksand, then the supporting force will not equal your weight, and you will continue to sink.

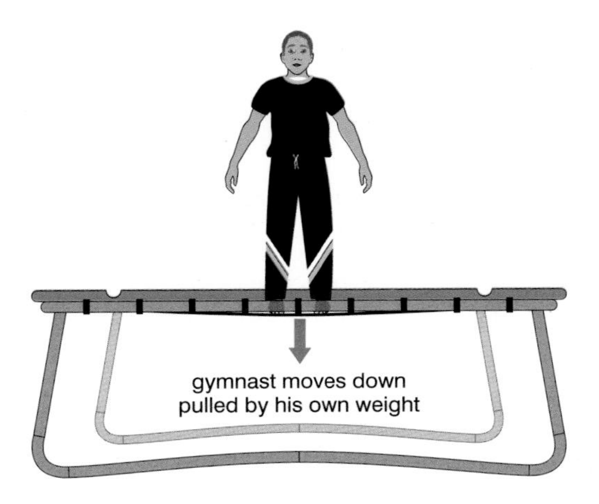

Δ Fig. 1.24 A trampoline stretches until it supports the weight on it.

QUESTIONS

1. Describe the motion of an object if the forces on it are balanced.
2. Describe the motion of an object if the forces on it are unbalanced.
3. If a gymnast is standing on a trampoline, what must the supporting force of the trampoline be equal to?

HOW ARE MASS, FORCE AND ACCELERATION RELATED?

The acceleration of an object depends on its mass and the force that is applied to it. The relationship between these factors is given by the formula:

force = mass × acceleration

$F = ma$

where F = force in newtons, m = mass in kg and a = acceleration in m/s^2

This equation comes from **Newton's second law of motion**, which links acceleration (the rate of change of velocity) to the resultant force that is causing the change. In this form we can only use the equation if the mass stays constant. A slightly different version is used later in this section (momentum).

The equation explains the definition of the newton: 'One newton is the force that will accelerate a mass of 1 kg at 1 m/s^2.'

The equation is perhaps easier to understand in the form $a = \frac{F}{m}$. This shows that if we use a big force we will get a larger acceleration, but if the object has more mass then we get a smaller acceleration.

Fig. 1.25 shows that, as mass increases, acceleration decreases (they are **inversely proportional**). Fig. 1.26 shows that as acceleration increases, force increases (they are **directly proportional**).

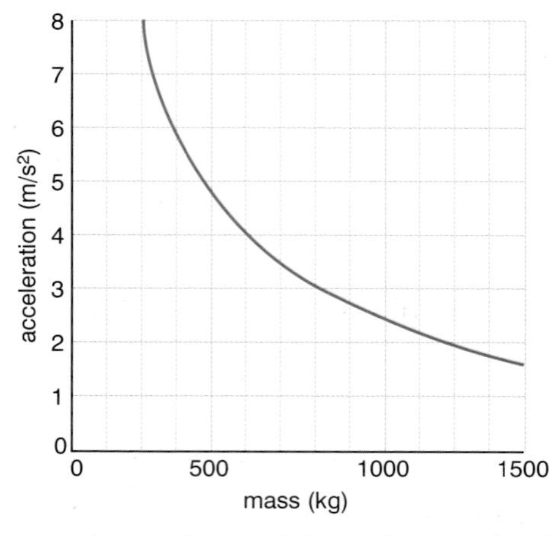

Δ Fig. 1.25 Graph showing that acceleration is inversely proportional to mass.

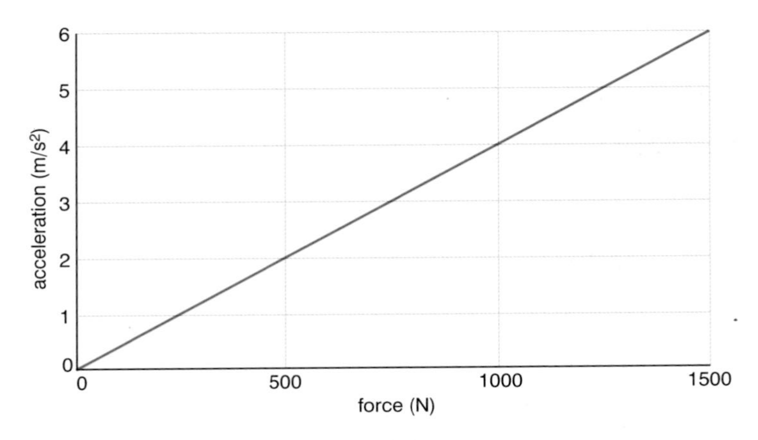

△ Fig. 1.26 Acceleration is directly proportional to force.

So a light object with a large force applied to it will have a large acceleration. For example, an athlete with a racing bicycle applies a large force to the pedals of the light bicycle, and so the bicycle will have a large acceleration. In contrast, a massive object with a small force applied to it will have a small acceleration. For example, a small child trying to pedal a large bicycle rickshaw will only be able to apply a small force to the pedals of the heavy rickshaw, and so the acceleration will be small.

REMEMBER

The equation $F = ma$ shows that the acceleration of an object is directly proportional to the force acting (if its mass is constant) and is inversely proportional to its mass (if the force is constant).
The gradient of a force–acceleration graph gives the mass of the object.

WORKED EXAMPLES

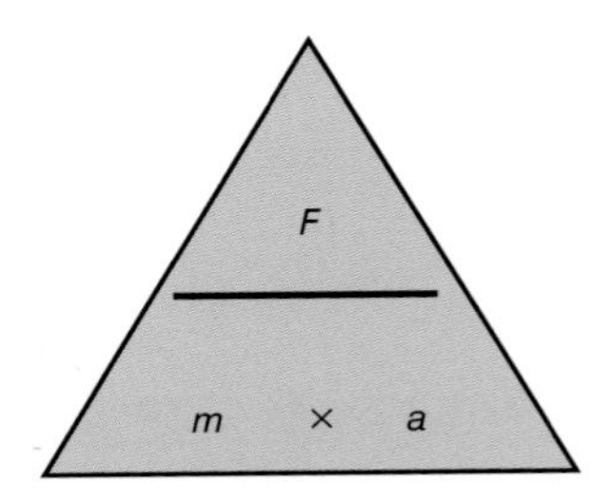

△ Fig. 1.27 Force, mass and acceleration triangle.

1. What force would be required to give a mass of 5 kg an acceleration of 10 m/s²?

Write down the formula: $F = ma$

Substitute the values for m and a: $F = 5 \times 10$

Work out the answer and write down the units: $F = 50$ N

3. A car has a resultant driving force of 6000 N and a mass of 1200 kg. Calculate the car's initial acceleration.

Write down the formula in terms of a: $a = \frac{F}{m}$

Substitute the values for F and m: $a = \frac{6000}{1200}$

Work out the answer and write down the units: $a = 5 \text{ m/s}^2$

QUESTIONS

1. Write down the equation linking force, mass and acceleration.
2. For the equation to apply, what conditions must the 'force' and 'mass' meet?
3. What force is acting if a 60 kg bungee jumper is accelerating at 10 m/s²?
4. The resultant force on an object of mass 3.2 kg is 2560 N. What is the acceleration of the object?

MASS AND WEIGHT

Scientists use the words **mass** and **weight** with special meanings. The 'mass' of an object means how much material is present in it. It is measured in kilograms (kg).

Weight is the force on the object due to gravitational attraction. It is measured in newtons (N). The weight of an object depends on its mass and **gravitational field strength**. Any mass near the Earth has weight due to the Earth's gravitational pull.

Weight is calculated using the equation:

weight = mass × gravitational field strength

$W = mg$

Scientists often use the word 'field'. We say that there is a 'gravitational field' around the Earth, and that any object that enters this field will be attracted to the Earth.

The value of the gravitational field strength on Earth is 9.8 N/kg, though we usually round it up to 10 N/kg to make the calculations easier. A gravitational force of 10 N acts on an object of mass 1 kg mass on the Earth's surface.

As mentioned earlier, gravity does not stop suddenly as you leave the Earth. Satellites go round the Earth and cannot escape because the Earth is still pulling them. The Earth is pulling the Moon, which is why the Moon orbits the Earth once per month. And the Earth goes round the Sun because it is pulled by the Sun's *much greater* gravity.

WORKED EXAMPLE

On Earth the gravitational field strength is 9.8 N/kg. Calculate the mass of a Mars exploration rover that weighs 1500 N on the Earth.

Write down the formula: $W = mg$

Rearrange the formula: $m = W \div g$

Substitute the values for W and g: $m = 1500 \div 9.8$

Work out the answer and write down the units: $m = 153.061\,224\,489\,8$ kg.

This answer has 10 decimal places. But the gravitational field strength, g is given as 9.8 N/kg which has only two **significant figures**. The weight is 1500 N which also has two significant figures. It would not be correct therefore to give an answer that was correct to 10 decimal places. The answer to a calculation can only have the same number of significant figures as the data provided. This answer needs to be rounded to two significant figures, 150 kg.

QUESTIONS

1. What is the difference between mass and weight?
2. What is the weight of someone whose mass is 60 kg?
3. What is the mass of someone whose weight is 500 N?

Gravity on the Moon

If you stood on the Moon you would feel the gravity of the Moon pulling you down. Your mass would be the same as on Earth, but your weight would be less. This is because the gravitational field strength on the Moon is about one-sixth of that on the Earth, and so the force of attraction of an object to the Moon is about one-sixth of that on the Earth. The gravitational field strength on the Moon is 1.6 N/kg and so a force of 1.6 N is needed to lift a 1 kg mass.

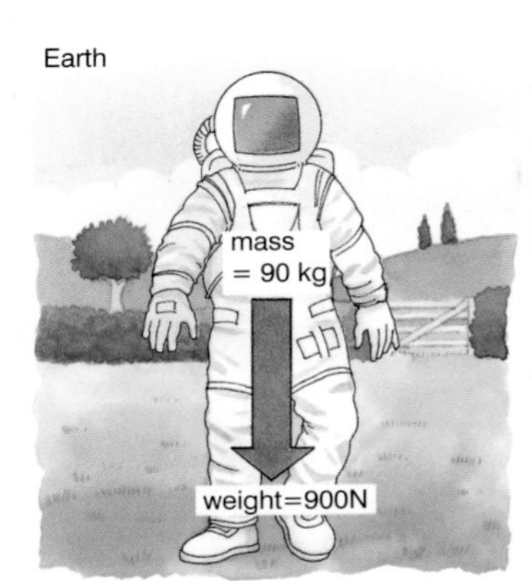

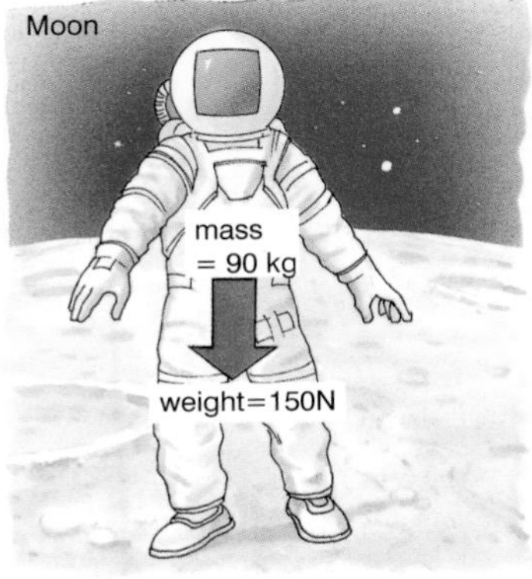

Δ Fig. 1.28 Though your mass remains the same, your weight is greater on Earth than it would be on the Moon.

If two astronauts played football on the Moon and crashed into each other, it would hurt just as much as it would on Earth. This is because the accelerations would be the same, the forces would be the same, and so from $F = ma$ the forces would be the same. It is the forces that hurt! A pressurised space suit would help to reduce the effects, just like an airbag in a car.

It is harder to get a massive object moving, and it is harder to stop it once it is moving. A supertanker, laden with oil, and travelling at 18 km/h takes over 12 km to stop. In contrast, a speedboat, which is very much lighter, takes less than 100 m to stop if it is travelling at the same speed as the supertanker.

△ Fig. 1.29 A supertanker will take very much longer to stop than a speedboat travelling at the same speed.

How do you weigh something?

The balance is level when the forces pulling down both sides are the same. In the balance shown, the forces of 10 N and 20 N on the one side balance the force of 30 N on the other side. The balance compares the weight of the objects on each side. If the balance is on the surface of the Earth, then the masses of these objects are 1 kg and 2 kg on one side, and 3 kg on the other. So the balance also allows you to compare masses.

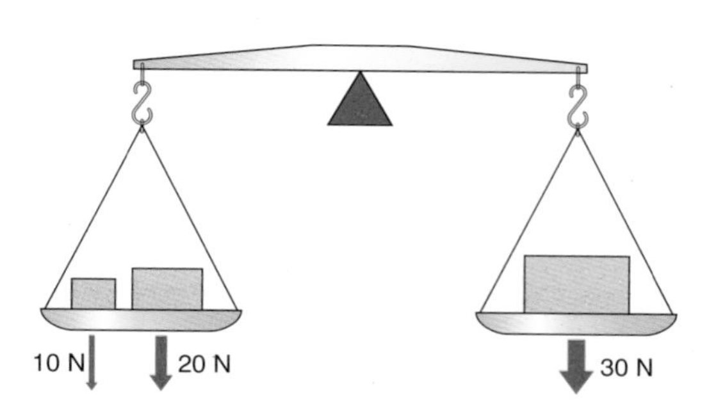

◁ Fig. 1.30 The total force is the same on each side of the balance.

QUESTIONS

1. Explain why your weight would change if you stood on the surface of different planets.
2. What would be the weight on the surface of the Moon of someone of mass 60 kg?
3. A person weighs 80 N on the Moon. What is their mass?
4. A balance has 30 N on the left hand side and 50 N on the right hand side. What weight must be added so that the sides are balanced, and to which side?

Falling objects and terminal velocity

As a skydiver jumps from a plane, the weight will be much greater than the opposing force caused by air resistance. Initially she will accelerate downwards at 10 m/s^2.

The skydiver's velocity will increase rapidly – and as it does, the force caused by the air resistance increases. Eventually it will exactly match the weight, the forces will be balanced and the velocity of the skydiver will remain constant. This speed is known as the **terminal velocity**, typically 180 km/h.

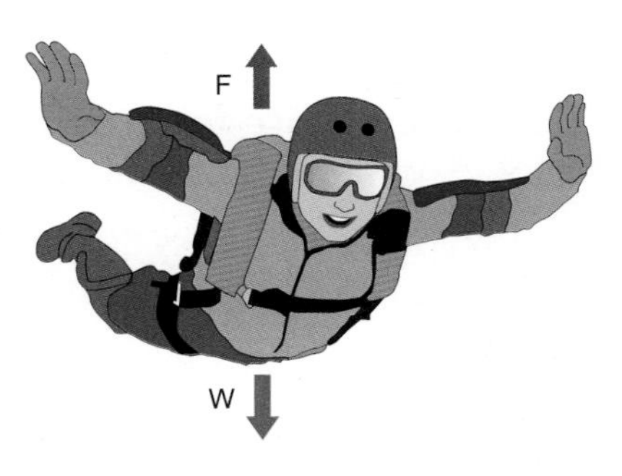

Δ Fig. 1.31 The forces on the skydiver are balanced and so the velocity is terminal velocity.

In Figure 1.31, the skydiver has her arms outstretched and the air resistance force is fairly high. If the skydiver makes herself streamlined by going headfirst with her arms by her side, then she will cut through the air more easily, and the air resistance force will go down. She will then accelerate again, until the force of air resistance increases again to equal her weight. She will now be going at almost 300 km/h.

A parachute has a very large surface, and produces a very large resistive force, so the terminal speed of a parachutist is quite low. This means that he or she can land relatively safely.

◁ Fig. 1.32 The terminal speed of this parachutist is quite low so he will be able to land relatively safely.

REMEMBER

You need to think carefully about the two opposing forces that lead to terminal velocity. The force causing the motion (such as gravity or the force from a car engine) usually remains constant. It is the drag force (such as air or water resistance) that increases as the velocity increases, until the two are balanced and the velocity stays constant.

QUESTIONS

1. What is terminal velocity?
2. A skydiver jumps from a plane. How fast will he be travelling after 1 s?
3. If the air resistance force on the skydiver in question 2 is $0.15 \times v^2$, what will its magnitude and direction be after 1 s?
4. If the skydiver has mass 60 kg, what will be the magnitude and direction of the resultant force on him after 1 s?
5. What is the magnitude and direction of the skydiver's acceleration after 1 s?

EXTENSION

Graphs can be drawn to show how the force, acceleration and velocity of a falling object vary with time. These three graphs show these variables, but the labels have been missed from the *y*-axis in each case.

1. What is the correct label for each *y*-axis?

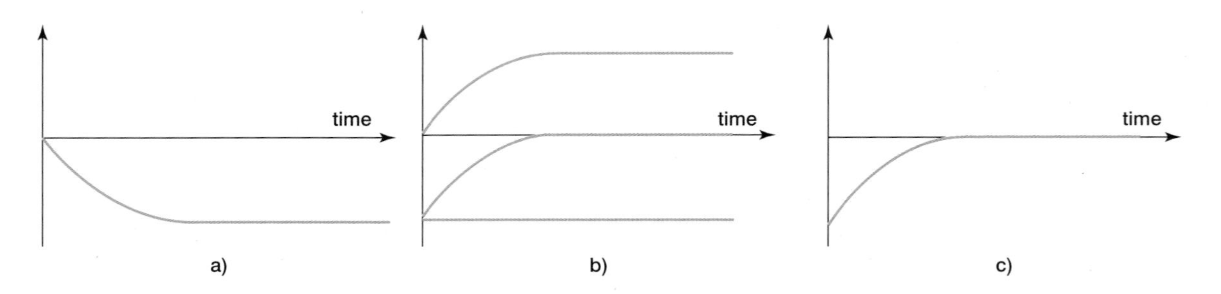

2. Explain the curve of each graph.

VEHICLE STOPPING DISTANCES

When a car driver sees a hazard (such as a child running out into the road), he or she takes time to react. This is called the **reaction time**. During this time the car will be travelling at its normal speed. The distance it travels in this time is called the **thinking distance**.

The driver then puts on the brakes. The distance the car travels while it is braking is called the **braking distance**. The overall **stopping distance** is made up of the thinking distance and the braking distance.

The thinking distance varies from person to person and from situation to situation. The braking distance can vary from car to car.

Factors affecting thinking distance are:

- speed
- reaction time, which is affected by tiredness; medication, drugs or alcohol consumption; level of concentration and distraction.

Factors affecting braking distance are:

- speed
- amount of tread on tyres
- friction from the brakes
- road conditions (dry, wet, icy, gravel, etc.)
- mass of the car.

QUESTIONS

1. List three factors that affect the thinking distance and three factors that affect the braking distance of a car when it stops.
2. At a certain speed, the thinking distance is 12 m and the braking distance is 24 m. What is the stopping distance?
3. The average car length is 4 m. At a given speed, the thinking distance is 18 metres and the braking distance is 55 m. Approximately how many car lengths is this?
4. You are travelling at 15 m/s and you are wide awake and alert. You see a hazard ahead. Your reaction time is 0.5 s. What is your thinking distance?
5. How would your answer to question 4 be affected if you were tired and distracted by your passengers?

REMEMBER

If you are asked about factors that affect stopping distance – (1) check whether you are asked for 'increase', 'decrease' or just change, (2) avoid vague phrases like 'road conditions' without adding details.

EXTENSION

A residential area in your town is considering changing the speed limit on the roads from 45 km/h to 30 km/h. There is a lot of disagreement over whether this should happen or not, and some residents have written to the local paper. Read the following two letters.

To the editor,

The local area desperately needs a lower speed limit. There are a lot of families in the area and lower speed increases the chance of a child surviving if they get hit after chasing their ball into the road. The difference in the stopping distances also means that the child would be much less likely to be hit anyway!

It has been shown that reducing speed limits to 45 km/h means a pedestrian being hit has an 80% chance of survival compared to a 90% chance of death at 60 km/h. Surely this would be the same kind of step change for going from 45 km/h to 30 km/h.

Anything we can do to reduce deaths on the roads should be done!

Yours,

Mr C. Trevithick

To the editor,

In response to Mr Trevithick's letter, I believe the limit should remain at 45 km/h on our roads as it is perfectly safe. The roads are wide and free from obstruction so any children playing in the area will be visible and drivers can be aware of them long before they 'chase their ball into the road'.

At busy times of the day the traffic lights in the area mean that the cars are often moving slower than 45 km/h anyway. To introduce a lower limit would just disadvantage drivers at the other times of the day.

Finally, the difference in moving from 45 km/h to 30 km/h on people survival and safety is far less than that from 60 km/h to 45 km/h - it's simply not worth the cost of changing all the signs.

Yours,

Mrs D. Simmons

1. Why would stopping distances be less at 30 km/h?

2. Would being able to see children clearly make a difference to the overall stopping distance?

3. The car's brakes can provide 6.5 kN of force, and the mass of the car is 1000 kg. What is the stopping distance for each speed?

TURNING EFFECT OF A FORCE

If you have used a spanner to tighten a nut, or you have turned the handle of a rotary beater, you have used a force. Turning also applies to less obvious examples, such as when you push the handle to close a door, or when a child sits on the end of a see-saw to push the end down.

The turning effect of a force is called the **moment** of the force.

The moment of a force depends on two things:

- the size of the force
- the distance between the line of the force and the turning point, which is called the pivot.

We calculate the moment of force using this formula:

moment of a force = force × perpendicular distance from **pivot**

moment = Fd

Moment is measured in newton metres (Nm).
F = force in newtons (N) and d = distance in metres (m).

distance, *d*
force, *F*
pivot

Δ Fig. 1.33 The moment of the force, *F*, is given by $F \times d$

WORKED EXAMPLE

Amy pushes open a door with a force of 20 N applied to the handle.
The door handle is at a distance of 0.8 m from the hinges.

Write down the formula:	moment = force × distance from pivot
Substitute the values for F and d:	moment = 20 N × 0.8 m
Work out the answer and write down the units:	moment = 16 N m

QUESTIONS

1. What two things does the moment of a force depend on?

2. What is the moment when a force of 4 N is 0.5 m from the pivot?

3. What is the moment when a force of 5 N is 0.25 m from a pivot?

4. A force of 4 N produces a moment of 1.6 N m about a pivot. How far is it from the pivot?

CENTRE OF GRAVITY

The **centre of gravity** is the point at which we can assume all the mass of the object is concentrated. This is a useful simplification because we can pretend gravity only acts at a single point in the object, so a single arrow on a diagram can represent the weight of an object. For this reason, the centre of gravity is sometimes called the **centre of mass**.

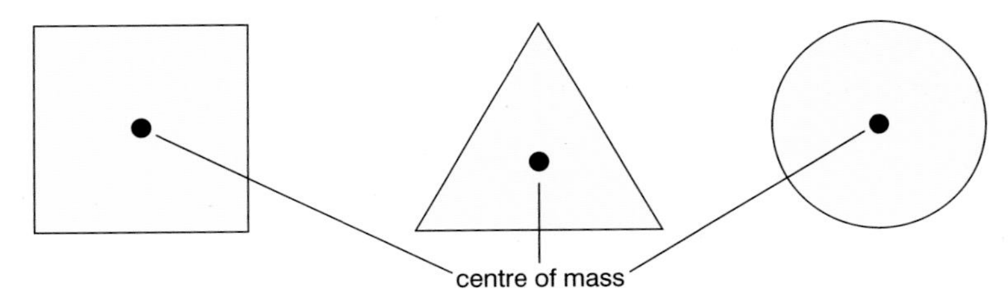

Δ Fig. 1.34 The centre of gravity for objects with a regular shape is in the centre.

What about irregular shapes?

To find the centre of gravity of simple objects, such as a piece of card, follow these steps:

1. Hang up the object.

2. Suspend a mass from the same place.

3. Mark the position of the thread.

4. The centre of mass is somewhere along the line of the thread.

5. Repeat steps 1 to 3 with the object suspended from a different place.

6. The centre of gravity is where the two lines meet.

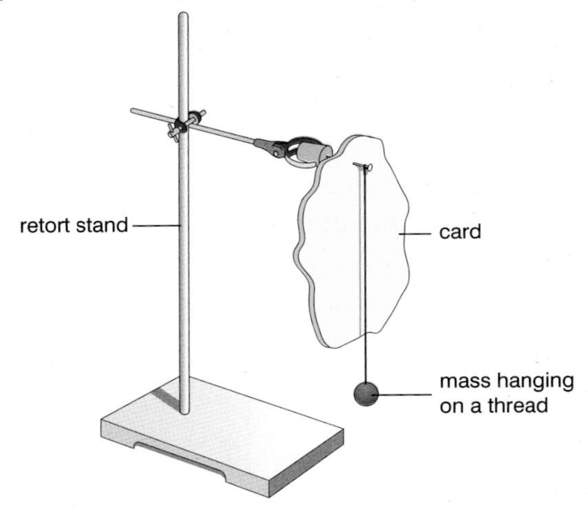

Δ Fig. 1.35 Finding the centre of gravity of a plane object.

CENTRE OF GRAVITY IN AIRCRAFT

The concept of centre of gravity is important in determining whether or not an aircraft is safe to fly. For an aircraft to be safe to fly, its centre of gravity must fall within limits which are set by the manufacturer. The area between these limits is called the CG range for the aircraft. The centre of gravity needs to be calculated before each flight and, if it is not within the CG range, weight must be removed, added (which is rare) or redistributed until the centre of gravity falls within the range.

Part of the calculation of the centre of gravity involves the weights within the aircraft. The weight of fixed parts of the aircraft such as engines and wings does not change and is provided by the manufacturer. The manufacturer will also provide information about the effect of different fuel loads. The operator is responsible for allowing for removable weight such as passengers and crew, and luggage, in the calculation.

Δ Fig. 1.36 An aeroplane taking off.

All aircraft have a maximum weight for flight. If this maximum is exceeded, then the aircraft may not be able to fly in a controlled, level flight. It may be impossible to take off with a given length of runway. Excess weight may make it impossible to climb beyond a particular altitude (so the aircraft may not be able to take its proper course determined by air traffic control).

QUESTIONS

1. What is the centre of gravity of an object?
2. Describe how to find the centre of gravity of an object.
3. If the centre of gravity of an object falls outside its pivot, will the object be stable?

PRINCIPLE OF MOMENTS

If an object is not turning, the sum of the clockwise moments about any pivot equals the sum of the anticlockwise moments about the same pivot. This is the **principle of moments**.

WORKED EXAMPLE

Phil and Ben are sitting on a see-saw. The see-saw is balanced on a pivot. Work out Phil's weight.

Ben is causing the clockwise moment of 400 N × 3 m.

Phil is causing the anticlockwise moment of $W \times 2$ m.

The see-saw is balanced, so, taking moments about the pivot

the sum of the clockwise moments = the sum of the anticlockwise moments

$400 \times 3 = W \times 2$

$W = 600$ N

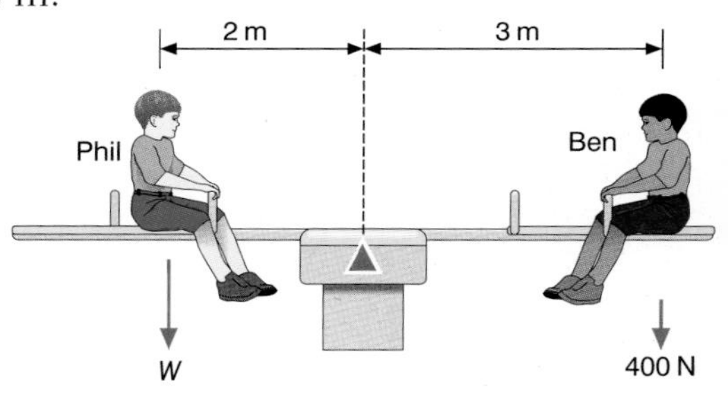

Δ Fig. 1.37 Forces on the see-saw.

CONDITIONS FOR EQUILIBRIUM

The word 'system' describes a collection of objects working together. So in the example of the see-saw, the two children and the see-saw form a system. If a system is not moving in any direction and it is not rotating then it is in **equilibrium**. We already know that for a system not to be moving, the forces on it must be equal and opposite. So:

For a system to be in equilibrium, there must be no resultant force and no resultant turning effect.

In the case of the balanced see-saw, there is no resultant turning effect on the see-saw because the clockwise and anticlockwise turning effects are equal and opposite. In addition, the downward weight of the two children on the see-saw is 1000 N, and the upward force on the see-saw from the pivot must also be 1000 N.

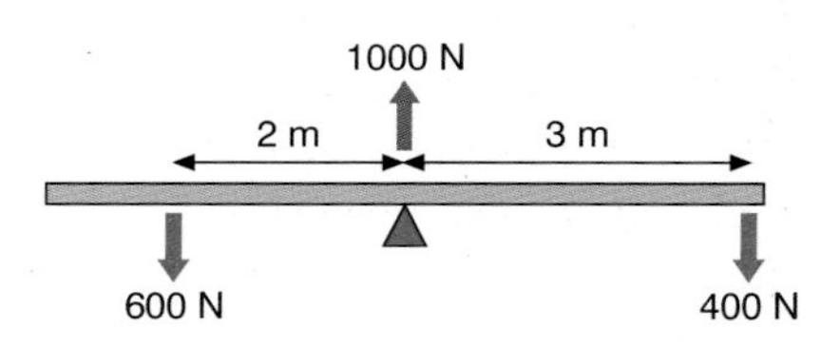

Δ Fig. 1.38 This beam is balanced because the moments about the pivot are equal and opposite, and the downwards and upwards forces are equal and opposite.

REMEMBER

Questions involving balancing can look quite difficult. Work out each moment in turn and add together the moments that turn in the same direction.

FORCES ON A BEAM

Let us look at the forces on a beam that is supporting a load. Scientists often make some approximations that cannot be exactly true, but which make it easier to understand a situation. In this case, assume that the beam is light compared with the load it is supporting, and ignore the weight of the beam completely.

If the beam is in equilibrium and is not moving, then the forces balance. We can immediately say that the downwards force on the beam from the 2 kg mass equals the value of the two upwards forces on the beam from the two supports added together:

$F_1 + F_2 = 20\text{ N}$

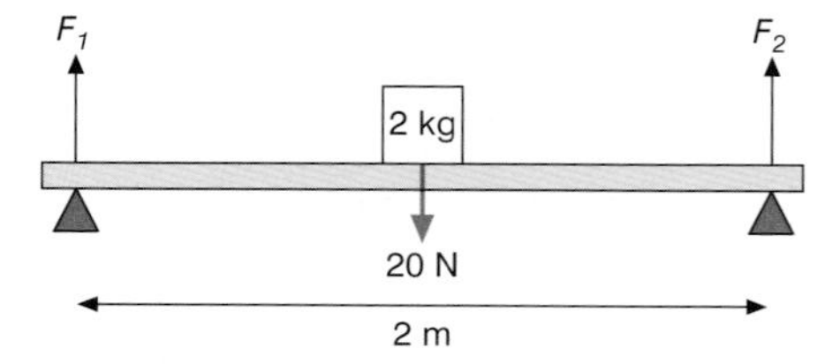

Δ Fig. 1.39 Forces on a beam loaded at its centre.

Note that the question is 'What is happening to the beam?' So we need to consider the three forces that are trying to move the beam. The 2 kg mass is trying to push it downwards, and the two supports are trying to push it upwards. If the mass is put in the middle of the beam, then it is obvious that the two forces are equal and that

$F_1 = F_2 = 10\text{ N}$

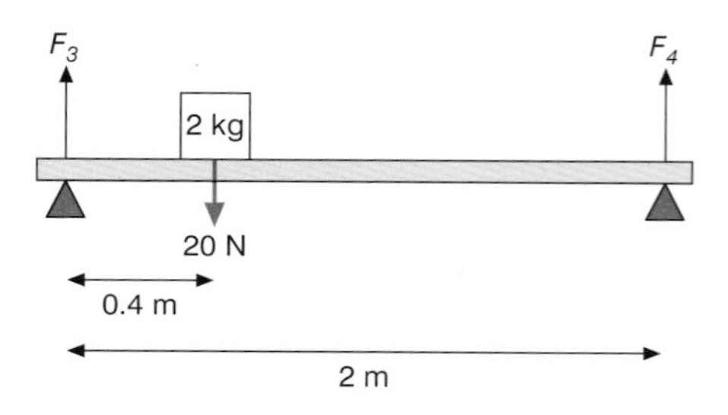

Δ Fig. 1.40 Forces on a beam loaded near one end.

If the mass is towards one end of the beam, then the forces F_3 and F_4 will not be equal. It is still true that $F_3 + F_4 = 20\text{ N}$, but how do we calculate the individual values? The easiest way is to note that the beam is not rotating around the left-hand pivot! That immediately tells us that the anticlockwise moment of F_4 about this pivot equals the clockwise moment of the 20 N about the pivot:

$2 \times F_4 = 0.4 \times 20\text{ N}$

$F_4 = (0.4 \times 20) \div 2 = 4\text{ N}$

because

$F_3 + F_4 = 20\text{ N}$

$F_3 = 16\text{ N}.$

QUESTIONS

1. Two children are sitting on a see-saw. The child to the left of the pivot is sitting x m from the pivot and has a weight of 400 N. The child on the right of the pivot is sitting 2 m from the pivot and has a weight of 300 N. What is the distance x?
2. In the diagram, what is the value of F_1 and F_2? The 4 kg mass is halfway along the beam.

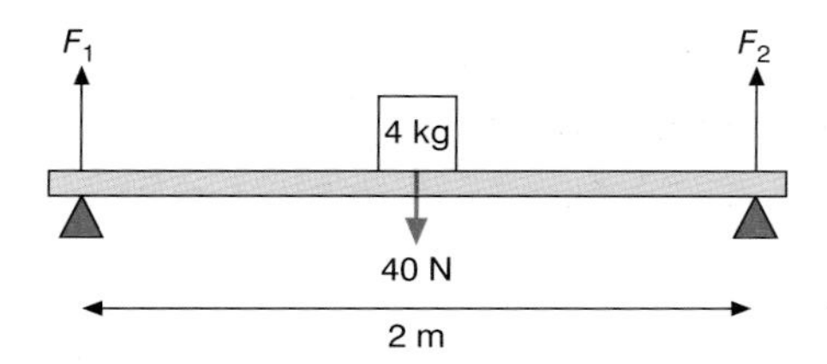

3. In the diagram, find the values of F_4 and F_3.

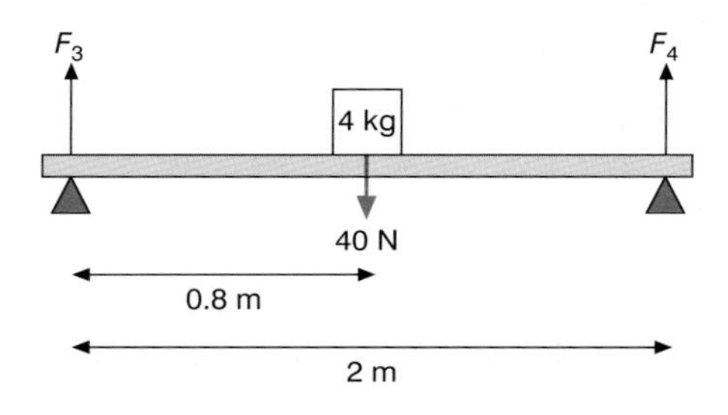

HOW ARE MATERIALS AFFECTED BY STRETCHING?

If masses are added to a length of wire, then the wire will stretch. The top graph in Figure 1.41 on the next page shows how the amount that the wire stretches (the **extension**) varies with the load attached to it. The wire will stretch in proportion to the load up to a certain point, which depends on the material from which the wire is made. Beyond this point, the extension is no longer proportional to the load, and so this point is called the **limit of proportionality**.

A wire on a musical instrument, such as a guitar string, will behave as shown in the graph for wire in Figure 1.41, but will break shortly after the limit of proportionality is reached. This means that, when tuning a string on a musical instrument, we need to take care that we do not tighten the string too much or we risk it breaking.

A piece of rubber stretches quite a lot for small forces. The long polymer molecules are being 'straightened out'. Once this is done, it becomes much stiffer and harder to extend further. However, unless it breaks, it behaves like elastic. This is shown in the graph for rubber.

A copper wire has a large plastic section, seen on the 'plastic flow' graph in Figure 1.41. This is between the limit of proportionality and where the wire breaks. The wire will no longer return to its original length when the load is removed. As it stretches, the wire becomes thinner and thinner until it finally breaks. This stretching is permanent, and the extension is caused by **plastic flow**. This is when planes of atoms slip past each other.

A strip of polythene will stretch relatively easily, but it will scarcely shorten at all when the load is removed. The 'polythene strip' graph shows the force–extension graph for this. This means that polythene stretches almost entirely by plastic flow. This was the original meaning of the word 'plastic'. When people started to invent new materials in the early 1900s, many of them stretched in a plastic way. The word plastic was then used to describe them.

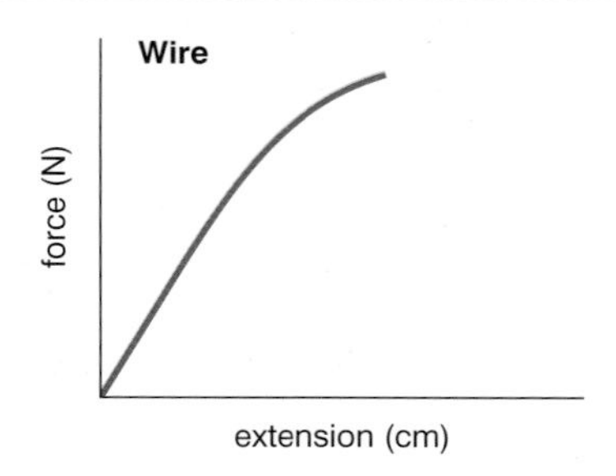

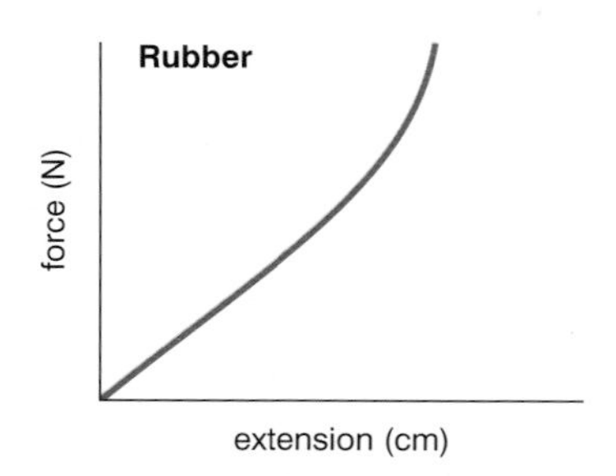

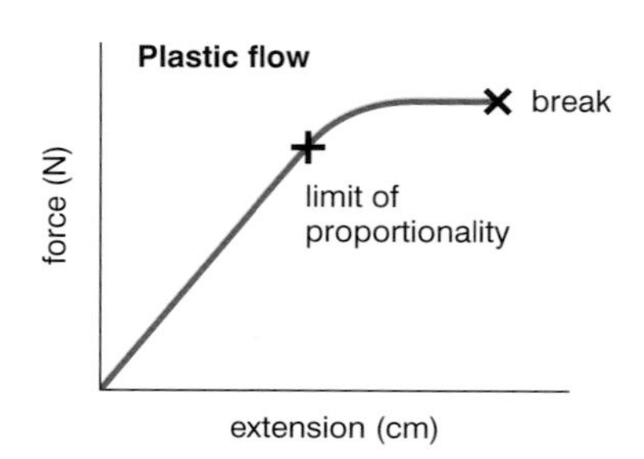

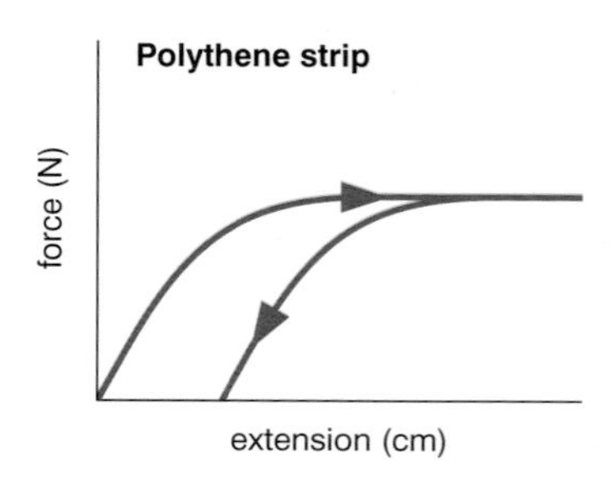

Δ Fig. 1.41 Force–extension graphs for a wire, and rubber, showing plastic flow in copper and in a polythene strip.

A material shows **elastic** behaviour if it returns to its original length when any deforming forces have been removed. During elastic behaviour, the particles in the material are pulled apart a little, so they return to their original positions when the forces are removed.
A material shows plastic behaviour if it remains deformed when a load is removed. During plastic behaviour the particles slide past each other and the structure of the material is changed permanently.

HOOKE'S LAW

On page 50 you saw that, for a wire, there is a section of the force–extension graph which is linear. Like a wire, when a spring stretches, the extension of the spring is proportional to the force stretching it, provided the elastic limit (the point where the spring returns to its original length after the load is removed) of the spring is not exceeded. This is **Hooke's law** and is shown by a straight line on a force–extension graph.

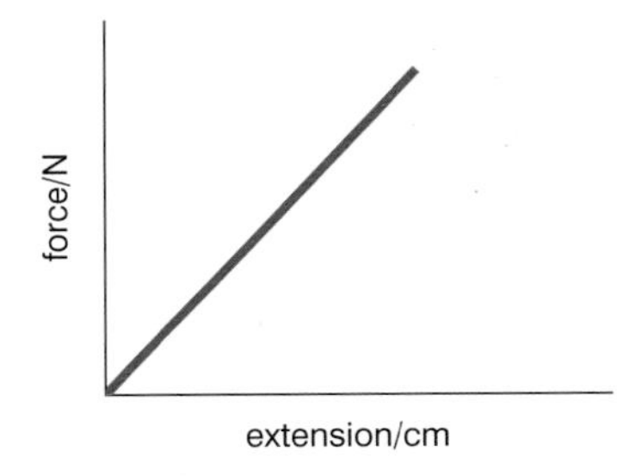

◁ Fig. 1.42 Hooke's law is shown by a straight line on a force-extension graph, for any object being stretched.

The gradient of the line is a measure of the stiffness of the spring.

An experiment to investigate how extension varies with applied force for a spring:

1. Assemble the apparatus (left) and allow the spring to hang down. Measure the starting position on the ruler.
2. Take the first mass, which consists of the hook and base plate, typically of mass 100 g (a weight of 1 N), and hang it on the spring. Measure the new position on the ruler. The difference in the readings is the extension of the spring.
3. Add masses one by one to the first one. Typically each mass is C-shaped, and adds an additional 100 g. Add the masses carefully so that the spring stretches slowly.
4. You should reverse the experiment to see what happens as the masses are removed.

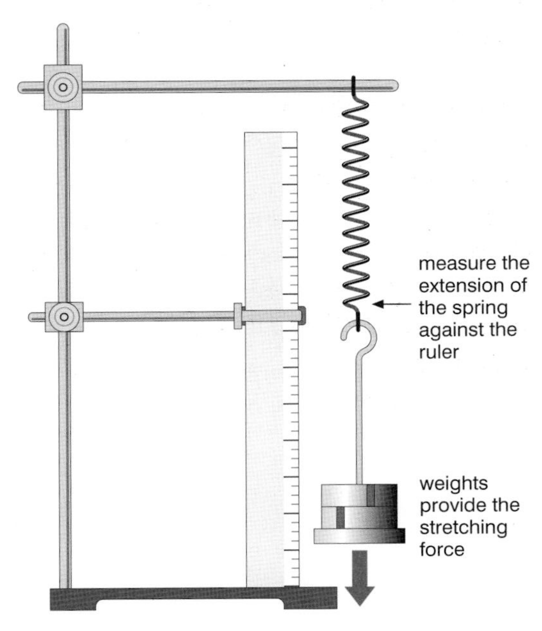

Δ Fig. 1.43 Apparatus to investigate Hooke's law.

Mass (g)	Force (N)	Reading (cm)	Calculate the extension (cm)	Extension (cm)
0	0	15.2	–	–
100	1.0	16.8	16.8 – 15.2	1.6
200	2.0	18.5	18.5 – 15.2	3.3
300	3.0	19.9	19.9 – 15.2	4.7
400	4.0	21.6	21.6 – 15.2	6.4
etc.				

Δ Table 1.1 Results of experiment

5. Calculate the extension, and plot a graph of force against extension.

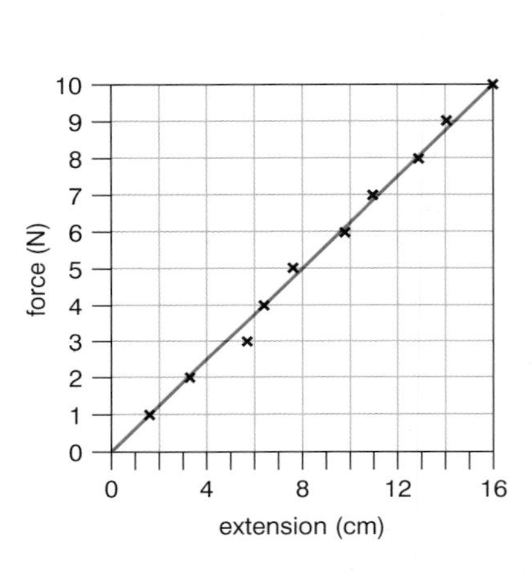

Δ Fig. 1.44 Graph of results.

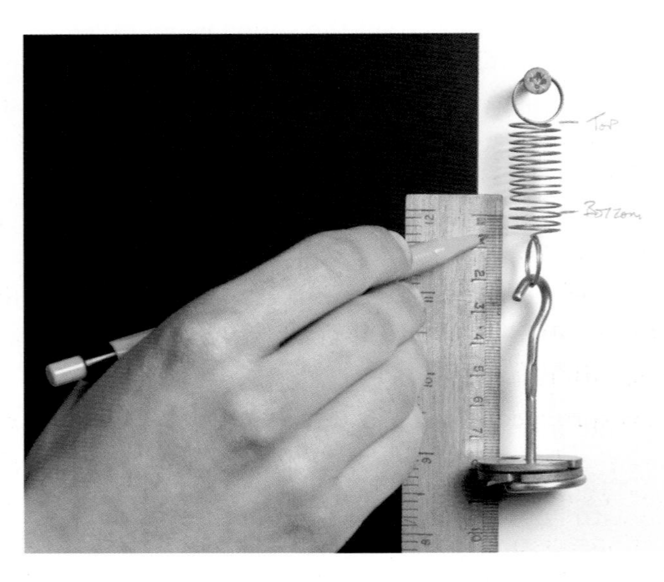

Δ Fig. 1.45 Carrying out the experiment.

A spring that obeys Hooke's law shows **'proportional' behaviour**: the extension of the spring increases in proportion to the load on the spring. It also shows **elastic behaviour** – when the force is removed, the spring returns to its original length.

This method can be used to investigate how extension varies with applied force for any object being stretched, such as a metal wire or a rubber band.

SPRINGS: HOOKE'S LAW IN ACTION

Hooke's law applies to springs that are both extended and compressed by a load, so whenever a spring is used it is an application of Hooke's law. For example, toys that use springs such as jack-in-the boxes, or trampolines which rely on springs returning to their original length and then stretching again to give the 'bounce' required. Anyone who sleeps on a mattress which contains springs also experiences Hooke's law in action on a nightly basis – a mattress whose springs did not return to their original length after being compressed would be rather uncomfortable to sleep on!

EQUATION FOR HOOKE'S LAW

If the spring is stretched too far, the line is no longer straight, and Hooke's law is no longer true. This point at the end of the straight line is known as the limit of proportionality.

If the spring does not stretch too far, it may be elastic and go back to its original length.

However, if stretched beyond the limit of proportionality, different materials can behave in widely different ways.

The equation for Hooke's law is:

force = spring constant × extension of spring or $F = kx$

Where F = force in newtons, k = spring constant in N/m and x = extension of the spring in m.

Note that you are allowed to use a spring constant in N/cm or N/mm, so long as the extension is measured in the same units. You can use this triangle to help you to rearrange the equation. Cover the quantity you want to find and the form of the other two will show you how to write the equation. For example, to find x, cover it and you will see that the equation should be written as $x = F/k$.

This equation works for springs that are being stretched or compressed. The value of k will be the same for both, but note that some springs cannot be compressed (if, for example, the turns of the spring are already in contact).

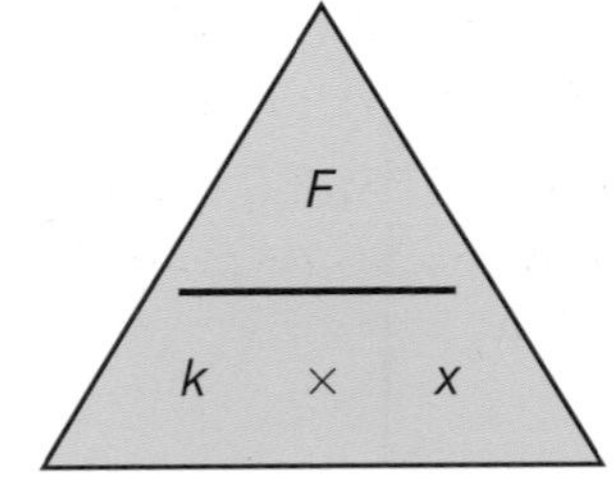

Δ Fig. 1.46 Use this triangle when answering problems on Hooke's law.

WORKED EXAMPLE

A motorbike has a single compression spring on the rear wheels. When the rider sits on the bike, he pushes on the rear wheel with 60 per cent of his weight. If his mass is 50 kg, and the spring constant is 60 N/cm, how much does the spring compress when he sits on the bike?

Write down the formula for the weight of the rider: $W = mg$

Substitute the values for m and g: $W = 50 \times 10$

Work out the answer and write down the units: $W = 500$ N

The force on the rear spring

$= 60$ per cent of 500 N

$= 0.6 \times 500$ N

$= 300$ N

Write down the formula for the compression of the spring: $x = \frac{F}{k}$

Substitute the values for F and k: $x = \frac{300}{60}$

Work out the answer and write down the units: $x = 5$ cm

The spring compresses by 5 cm.

QUESTIONS

1. What force is required to stretch a spring with spring constant 0.2 N/cm a distance of 5 cm?
2. A vertical spring stretches 5 cm under a load of 100 g. Determine the spring constant.
3. A force of 600 N compresses a spring with spring constant 30 N/cm. How far does the spring compress?

Developing investigative skills

A student assembled the apparatus as in Figure 1.44, allowing the spring to hang vertically. Wearing safety glasses, the student measured the initial length of the spring and then measured it again after hanging an additional 100 g onto it. The student continued adding 100 g masses, measuring the length of the spring after each one. The measurements are shown in the table.

Mass added (g)	Force (N)	Length of spring (cm)	Extension of spring (cm)
0		2.0	
100		6.0	
200		10	
300		14	
400		18	
500		22	
600		26	
700		30	
800		34	
900		38	
1000		42	
1100		46	
1200		52	
1300		59	
1400		77	

Devise and plan investigations

❶ Why should the student wear eye protection during this experiment?

❷ Describe how any other safety risks can be minimised.

❸ The student carried out a preliminary experiment before deciding to use 100g masses. Why is a preliminary experiment valuable?

Analyse and interpret

❹ The force stretching the spring is equal to the weight of the 100 g masses that have been added. Use the equation $W = mg$ to calculate the values for the 'force' column.

❺ Use the equation extension of spring = length – original length to calculate the values for the 'extension' column.

❻ Plot a graph of force (on the *y*-axis) against extension (on the *x*-axis)

❼ Use your graph to justify whether or not the spring obeyed Hooke's law.

Evaluate data and methods

❽ The student could not use this spring for a further test of Hooke's law. Explain why not.

❾ The student found it difficult to judge the 'end' of the spring. How could this be improved?

MOMENTUM

Momentum is a quantity of motion. It is a measure of how hard it is to change the motion of an object. Momentum is useful in analysing the motion of objects. It can be calculated (and is defined) using the formula:

momentum = mass × velocity m = mass in kg

momentum = mv v = velocity in m/s

Momentum is usually measured in kg m/s.

WORKED EXAMPLE

Calculate the momentum of a racing car of mass 600 kg travelling at 75m/s.

Write down the formula:	momentum = mass × velocity
Substitute the values for m and v:	momentum = 600 kg × 75 m/s
Work out the answer and write down the units:	momentum = 45 000 kg m/s

REMEMBER

The unit of momentum is quite a tricky one – usually kg m/s. It will often be given to you in the question, but make sure you learn it; if it isn't given in the answer space there will be a mark for it. In other words, if a question does not give a unit beside the space for the answer, then you must insert the unit (if the quantity has one) or you will lose a mark!

QUESTIONS

1. What is the momentum of a 58 g tennis ball travelling at 40 m/s?
2. What is the momentum of a 2000 kg car travelling at 25 m/s?
3. What is the mass of a car that is travelling at 20 m/s and has a momentum of 20 000 kg m/s?
4. What is the velocity of a car that has a mass of 1500 kg and a momentum of 37 500 kg m/s?

CHANGING MOMENTUM

In practical applications, such as safety in cars (see page 58) we are interested in the *change* in momentum. Since velocity is a vector quantity (it has a direction), momentum is also a vector quantity and has a direction. When working out problems, one direction is called the positive direction, for example, going to the right, with the opposite direction taken as the negative direction, for example, going to the left.

For example, the ball hitting a wall as shown in Figure 1.47 will have a positive momentum before it hits the wall and a negative momentum after it hits the wall.

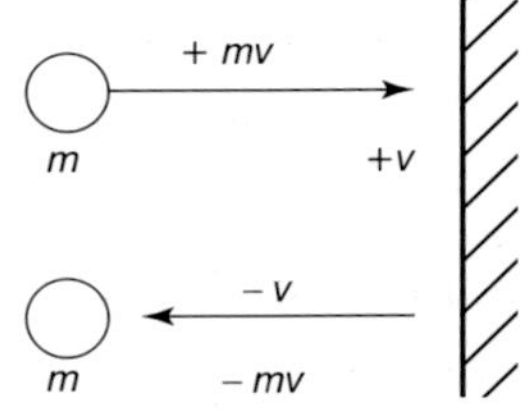

Δ Fig. 1.47 Momentum before and after collision with a wall.

The momentum of an object will change if:

- the mass changes
- the velocity changes
- the direction of the movement changes.

The momentum of an object changes when a force acts on it.

Momentum and Newton's second law of motion

If a net force acts on an object to change its motion, then we can apply Newton's second law. On page 36, you saw that **Newton's second law of motion** is usually given as:

force = mass × acceleration or $F = ma$

where F = resultant force, m = mass of object and a = acceleration

This version works if the mass of the object does not change.

A fuller version of this law is:

force = change in momentum ÷ time taken

$$F = \frac{(mv - mu)}{t}$$

where v = final velocity, u = starting, or initial, velocity, t = time for which the force is acting

This version works if the mass of the object changes.

The resultant force is proportional to how quickly the momentum changes, that is, the *rate of change of momentum*.

REMEMBER

There are a few situations where the mass does change. For example, when a rocket launches the fuel is burned and so the overall mass of the rocket decreases. In situations like this, use two separate values for the mass when you work out the starting momentum and final momentum.

Re-arranging the formula gives:

force = change in momentum ÷ time taken

$$F = \frac{m(v - u)}{t}$$

which leads back to $F = ma$, since acceleration $a = \frac{v - u}{t}$

Re-arranging the formula in a slightly different way gives:

$Ft = mv - mu$

force × time the force is acting = change in momentum

This formula helps to illustrate two important ideas:

1. For a particular force, there is a greater change in momentum if the force acts for a longer time. For example, a footballer who follows through when kicking (continues the movement of the leg after making contact with the ball) makes contact with the ball for longer, so there is a greater change in the ball's momentum and the ball flies faster.

2. For a particular change in momentum, the longer the change takes, the smaller the force will be. For example, when a parachutist lands, he bends his knees. This makes the momentum change take a longer time, so the force on the parachutist is smaller. This is also very important in car safety features such as crumple zones – the car is designed to crumple on impact, which makes the momentum change take place over a longer time and reduces the forces on the occupants.

How do seat belts work?

Momentum changes can be important when considering safety features in cars, such as seat belts. When a vehicle stops suddenly, its passengers tend to keep going until something stops them (which follows on from Newton's first law of motion). Without seat belts, they may stop very suddenly when they hit the windscreen or a passenger in front of them. The seat belt applies a force in the opposite direction to the direction of motion of the vehicle.

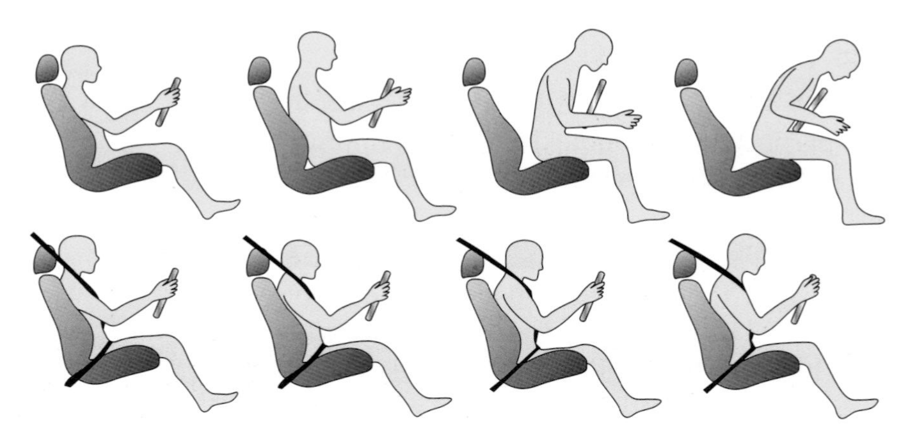

Δ Fig. 1.48 Seat belts are designed to prevent people in vehicles from continuing to move forwards in the vehicle.

Seat belts are designed to stretch a little during a crash. If they did not, they would hold the person in place too strongly, which would make the person stop too quickly and increase the forces. As the seat belt stretches, the momentum change is spread over a longer period of time, which reduces the force experienced by the person. However, because of this stretching, seat belts should be replaced after a collision, since the material will not be able to stretch again in a subsequent accident, so damaged seat belts could cause further injury.

In any collision, the total momentum before the collision is the same as the total momentum after the collision. This is called the **principle of conservation of momentum** and is explored in the following examples.

WORKED EXAMPLES

1. In an experiment, a 2 kg trolley travelling at 0.5m/s crashes into, and sticks to, a 3 kg trolley that is stationary before the crash. Work out the velocity of the pair after the crash.

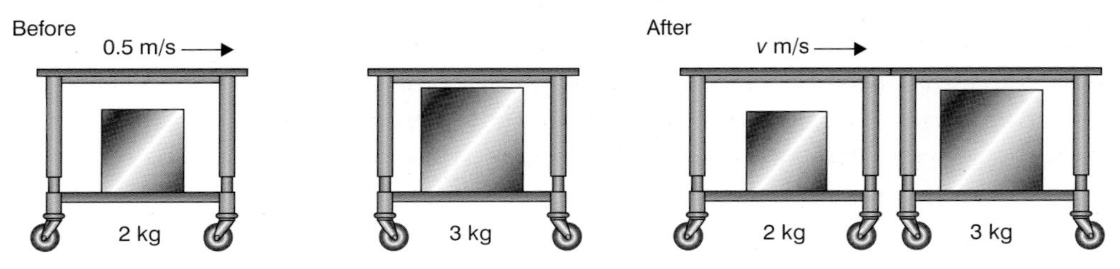

Δ Fig. 1.49 Lab trolleys before and after collision.

Let v = final velocity

Total momentum before collision = total momentum after collision

$(2 \times 0.5) + (3 \times 0) = (5 \times v)$

$1 + 0 = 5v$

Final velocity = 0.2 m/s

2. Two cars collide head-on as shown in the diagram. After the collision both cars are stationary. Calculate the velocity of the 750 kg car before the collision.

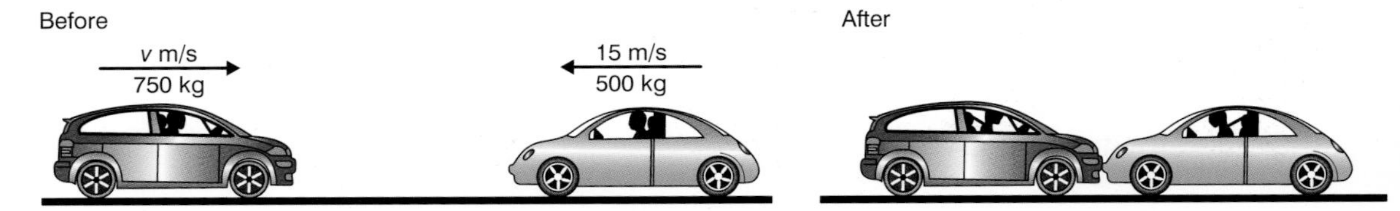

Δ Fig. 1.50 Cars before and after collision.

Let v = required velocity

total momentum before collision = total momentum after collision

$(750 \times v) + (500 \times -15) = (750 \times 0) + (500 \times 0)$

$750v - 7500 = 0$

$750v = 7500$

$v = 10$ m/s

REMEMBER

Newton's second law can make your explanations very specific and clear – it avoids using vague phrases like 'absorbs the impact'.

QUESTIONS

1. Describe how to calculate momentum. Is it a scalar quantity or a vector quantity?

2. Use ideas about change in momentum to explain why a parachutist should bend their knees when they land.

3. Consider two trolleys. Both have a mass of 1 kg. The first trolley has a velocity of 5 m/s and the second trolley is stationary before the collision. They collide and move off together after the collision as shown in the diagram.

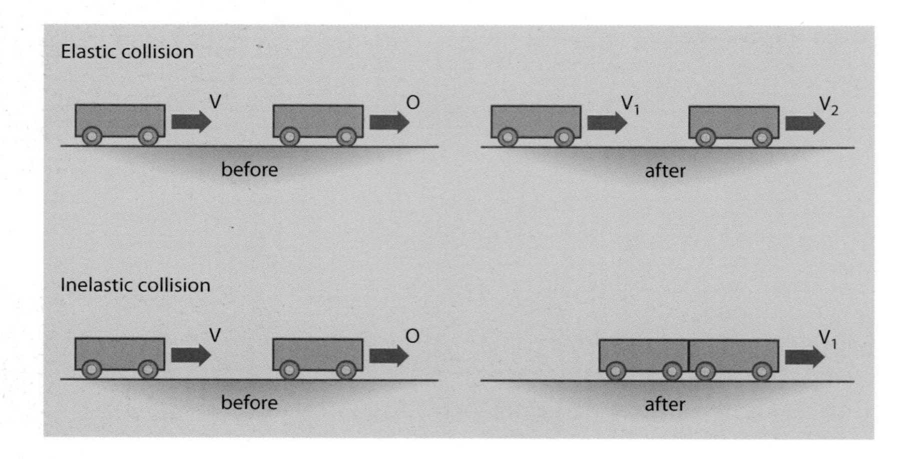

What is the velocity of the two trolleys after the collision?

4. A skater of mass 60 kg moving at 6 m/s collides with, and holds on to, another stationary skater of mass 30 kg. The pair moves off at a velocity *v*. Find *v*.

5. A lorry of mass 3000 kg is travelling at 30 m/s. It collides with a car of mass 1000 kg. They move off together with a velocity of 25 m/s.

a) What is the momentum of the vehicles after the collision?

b) What was the momentum of the vehicles just before the collision?

c) What was the speed of the sports car just before the collision?

6. Goalkeepers in hockey wear extra padding to protect themselves from being hit by fast hockey balls. Explain how the padding reduces injuries from hockey balls. Use ideas about force, momentum and time in your answer.

NEWTON'S THIRD LAW OF MOTION

Newton's third law is quite simple to use. It says that forces always come in pairs that are equal in size and opposite in direction, and act on different objects. For example, a stationary book on a table exerts a force on the table which is equal in size but opposite in direction to the force exerted by the table on the book.

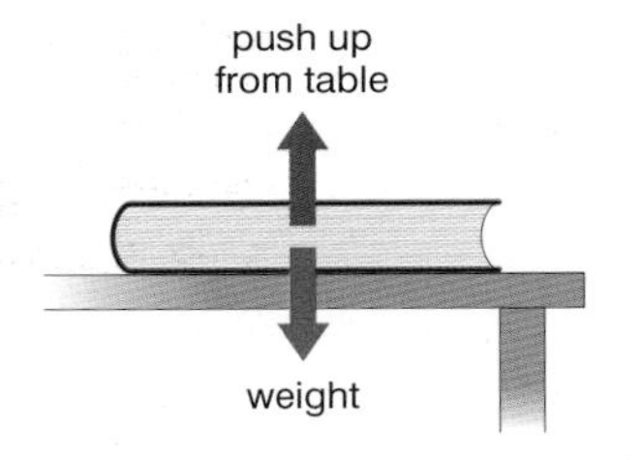

Δ Fig. 1.51 The forces a book.

QUESTION

1. Two ice skaters have different masses and are stationary on the ice. One skater pushes the other using a 200 N force. The skaters move apart with a different acceleration.

   Explain why the skaters move apart with a different acceleration.

ROCKETS: NEWTON'S SECOND AND THIRD LAWS

Rocket launches make use of Newton's second and third laws. Fuel is burned to provide a backwards force on the exhaust gases (force F_E in Figure 1.52), which, in turn, from Newton's third law of motion, which states that every force has an equal and opposite force, produces a forwards force on the rocket itself (force F_F in Figure 1.52).

If the force is constant (from burning the fuel at a constant rate), then the rate of change of momentum will be constant. Since the mass of the rocket is decreasing because the fuel is being burned, then the velocity of the rocket will increase at an increasing rate. So the rocket will not only accelerate, its acceleration will increase as it gets faster, which is exactly what would be predicted from Newton's second law:
force = mass × acceleration

Δ Fig. 1.52 Forces on a rocket.

KEEPING A TRAIN ON THE RAILWAY TRACK

How does a train stay on the track? Your first answer may involve the flanges on the inside of the wheels. Think again, though, and you'll realize this can't be right – there would be a huge amount of friction (and noise!), and the train would have great difficulty going around bends.

The answer lies in the coned shape of the wheels. Imagine the train moves a little way from the centre. The wheel on one side will have moved so that it is rolling on a slightly wider part of the wheel. Similarly, the other wheel will be running on a smaller radius. Since the wheels are rotating at the same speed (they are connected together), this produces a sideways force which pushes the train back towards the centre.

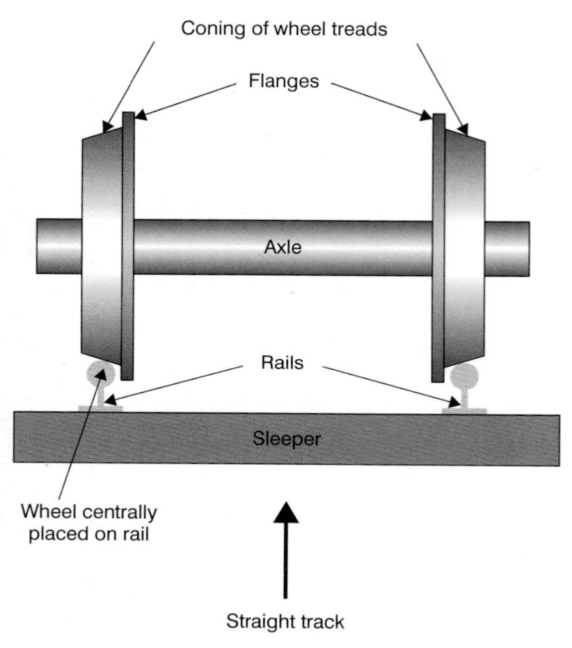

Δ Fig. 1.53 How does a train stay on the track?

When the train goes around a curve, sideways force causes the train to move. This is why train tracks need to move in gentle curves and why straight tracks are needed for high-speed trains such as the *Hayabusa* in Japan and the Beijing–Shanghai link in China.

But how does the train move forward? The track is smooth and the wheels are smooth, but there is still enough friction between the two. As the wheels turn, they try to push the track backwards, causing the train to be pushed forward – this demonstrates Newton's third law. However, using smooth surfaces means that trains cannot climb very steep gradients, and if the track is slippery (with fallen leaves for example) it is difficult to produce enough friction.

Developing investigative skills

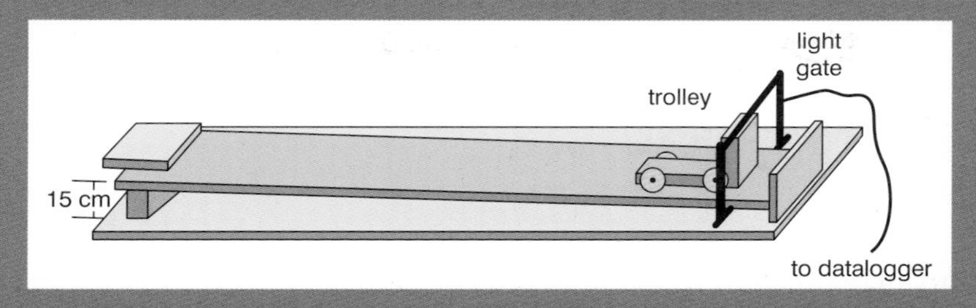

Δ Fig. 1.54 Apparatus for experiment.

A student investigates the motion of a trolley as it rolls down a ramp. They want to find out if there is a link between the starting height of the trolley and its momentum when it reaches the bottom of the ramp. To find the momentum, the student uses a light gate (a beam of light which is broken when an object moves through it) connected to a data logger (a piece of equipment which records measurements (data) automatically). When the trolley passes through the light gate an infrared beam is broken. The data logger records the time the beam was broken for and uses this to calculate the speed of the trolley. The student's data is shown below.

Height of ramp (cm)	Momentum of trolley at the bottom (kg m/s)
15	1.1
20	1.3
25	1.5
30	1.6
35	1.8
40	2.0

Devise and plan investigations

❶ How could the student measure the starting height of the ramp as accurately as possible using a ruler marked in cm?

❷ To calculate the momentum, the student needed to make another measurement of the trolley. State what that measurement was and describe how it could have been measured.

❸ What was the independent variable in this investigation? What was the dependent variable?

Analyse and interpret

❹ Draw a graph of the student's results.

❺ Describe the pattern (if any) shown by the graph.

Evaluate data and methods

❻ Do the data points of the graph suggest that the line should go through the origin?

❼ Using ideas about momentum, explain whether you would expect the line to go through the origin.

❽ Suggest what further measurements the student should make. Explain your answer.

End of topic checklist

Resultant **force** = mass × acceleration

Weight = mass × gravitational field strength

Momentum = mass × velocity

Force = change in momentum/time taken

Moment = force × perpendicular distance from the pivot

The facts and ideas that you should understand by studying this topic:

- ❍ Know the effects that forces have on the motion or shape of objects.
- ❍ Know the difference between vector quantities and scalar quantities.
- ❍ Know and use the relationship between resultant force, mass and acceleration.
- ❍ Know and use the relationship between weight, mass and gravitational field strength.
- ❍ Know and use the relationship between momentum, mass and velocity.
- ❍ Apply ideas about rate of change of momentum to safety features.
- ❍ Use momentum calculations to predict what happens in a collision.
- ❍ Know and use the relationship between moment, force and perpendicular distance from the pivot.
- ❍ Know and use the principle of moments for a simple system of parallel forces.
- ❍ Understand how to calculate how the upwards forces on a beam, supported at both ends, vary with the position of the object on it.
- ❍ Know that the weight of an object acts through the object's centre of gravity.
- ❍ Describe the forces that affect falling objects.
- ❍ Describe factors that affect the stopping distance of vehicles, including speed, mass, road conditions and reaction time.
- ❍ Be able to investigate how extension varies with force for springs, metal wires and rubber bands.
- ❍ Understand that force is a vector quantity.
- ❍ Understand that friction is a force that opposes motion.
- ❍ Understand how falling objects reach a terminal velocity.

- ❍ Understand the concept of centre of gravity and how it can be used in situations involving the turning effect of forces.
- ❍ Understand how Newton's third law can be applied to motion, such as of rockets taking off.
- ❍ Understand how Hooke's law is related to the linear region of a force-extension graph.
- ❍ Describe what is meant by elastic behaviour.

End of topic questions

1. An astronaut has a mass of 60 kg when she gets into her spacecraft on Earth.

a) What is her weight on Earth? **(1 mark)**

The astronaut travels to Mars. Note that the gravitational field strength on the surface of Mars is 3.8 N/kg.

b) What is her mass on the surface of Mars? **(1 mark)**

c) What is her weight on the surface of Mars? **(1 mark)**

d) If she stands in one pan of a large balance, what masses would be needed in the other pan to balance her? **(1 mark)**

e) If she stands on bathroom scales (which are a type of spring balance) what would be the reading in newtons? **(1 mark)**

2. The height that you can jump has an inverse relationship to gravitational field strength. So if the field strength doubles, the height halves. If the Olympic Games were held on Mars in a large dome to provide air to breathe, what would happen to the records for:

a) weightlifting (weight in N) **(2 marks)**

b) high jump (height) **(2 marks)**

c) pole vault (height) **(2 marks)**

d) throwing the javelin (distance) **(2 marks)**

e) the 100 m race (time)? **(2 marks)**

In every case, describe what is likely to happen to the record, choosing between:

i) increase, ii) stay similar or iii) decrease and explain your choice.

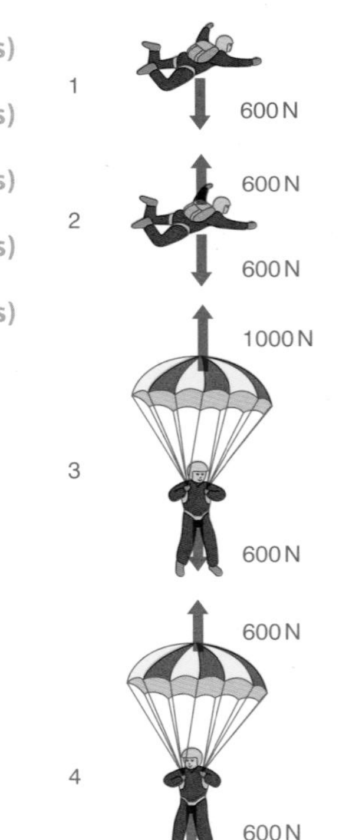

3. The diagram shows the stages in the descent of a skydiver.

a) Describe and explain the motion of the skydiver in each case. **(4 marks)**

b) In stage 5 explain why the parachutist does not sink into the ground. **(1 mark)**

4. Masood performed an experiment stretching a spring. He loaded masses onto the spring and measured its extension. Here are his results.

Extension (cm)	0	4	8	12	16	20	24
Load (N)	0	2.0	4.0	6.0	7.5	8.3	8.6

a) On graph paper, plot a graph of load (vertical axis) against extension (horizontal axis). Draw a suitable line through your points. **(2 marks)**

b) Mark on the graph the region where Hooke's law is obeyed. **(1 mark)**

c) How does Masood check whether the spring, after being loaded with 8.6 N, has shown purely elastic behaviour? **(1 mark)**

5. The manufacturer of a car gave the following information:

Mass of car 1000 kg. The car will accelerate from 0 to 30 m/s in 12 seconds.

a) Calculate the average acceleration of the car during the 12 seconds. **(1 mark)**

b) Calculate the force needed to produce this acceleration. **(1 mark)**

6. A rocket is lit and the fuel burns inside. It sends out hot gas downwards from the bottom of the rocket. Use Newton's third law to describe and explain the motion of the rocket until all the fuel is burned. **(4 marks)**

7. A flag is being blown by the wind. The force on the flag is 100 N and the flagpole is 8 m tall. Calculate the moment of the force about the base of the flagpole. **(1 mark)**

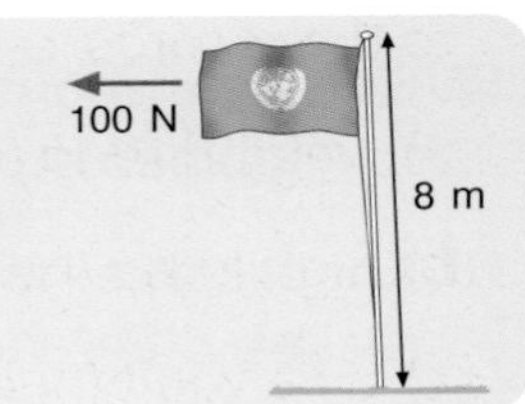

8. Seatbelts in cars, use a material that stretches in a crash.

a) How does this material help reduce injuries in a crash? **(2 marks)**

b) Why do seatbelts have to be changed after a crash? **(1 mark)**

c) Name two other car safety features that work in the same way as seatbelts. **(2 marks)**

9. Rod and Ahmed are sitting on a see-saw. The see-saw is not balanced. Rod and Ahmed have a friend, Freddy, who weighs 300N. Where should Freddy sit in order to balance the see-saw? **(2 marks)**

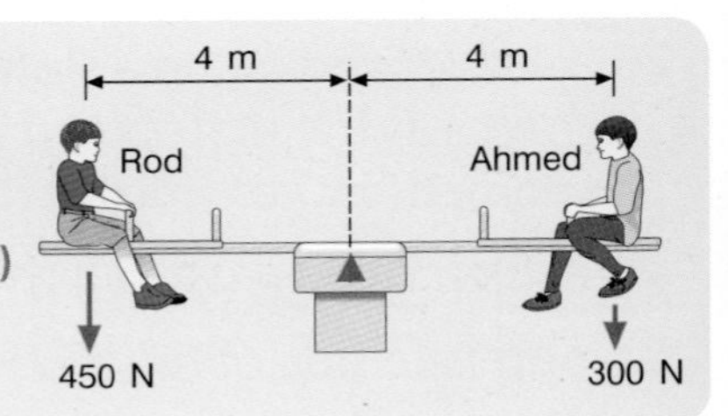

10. Velocity is a vector quantity – it has a magnitude and a direction. So your velocity changes if you change speed and/or direction. Explain why acceleration must also be a vector quantity. **(1 mark)**

End of topic questions continued

11. Crash dummies are used in car crashes to study the forces involved in car safety. A test car hits a wall at 35 kmph. The crash dummy is not wearing a seatbelt. Explain using ideas about force and motion, what happens to the dummy in the crash. **(3 marks)**

12. A hot air balloon rises vertically at a steady speed. Are the forces on it balanced or unbalanced? Explain your answer. **(2 marks)**

13. A football slows down as it rolls along on some grass. Are the forces on it balanced or unbalanced? Explain your answer. **(2 marks)**

14. A car of mass 1500 kg sets off from rest and reaches 20 m/s in 10s. Calculate the resultant force on the car. Why is this force an average? **(2 marks)**

15. In question 14, if the average frictional forces (from the road and the air) are 2000 N, find the force produced by the engine. **(1 mark)**

16. During the Apollo missions, some astronauts took golf clubs to the Moon and hit a golf ball. Explain carefully why the ball flew further than it would on Earth. **(3 marks)**

17. Use ideas about terminal velocity to explain why raindrops do not cause serious injury when they land on you. **(3 marks)**

18. Choose three factors that affect the stopping distance of a car. Explain why they make a difference. **(3 marks)**

19. A car of mass 500 kg accelerates from 15 m/s to 30 m/s.

a) Calculate the change in momentum of the car. **(1 mark)**

b) If the car takes 5 seconds to make this change, calculate the net force required. **(1 mark)**

c) The actual force provided by the car engine will be larger than the value calculated in part **b)**. Explain why. **(1 mark)**

20. Jack is standing on a stationary skateboard. When he steps off the skateboard forwards the skateboard moves backwards. Use the principle of conservation of momentum to explain why this happens. **(2 marks)**

21. Explain why it is easier to push open a door when you push on a point at the opposite side of the hinge. **(2 marks)**

22. Look at the 2016 world records for different races.

Distance (m)	Time (s)	Average speed in m/s
100	9.58	
200	19.19	
400	43.03	
800	100.91	
1000	131.96	

a) Complete the table by calculating the average speeds in m/s. **(5 marks)**

b) Suggest why the average speeds decrease as the races get longer. **(2 marks)**

23. Does a piece of rubber follow Hooke's law when it is being stretched? Explain your answer. **(2 marks)**

24. If you were choosing the springs for the suspension of a sports car, would you choose very stiff springs or very 'soft' springs? Explain your answer. **(2 marks)**

25. A toy car moves across the floor. Look at the speed–time graph for the car as it moves across the floor. Describe in detail the motion of the car and the distance it moves. **(6 marks)**

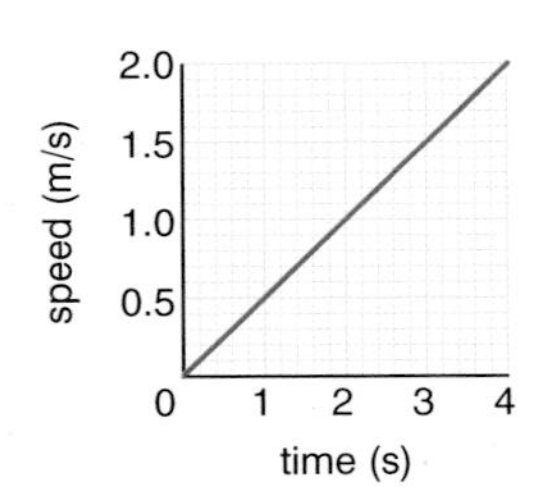

Exam-style questions

Sample student answers

Question 1

The diagram shows a car travelling to the right.

The arrows represent four forces on the car as it moves.

The arrows are **not** drawn to scale.

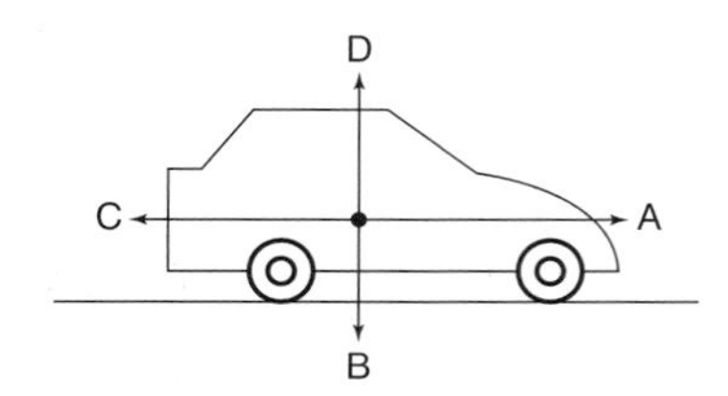

a) i) Which arrow represents the weight of the car?

(1)

ii) Which arrow represents the friction acting on the car?

C ✓ (1) **(1)**

iii) Describe **two** causes of friction on the car as it moves.

1. *the drag on the car as it moves* ✗ **(2)**

2. *air resistance as the car pushes through the air* ✓ (1)

b) The horizontal forces on the car are unbalanced.

i) State the equation linking unbalanced force, mass and acceleration.

F = ma ✓ (1) **(1)**

ii) The car accelerates at 2 m/s^2 and it has a mass of 1500 kg.

Calculate the magnitude of the unbalanced force on the car.

$1500 \times 2 = 3000$

force = 3000 N ✓ ✓ (2) **(2)**

iii) Which direction does this force act in?

Horizontal ✗ **(1)**

EXAMINER'S COMMENTS

1) a) i) and **ii)** The first two questions are straightforward, recalling and applying simple ideas in a familiar situation. Two marks given.

iii) Only one mark is awarded. The first answer given is mostly a copy of the question ('on the car as it moves') so the only new knowledge is the word 'drag'. On its own, this does not really answer the question – it gives a name, but it does not really give the cause of the friction: that is, what is in contact with what. 'Friction between the wheels and the road' would have gained the second mark.

b) i) and **ii)** The student has handled the mathematical parts of the question correctly.

iii) The student almost certainly knew which direction the force was acting in (forwards) but loses the mark because 'horizontally' has two possibilities – forwards or backwards. This is an easy slip to make, and the student might have noticed the mistake if they had time to check through their answers.

iv) The student has missed the point of the question. It is still in part b) and that is the clue that the answer is still related to $F = ma$. The question asks about forces, so the answer should also refer to forces (the calculated value is a resultant force made up from the engine force minus the friction. Instead, the student talks about 'speeding up' and 'energy losses' which are too vague to gain credit here.

c) 'Bigger' is too vague to gain credit. The examiner was looking for 'larger mass' – the question is still linked to $F = ma$. The student does then go on to link force to acceleration correctly so gains the second mark. The third mark is also gained because the idea of more air resistance is a sensible suggestion (and it is a 'suggest' question) – it also makes it clearer what the student meant by writing 'bigger' at the start of the sentence. Overall, however, the examiner has to link the student's ideas together more than they should need to.

Exam-style questions continued

iv) The force provided by the car engine is larger than the value you have calculated. Explain why.

The car is speeding up and energy is lost to the surroundings, so the engine has to give a bigger force. ✗ ✗ **(2)**

c) A truck tries to keep up with the car as it accelerates.

The truck falls behind, even though the engine in the truck provides a bigger force than the engine in the car.

Suggest why the truck falls behind.

The truck is bigger, so it has to have a bigger force to get the same acceleration because there will be more air resistance. ✗ ✓ ✓ **(3)**

(Total 13 marks)

Exam-style questions continued

Question 2

In 2009, Usain Bolt set the world record for 100 m at 9.58 s.

a) i) State the equation linking average speed, distance moved and time taken. **(1)**

ii) Calculate the average speed for Usain Bolt's run.

Give your answer to an appropriate number of significant figures. **(3)**

iii) At some part of the race, Usain Bolt must have run faster than this.

Explain why. **(3)**

b) At the start of the race, it is counted as a false start if a runner moves before 0.10 s ***after*** the starter has fired the starting gun.

Suggest why. **(3)**

(Total 10 marks)

Question 3

The graph shows how the velocity of a cyclist varies.

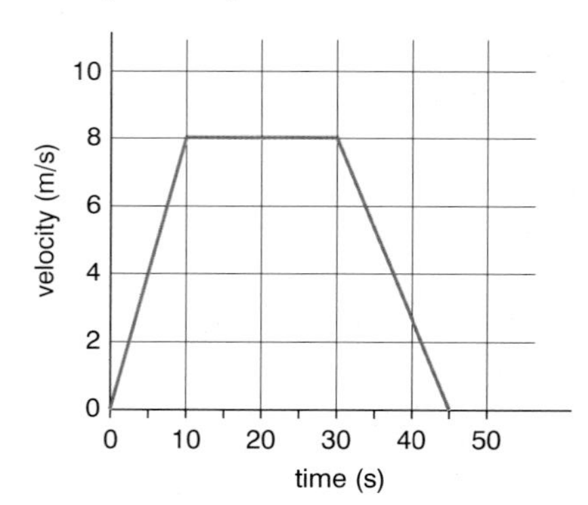

a) i) Between what times was the cyclist travelling at a constant speed? **(1)**

ii) How can you tell from the graph? **(1)**

b) Calculate the acceleration of the cyclist during the first 10 s. **(3)**

c) Use the graph to find the total distance travelled by the cyclist. **(3)**

(Total 8 marks)

Exam-style questions continued

Question 4

A student is planning to investigate the speeds of cars going past their school. The speed limit outside the school is 14 m/s. She thinks that more than 50% of these cars are breaking the speed limit.

Describe how the student should carry out her investigation. You should include:

a) A list of the equipment she should use **(2 marks)**

b) One safety precautions needed for the investigation **(1 mark)**

c) What measurements she should take and how she should take them. **(4 marks)**

d) A blank results table with headings and units. **(3 marks)**

e) How she can use her measurements to confidently draw a conclusion and test her prediction. **(4 marks)**

(Total 14 marks)

Question 5

Tom and Alex are standing still facing each other.

They are both wearing roller skates.

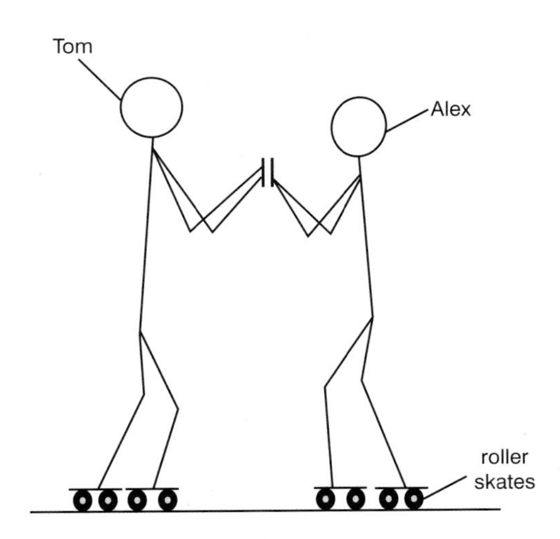

a) Tom pushes Alex away from him.

Tom is surprised to find that he moves backwards.

Use ideas about the conservation of momentum to explain why Tom moves backwards. **(4)**

b) Alex has a mass of 50 kg.

He moves away from Tom at 1.5 m/s.

i) State the equation linking momentum, mass and velocity. **(1)**

ii) Calculate Alex's momentum after he has been pushed.

State the correct unit. **(3)**

iii) What is Tom's momentum at the same time? **(2)**

iv) Tom moves backwards at 1.2 m/s.

Calculate Tom's mass. **(2)**

c) Tom moves because there is a force on him.

Use Newton's third law to explain how the force on Tom is linked to the force on Alex. **(2)**

(Total 14 marks)

Question 6

A crane at a building site is lifting a container.

The point labelled C is the centre of gravity of the container.

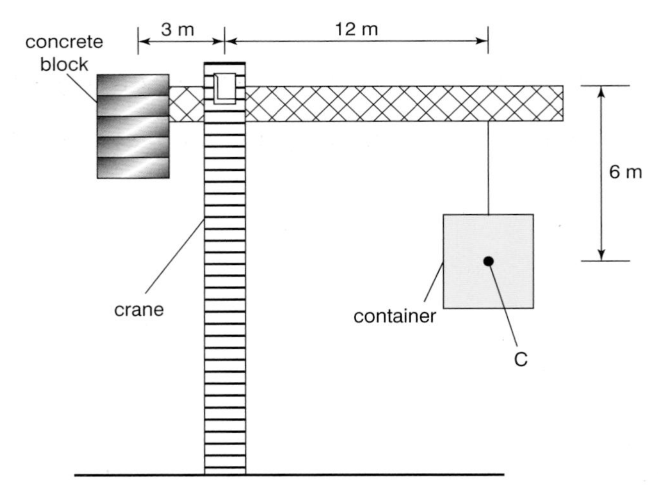

a) The container has a mass of 2000 kg.

i) State the equation linking weight, mass and g. **(1)**

ii) Calculate the weight of the container. **(2)**

b) i) Use the principle of moments to explain the purpose of the concrete block. **(3)**

ii) In the diagram, the crane is balanced.

Calculate a suitable value for the mass of the concrete block. **(4)**

(Total 10 marks)

Exam-style questions continued

Question 7

A student investigates stretching a spring.

The student hangs 100 g masses onto the spring one at a time and measures the length of the spring.

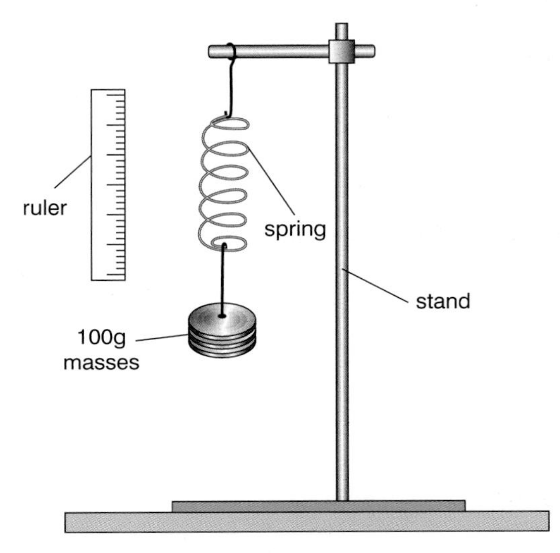

a) Describe how the student should use the ruler to obtain accurate measurements for the length of the spring. **(3)**

b) Suggest one safety precaution the student should take during the investigation. **(1)**

c) The student calculates the extension in the spring for each weight attached.

The table shows the results.

Weight (N)	Extension (cm)
0	0.0
1	3.1
2	6.0
3	8.8
4	13.7
5	14.9

i) Draw a graph of these results. **(5)**

ii) Is there any evidence of an anomalous measurement?

Explain your answer. **(2)**

iii) Does the evidence from this experiment support Hooke's Law?

Explain your answer. **(3)**

(Total 14 marks)

Electricity is something that many people use every day. It is used for lighting, heating, cooking and powering many different items of equipment. However, it can be dangerous. This section begins by looking at how to stay safe when using electricity. You will then learn the mathematical relationships behind electricity, which will be useful when you have your own home and have electricity bills to pay! You will also learn that the electricity supplied by electricity companies has one form (alternating current) while most applications of electricity that you will use need direct current.

You will then move on to find out about electrical circuits and how it is possible to calculate one quantity in a circuit if you know others. Finally, you will learn about electric charge, without which there would be no electricity.

You will already have explored different models to explain the flow of current and the transfer of electrical energy, and will have evaluated these models. You will also have investigated current and voltage in circuits and drawn conclusions from data. You should also be able to explain, using data and a simple model, the differences between series and parallel circuits.

STARTING POINTS

1. Two lamps, A and B, are connected in series with a battery. Lamp A is unscrewed from its holder, which causes a break in the circuit. What will happen to lamp B? Why?
2. Two lamps, A and B, are connected in parallel with a battery. Lamp A is unscrewed from its holder. What will happen to lamp B? Why?
3. What happens to electric current as it goes round a circuit?
4. A series circuit consists of two ammeters, A1 and A2, and two bulbs connected to a battery. Which ammeter will have the biggest reading? Why?
5. Name some safety devices used with mains electricity. Can you explain how each one works?
6. Suggest why different electrical appliances in the home use different voltages and make a list of some common examples and their voltages.

CONTENTS

2
Electricity

Δ Electricity is now an essential part of our lives - we rely on it to operate many devices.

Units

INTRODUCTION

For the topic of electricity, you will need to be familiar with:

Quantity	Unit	Symbol
charge	coulomb	C
time	second	s
current	ampere, amp	A
energy	joule	J
potential difference (voltage)	volt	V
power	watt	W
resistance	ohm	Ω

Some of these you have met previously but the new units will be defined when you meet them for the first time.

LEARNING OBJECTIVES

✓ Use the following units: ampere (A), coulomb (C), joule (J), ohm (Ω), second (s), volt (V) and watt (W).

Mains electricity

INTRODUCTION

Mains electricity is a clean and effective method of generating heat and movement. It is incredibly useful if used properly. When a domestic appliance (such as a washing machine or a fridge) is switched on, a circuit is completed between the local substation and the appliance. Electrical energy travels from the substation to the appliance through the 'live' and 'neutral' wires. Some appliances have a third wire, the 'earth' wire. You will also meet the American word 'ground' instead of 'earth'. This wire does not normally carry any current, but it is there for safety.

Δ Fig. 2.1 Electricity travels from power stations to our homes on pylons like these.

KNOWLEDGE CHECK

✓ Know some advantages of using mains electricity to transfer energy.
✓ Know some safety precautions to take when dealing with mains electricity.

LEARNING OBJECTIVES

✓ Understand how insulation, double insulation, earthing, fuses and circuit breakers protect a circuit or device in a range of domestic appliances.
✓ Understand how electrical resistance leads to a heating effect, and describe how this can be used.
✓ Understand the difference between a d.c. supply and an a.c. supply.
✓ Know and use the relationship between power, current and voltage: $P = I\,V$ and apply the relationship to choosing fuses.
✓ Be able to calculate how much energy is transferred in a given time, using the relationship $E = I\,V\,t$.

If there is a fault in an electrical appliance, it could take too much electrical **current**. This might make the appliance itself dangerous, or it could cause the flex between the appliance and the wall to become too hot and start a fire.

INSULATION, FUSES AND CIRCUIT BREAKERS

There are several ways to make appliances safer to use, and protect the device or user if a fault should develop. In some countries the **fuse** is fitted into the plug of the appliance. The fuse fits between the live brown wire and the pin. The brown live wire and the blue neutral wire carry the current. The green and yellow striped earth wire is needed to make metal appliances safer, as we shall see later.

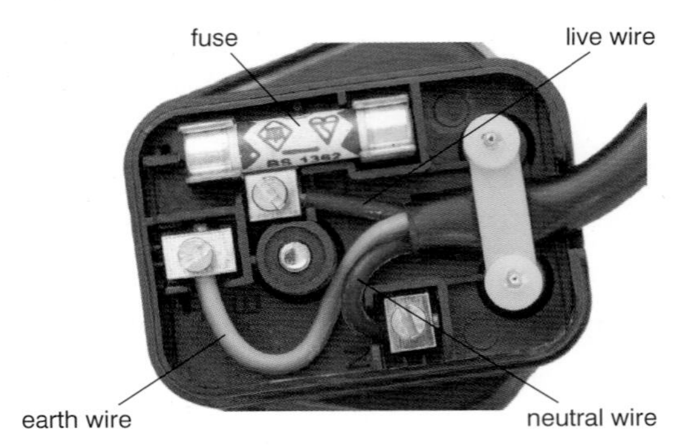

∆ Fig. 2.2 A three pin plug has a built-in safety device, the fuse.

The laws for the safe use of electricity are constantly being improved by governments, and electricians learn to work to the latest standards. The most important aids to the safe use of electricity are **insulation** and fuses or **circuit breakers**.

Insulation these days is usually a plastic such as PVC, which is used to cover the copper wires. This prevents them from touching each other, and also prevents the operator from touching them. In parts of objects where the temperature goes above 100 °C, other plastics, glass or ceramic are used.

The electric current usually has to pass through a fuse or circuit breaker before it reaches the appliance. If there is a sudden surge in the current, the wire in the fuse will heat up and melt – it 'blows'. This breaks the circuit and stops the current. If a circuit breaker is used, then the circuit breaker springs open (trips) a switch if there is an increase in current in the circuit. This can be reset easily after the fault in the circuit has been corrected.

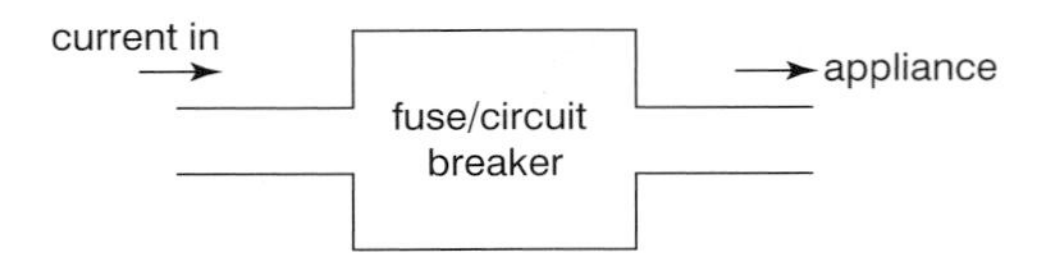

∆ Fig. 2.3 How a fuse or circuit breaker protects an appliance.

REMEMBER

The fuse does not 'provide' current, nor does it 'allow' a certain amount of current to go through. It is just a wire that melts if the current gets too high.

In all houses with mains electricity, there is a distribution box that takes all of the electricity for the house and sends it to the different rooms. In old houses this box may still use fuses, but in modern installations the box uses miniature circuit breakers, often known as MCBs.

Where a fuse is fitted to the plug, it must have a higher current rating than the appliance needs, but it should have the smallest current rating available above this. For fuses in the plug, the most common fuses are rated at 3 A, 5 A and 13 A. Any electrical appliance with a heating element in it should be fitted with at least a 13 A fuse. An appliance working at 3.5 A should have a 5 A fuse.

Metal-cased appliances, such as washing machines or electric cookers, must have an **earth wire** as well as a fuse. If the live wire worked loose and came into contact with the metal casing, the casing would become live (at a high voltage). If the user touched the live case, current could pass through their body and they could be electrocuted. The earth wire is connected to the ground outside and is at 0 V. It has a much lower **resistance** than the human body, so the current will go directly to earth via the earth wire rather than via the user.

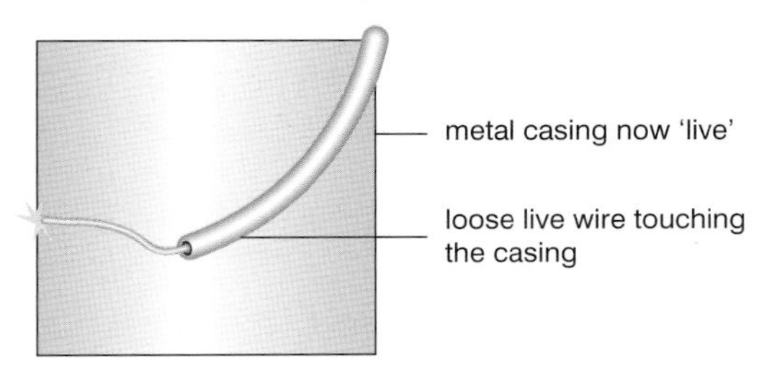

Δ Fig. 2.4 The earth wire provides a very low resistance route to the 0 V earth – usually water pipes buried deep underground.

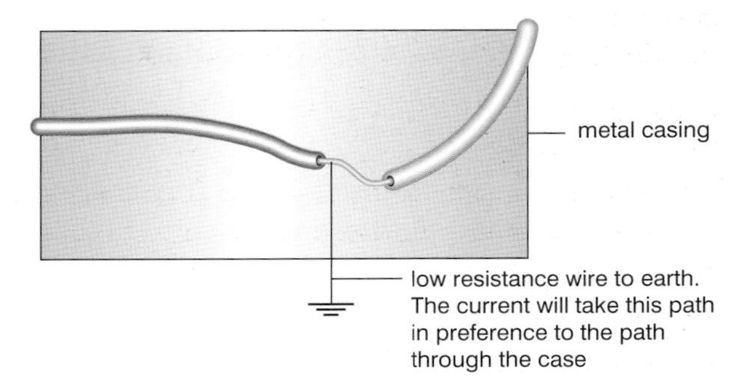

Δ Fig. 2.5 An earth wire provides a path for current to ground.

This low resistance means that a large current passes from the live wire to earth, causing the fuse to melt and break the circuit. If the earth wire is not fitted correctly, or if it has broken, the appliance will be extremely dangerous! If there is any doubt about the earthing of the appliance, or of the whole house, it must be checked by an electrician.

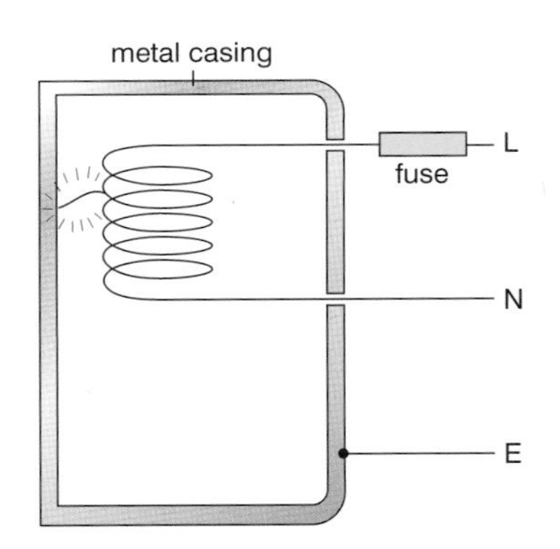

◁ Fig. 2.6 The earth wire and fuse work together to make sure that the metal outer casing of this appliance can never become live and electrocute someone.

Appliances that are made with plastic casing such as toasters do not need an earth wire. The plastic is an **electrical insulator** and so can never become live. Appliances like this are said to be **double insulated**.

In some situations people may be unexpectedly exposed to electricity: for example, using an electric drill, especially drilling into a wall with hidden power cables, or using power tools out of doors, perhaps in wet conditions). In these cases, a residual current circuit breaker (RCCB) must be used in the power socket on the wall. If any of the electricity starts to leak out, the RCCB will turn off the power in 30 ms or less. The RCCB cannot be guaranteed to save the user's life, but it gives a much better chance of surviving.

QUESTIONS

1. Describe the wiring of a three-pin plug. You should explain what each of the wires in the plug does, and the colour of the insulation.
2. Explain the function of a fuse.
3. A student wants to run an appliance that requires a current of 6 A. They choose a fuse of 5 A 'because it's the nearest available'. Explain why this is not a good choice.
4. The earth wire connection to the ground is usually a thick piece of copper wire. Explain why.
5. Explain why appliances with plastic casing do not need to be earthed.

ELECTRICAL HEATING AND CURRENT IN A RESISTOR

Many household appliances consist of an electrically heated **resistor**. A resistor is a device that opposes the current. As it does so, its temperature increases. Examples of appliances that have such a device are electric kettles, electric fires, light bulbs, domestic irons and electric ovens. Even the washing machine, the dishwasher, the tumble drier and the hairdryer consist of an electric heater with an electric motor added.

If you touch the electric flex (the plastic-coated wire that comes out of the device and has the plug on the end of it) to any of these electric heaters when it is switched on, you will notice that the flex is either at room temperature or, perhaps, it will be slightly warm to the touch. The electric current is increasing the temperature of the heater, by giving it energy, but it is not having the same effect on the flex.

The reason is that the heater has a higher **resistance**, to make it difficult for the charges to flow through it. The flex contains copper wires to feed the electricity to and from the heater. Copper has a very low resistance – only silver is lower, and silver is seldom used for the obvious reason that it is expensive!

So when an electric current passes through a heater (which is a resistor), there is an electrical transfer of energy that increases the temperature of the resistor.

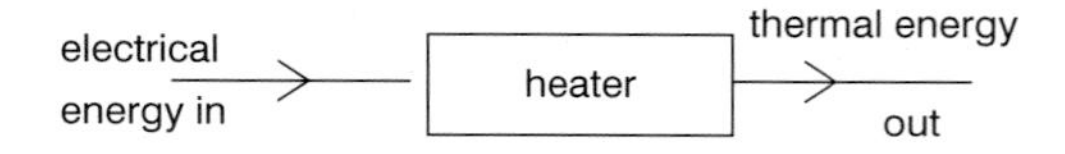

Δ Fig. 2.7 Energy transfer in a heater.

Note that if two heaters are each connected to the mains power, the one with the lower resistance will allow a greater current, and will become hotter than the other one. This is why it is so important that the live wire and the neutral wire in the flex do not touch. If they do touch, they make a very low resistance circuit, and the flex may burst into flames if the correct fuse is not fitted to break the circuit.

QUESTIONS

1. What is a resistor?
2. What happens to the energy store of a resistor when current passes through it?
3. What energy transfer takes place in a heater?

POWER, CURRENT AND VOLTAGE

The **voltage** ('pressure') with which the power station tries to drive electricity through the household appliances is measured in **volts** (V). The voltage between the live socket and the neutral socket on the wall varies from region to region. The most common options are 230 V and 110 V, though there are many other standards depending on what country you are in such as 200 V, 127 V and 100 V. Some equipment can adapt to run on any voltage, but some will be destroyed if it is connected to the wrong voltage, especially if it is too high.

All electrical equipment has a **power rating**, which indicates how many **joules** (J) of energy are supplied each second. The unit of power used is the **watt** (W). Light bulbs often have power ratings of 60 W or 100 W. Electric kettles have ratings of about 2 kilowatts (2 kW = 2000 W). A 2 kW kettle supplies 2000 J of energy each second.

The power of a piece of electrical equipment depends on the voltage and the current.

The units watt, volt and amp, are defined as follows.
1 watt = 1 J/s

1 V = 1 J/C

1 A = 1 C/s

From these definitions, we can see that 1 W = 1 V × 1 A.

In other words,

power = voltage × current or $P = I \times V$

where P = power in watts (W), I = current in amps (A) and V = voltage in volts (V)

You can use the following triangle to help you to rearrange this equation.

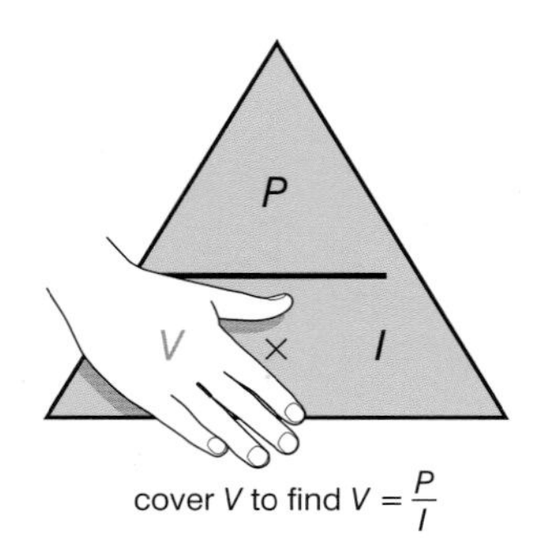

◁ Fig. 2.8 Power, current and voltage triangle to help with rearranging the equation.

CHOOSING THE RIGHT FUSE

Choosing the correct fuse is important when fitting one to an electrical appliance. The fuse is there to protect the user from electric shock, and to protect the appliance from damage.

For this task, take the mains supply to be 230 V.

So how do you select the correct fuse? You need to understand the relationship between the power supplied and the voltage across the appliance.

power = current × voltage

An example of the fuse required for a lawnmower that has a 500 W rating and is connected to the 230 V mains would be:

power = current × voltage

500 = current × 230

current = 2.2 A

Therefore, a 3A fuse should be fitted to the lawnmower.

WORKED EXAMPLES

1. What is the power of an appliance if a current of 7 A is obtained from a 230V supply?

Write down the formula in terms of P:	$P = V \times I$
Substitute the values:	$P = 230 \times 7$
Work out the answer and write down the unit:	$P = 1610$ W

2. An electric oven has a power rating of 2 kW. What will the current be when the oven is used with a 230V supply?

Write down the formula in terms of I:	$I = \frac{P}{V}$
Substitute the values:	$I = \frac{2000}{230}$
Work out the answer and write down the unit:	$I = 8.7$ A

3. What fuse should be fitted in the plug of a 2.2 kW electric kettle used with a supply voltage of 230 V?

Calculate the normal current:

$$I = \frac{P}{V}$$
$$= \frac{2200\ \text{W}}{230\ \text{V}}$$
$$= 9.6\ \text{A}$$

Choose the fuse with the smallest rating bigger than the normal current: the fuse must be 13 A (3 A, 5 A and 13 A fuses are available).

4. What fuse should be fitted to the plug of a reading lamp which has a 60 W lamp and a supply of 230 V?

Calculate the normal current:

$$I = \frac{P}{V}$$
$$= \frac{60\ \text{W}}{230\ \text{V}}$$
$$= 0.26\ \text{A}$$

Choose the fuse with the smallest rating bigger than the normal current: the fuse must be 3 A.

QUESTIONS

1. An appliance has 2 A of current through it and operates at 110 V. What is its power rating?

2. A 60 W lamp has a current of 5 A through it. What voltage is it operating at?

3. A 25 W lamp is designed to be used with a voltage of 230 V. What will be the current through it?

4. Calculate the fuses required for the appliances listed below:

a) a vacuum cleaner of 360 W

b) a television of 0.8 kW (Hint: 1 kW = 1000 W)

c) a table lamp of 100 W

d) a kettle of 2100 W

e) an iron of 900 W.

Assume that the mains supply is 230 V.

Use a spreadsheet to produce a printed table for your teacher to mark. For each appliance you should show:

- the power rating
- the voltage supplied

- the current produced
- the size of fuse required (use the internet or textbook to identify suitable fuses).

ENERGY, CURRENT, VOLTAGE AND TIME

If you switch on an electric kettle for a minute or a room heater for five minutes, then you can measure the temperature increase of the water in the kettle or of the room. The temperature will increase because energy has been given to the water or to the room. Energy is measured in joules (J). A heater rated at 1 watt will give out 1 joule of heat each second. A 2 kW heater will give out 2000 joules per second, and if the heater is switched on for 4 s, it will give out 8000 J.

energy = current × voltage × time
or $E = I \times V \times t$

where E = the energy transferred in joules (J), I = current in amperes (A), V = voltage in volts (V) and t = time in seconds (s)

WORKED EXAMPLE

Calculate the energy transferred when a 12 V motor, running at a current of 0.5 A, is left on for 5 minutes.

Write down the formula: $E = I \times V \times t$

Substitute the values: (remember the time must be in seconds) $E = 0.5 \times 12 \times 300$

Work out the answer and write down the unit: $E = 1800$ J

QUESTIONS

1. A laptop charger is designed for a country where the mains voltage is 230 V. If the owner takes the charger to a country where the mains voltage is 110 V, will they be able to charge their laptop?
2. What if someone takes a 110 V charger to a country with 230 V mains electricity?
3. How much energy is transferred when there is a current of 3 A in a circuit with a voltage of 12 V for 1 minute?
4. A heater which runs on 12 V transfers 4800 J of energy in 2 minutes. What is the current in the heater?

5. A bulb runs on a voltage of 110 V and has a current of 0.1 A through it. It transfers 2400 J of energy in a given time. What is this time?

6. A lamp transfers 24 J of energy and draws a current of 2 A for 1 s. What voltage is it operating at?

ALTERNATING CURRENT AND DIRECT CURRENT

A battery produces a steady current. The electrons are constantly flowing from the negative terminal of the battery round the circuit and back to the positive terminal. This produces a **direct current** (d.c.). The voltage stays constant.

The mains electricity used in the home is quite different. The electrons in the circuit move backwards and forwards. This kind of current is called **alternating current** (a.c.). Mains electricity moves forwards and backwards 50 times each second. It has a **frequency** of 50 hertz (Hz). Figure 2.9 shows how the voltage changes with time for alternating current. The frequency and voltage chosen varies from country to country.

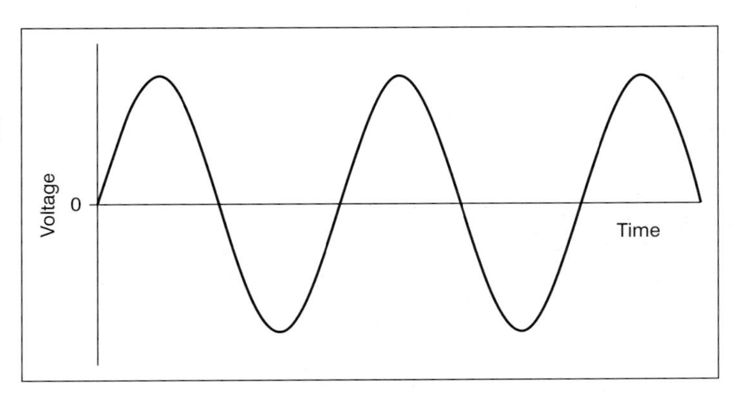

Δ Fig 2.9 How voltage varies with time for an alternating current.

The advantage of using an a.c. source of electricity rather than a d.c. source is that it can be transmitted from power stations to the home at very high voltages, which reduces the amount of energy that is lost in the overhead cables.

DEVELOPMENT OF MAINS ELECTRICITY

Electrical discoveries and techniques began as pure science in the late 18th century and continued throughout the 19th century, but the use of electricity in wider society developed only slowly. From the late 1830s onwards electricity was used for communications, but this only required the low power that batteries could provide. The invention of the incandescent light bulb in the 1870s was an important change, as it led to the possibility of electric street lighting. This meant that electrical supplies would need to be sent across larger distances.

As local companies and authorities developed electricity supply systems, sometimes a direct current (d.c.) system was used, sometimes an alternating current (a.c.) system. As distances became greater, however, the power losses in the cables became more significant and it became clear that higher voltages would be more economic. Since d.c. systems had no straightforward way to convert to higher voltages, a.c. became the standard that developed into the systems we use today.

Δ Fig. 2.10 A modern incandescent light bulb.

Different countries have different standards for mains supply, reflecting the historical development of the networks across the world. For example, the UK, Sri Lanka, India and Australia use a 230 V supply, China uses 220 V and the USA uses 110 V.

But what does the voltage refer to in an a.c. supply?

The voltage alternates between positive and negative. Where on the voltage axis (if we refer to the UK value) should the '230' go? The mean value of the voltage is 0 (equally spread in positive and negative values). If we label the highest value (called the 'peak' value), that doesn't seem very representative of the voltage, since the supply would only be at that value for very short times.

A value called the root mean square (RMS) value is calculated in three stages. Firstly, all values of voltage are *squared* – this makes all the values on the graph positive. Secondly, now that all the values are positive, a *mean* is calculated – this will not be zero. Finally, since this mean value is in (volts)^2, the *square root* is taken to give a value in volts. It is this final value, the RMS value, that is quoted as 230 V.

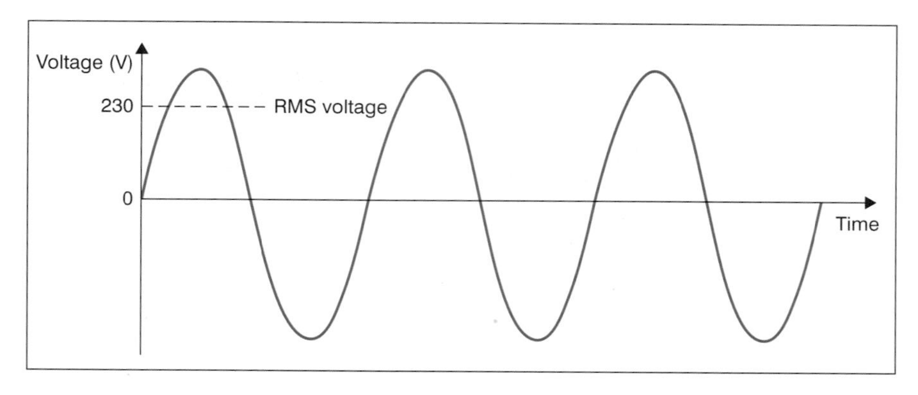

Δ Fig. 2.11 RMS voltage

The particular advantage of using the RMS value is that an a.c. supply of 230 V RMS would provide the same *heating effect* as a 230 V d.c. battery, so the two systems can be compared easily.

End of topic checklist

Power = voltage × current

Energy transferred = voltage × current × time (in seconds)

A number of features, such as double insulation, earthing and fuses, are designed to protect a device or the user against shocks and fire hazards.

Electrical resistance results in a heating effect.

A **circuit breaker** is a device that breaks a circuit when there is an increase in current.

An **earth wire** is a wire connecting the case of an electrical appliance, through the earth pin on a three-pin plug, to earth.

When a device is **double-insulated**, it has a casing that is made of an insulator and does not need an earth wire.

The **power rating** is a measure of how fast an electrical appliance transfers energy supplied as an electrical current.

Direct current is electric current where the direction of the current stays constant, as in cells and batteries.

Alternating current is electric current where the direction of the current constantly reverses, as in mains electricity.

The facts and ideas that you should understand by studying this topic:

- ❍ Know and use the relationship between power, current and voltage and use this to select an appropriate fuse.
- ❍ Use the relationship between energy transferred, current, voltage and time.

End of topic checklist continued

- ❍ Understand some of the operating features of mains electricity, including
  - insulation
  - double insulation
  - earthing
  - fuses
  - circuit breakers.
- ❍ Understand how the current in a resistor leads to a heating effect due to the electrical transfer of energy.
- ❍ Understand the difference between a d.c. supply and an a.c. supply.

End of topic questions

1. Suggest why the diameter of a cable must be suitable for the current it has to carry. Hint: current has a heating effect. **(2 marks)**

2. Why do you think the fuse must always be connected in series with the live wire? **(1 mark)**

3. a) A hairdryer works on mains electricity of 230 V and takes a current of 4 A. Calculate the power of the hairdryer. **(2 marks)**

b) In some countries it is illegal to have power sockets in a bathroom, to stop you using electrical devices such as hairdryers near the sink. Why would it be unadvisable to use a hairdryer near a sink? **(2 marks)**

4. An appliance has a power rating of 1400 W. The potential difference of the mains is 230 V. Calculate the approximate current. Explain what size standard fuse you would use. **(3 marks)**

5. In her living room, Felicity has the following items:

- three 100 W lamps
- a TV that takes 2 A
- an audio system that takes 1 A
- a 2 kW electric heater
- a 3 kW air conditioning unit.

The whole room is supplied from a 220 V AC power supply through one miniature circuit breaker (MCB).

a) What rating of MCB should you fit, if values of 10 A, 20 A, 30 A, 40 A, 50 A and 60 A are available? **(2 marks)**

b) How would your answer change if the supply were 110 V AC? **(2 marks)**

6. The potential difference across a bulb is 5 V and the current through the bulb is 3 A. In one minute, how much energy will be transferred by charge passing through the bulb? **(2 marks)**

7. What size fuse would you use in a microwave of power rating 800 W with **a)** 240 V **b)** 120 V mains supply? **(4 marks)**

8. What potential problems may there be with the fuse in question 7? **(2 marks)**

9. A lamp has a power rating of 11 W and runs from a supply of 230 V. What is the current in the lamp? **(2 marks)**

10. An appliance runs from a 110 V supply. It has a current of 3.2 A through it. What is its power rating? **(2 marks)**

11. An appliance has a current of 2.7 A through it and has a power rating of 300 W. What is the voltage of the supply? **(2 marks)**

∆ Fig. 2.12 You can investigate electrical circuits in the classroom.

Energy and voltage in circuits

INTRODUCTION

Whenever you use an electrical appliance, electrical circuits operate. Some are visible to the eye but some have been etched on to microchips. The basic operation of all circuits relies on connecting components, the nature of the components and the energy supplied to the circuit. In this section you will learn about different ways of connecting components in circuits. You will also learn how to draw circuit diagrams that can be followed by anyone anywhere in the world, and how to carry out calculations to choose the right values for the components in a circuit.

KNOWLEDGE CHECK

✓ Know that a complete circuit is needed for an electric current.
✓ Know the difference between connecting components in series and in parallel.
✓ Be able to use an ammeter and voltmeter.
✓ Be able to draw and interpret circuit diagrams.

LEARNING OBJECTIVES

✓ Be able to explain the advantages of a series or parallel circuit for particular applications.
✓ Describe how changing resistance affects the current in a circuit.
✓ Describe how current varies with voltage in wires, resistors, metal filament lamps and diodes.
✓ Describe how the resistance of LDRs and thermistors vary.
✓ Know and use the relationship between voltage, current and resistance: $V = IR$.
✓ Know that a current is a rate of flow of charge.
✓ Know and use the relationship between charge, current and time: $Q = It$.
✓ Know that electric currents in metals are due to the motion of electrons.
✓ Understand why current is conserved at a junction in a circuit.
✓ Know that the voltage across two components connected in parallel is the same.
✓ Calculate the currents, voltages and resistances of two resistive components connected in a series circuit.
✓ Know that voltage is the energy transferred per unit charge passed, and 1 volt is 1 joule per coulomb.
✓ Know and use the relationship between energy transferred, charge and voltage: $E = QV$.

CIRCUIT DIAGRAMS

When people started using electricity, they quickly found that it was not convenient to draw accurate pictures of the circuits that they made. It was much easier to understand how the circuit worked, and to correct any faults, if they used standard symbols for the parts. It was also much easier if the wires were drawn in straight lines, rather than trying to copy the exact route taken.

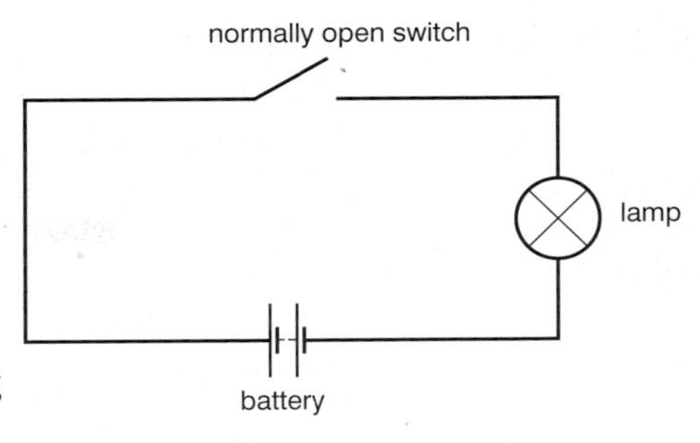

Δ Fig. 2.13 A circuit for a torch.

Study the circuits used in this chapter, and learn the symbols and what they represent.

Lamps are often used in electrical experiments. They light up when current is present and so they are a simple way of showing what is happening in a circuit. LEDs can also be used.

This simple circuit diagram shows how a torch is powered by a battery consisting of three 1.5 V cells, giving a total of 4.5 V. In the case of a torch, the cells are put in separately, but in the case of a 9 V battery, for example, the six cells are pre-assembled by the manufacturer. The word 'battery' means an assembly of several cells, but people often use the word to refer to a single cell.

The '+' terminal of the cell or power supply is indicated by the long thin line, and the '–' terminal by the short thick line. To help you remember, imagine yourself cutting the long thin line into two shorter pieces and turning them into a + sign.

The other symbols in the circuit are the normally open switch, and the lamp.

SERIES AND PARALLEL CIRCUITS

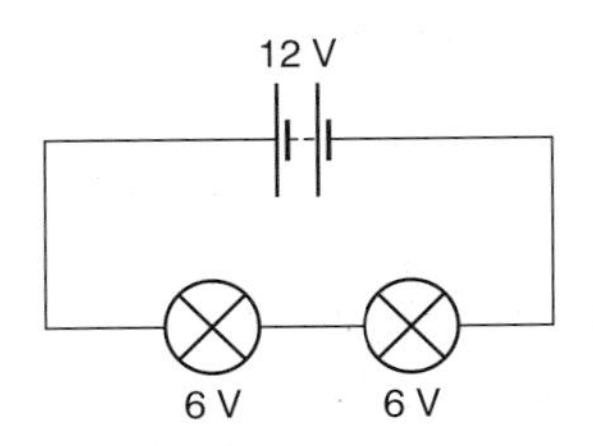

Δ Fig. 2.14 A simple series circuit.

There are two different ways of connecting two lamps (or other components such as LEDs) to the same battery (or other power source). Two very different kinds of circuit can be made. These circuits are called **series** and **parallel circuits**.

If the components are in series, then exactly the same electric current passes through each of the components in the circuit. The voltage of the supply is shared between the units in the circuit. Thus it is possible to join in series two identical lamps designed for 6 V and then to connect them to a 12 V battery.

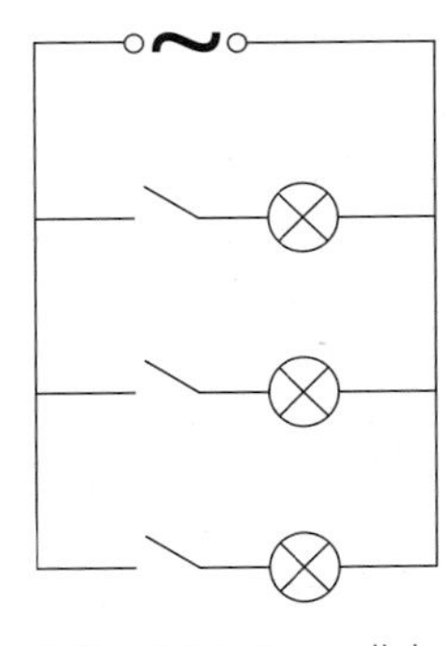

Δ Fig. 2.15 A parallel circuit

In a parallel circuit, there is more than one path for the current so the current splits. The voltage across two components connected in parallel is the same. All of the appliances in a house are connected in parallel to the mains supply, and each one sees the full 110 V or 230 V of the mains supply when it is switched on. The two great advantages of the parallel arrangement are that each appliance can be designed to work with the mains voltage supply, and that the appliances can be switched on and off individually.

◁ Fig. 2.16 All of these lights are in parallel. If they were in series then they would all go off if any one of them was switched off.

Here is a comparison of series and parallel circuits for two identical lamps.

	Series	**Parallel**
Circuit diagram	Δ Fig. 2.17	Δ Fig. 2.18
Appearance of the lamps	Both lamps have the same brightness, both lamps are dim.	Both lamps have the same brightness, both lamps are bright.
Battery	The battery is having a hard time pushing the same charge first through one bulb, then another. This means less charge flows each second, so there is a low current and energy is transferred slowly from the battery.	The battery pushes the charge along two alternative paths. This means more charge can flow around the circuit each second, so energy is transferred quickly from the battery.
Switches	The lamps cannot be switched on and off independently.	The lamps can be switched on and off independently by putting switches in the parallel branches.
Advantages/ disadvantages	A very simple circuit to make. The battery will last longer. If one lamp 'blows' then the circuit is broken so the other one goes out too.	The battery will not last as long. If one lamp 'blows' the other one will keep working.
Examples	Christmas tree lights are often connected in series.	Electric lights in the home are connected in parallel.

Δ Table 2.2 Comparison of series and parellel circuits.

QUESTIONS

1. If one bulb in a string of Christmas tree lights does not work, why does this mean that the whole string does not work?
2. A battery running two bulbs in parallel runs out before a battery running two bulbs in series. Explain why.
3. Why are electric lights in the home connected in parallel?
4. What is the difference in brightness in two bulbs connected a) in series b) in parallel with a given battery? Give a reason for your answer.

Current in a series or parallel circuit

The current in a circuit can be measured using an **ammeter**. If you want to measure the current through a particular component, such as a lamp or motor, the ammeter must be connected in series with the component. Figure 2.19 shows an ammeter in series with a motor. In a series circuit, the current is the same no matter where the ammeter is put. This is not the case with a parallel circuit.

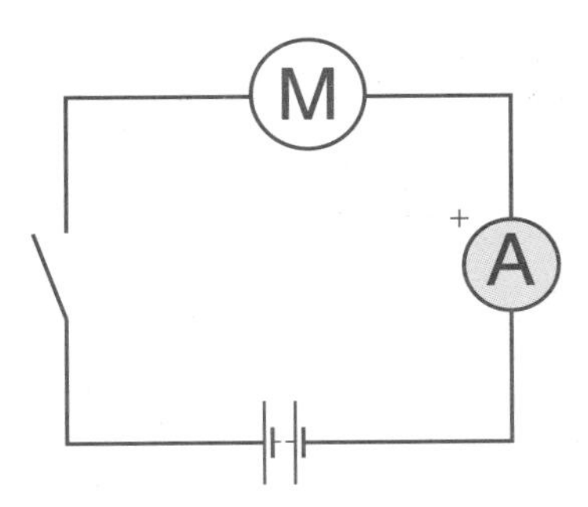

Δ Fig. 2.19 The circuit has to be broken to include the ammeter.

In a parallel circuit the current splits at a junction, with part of it going through each branch. The sum of all the currents in all the branches is equal to the current from the electrical supply.

The current from the power supply in this circuit is 3 A. The three motors are identical. There is a greater current (2 A) through the single motor. This motor has the full share of the 12 V supply. A smaller current (1 A) passes through the pair of motors. These two motors, which have equal resistance, will share the 12 V, so the voltage across each is 6 V. These two motors will rotate more slowly than the single motor, because together they have more resistance. The current recombines (2 A + 1 A = 3 A) before returning to the power supply.

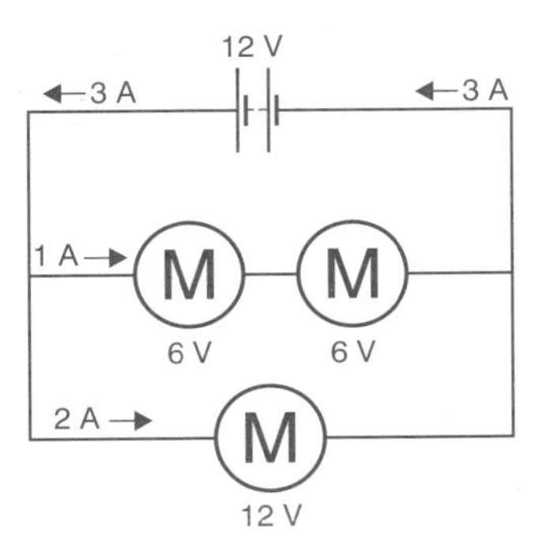

Δ Fig. 2.20 When two or more components are connected in parallel, the total current in the circuit is equal to the sum of the currents in each branch.

How much current in the circuit?

For a series circuit, the current in the circuit depends on the applied voltage and on the number and the nature of the components in the circuit. For example, it is possible to buy strings of lights designed to decorate trees. Some of these use lamps that are designed to be connected to 12 V and to pass 1 A of current. So the lamps are 12 W each. If the string of lights connects 20 of these in series for connection to a 240 V supply, as in Figure 2.21, then each lamp will have 12 V across it, and the current in the string will be 1 A, with exactly the same current in each lamp.

If any one lamp is removed or breaks, all of the lamps will go off. Treat this string very carefully if you are trying to check which lamp is faulty. The lamps may only be 12 V, but when you remove the faulty bulb, the voltage on one contact inside the socket will be 240 V!

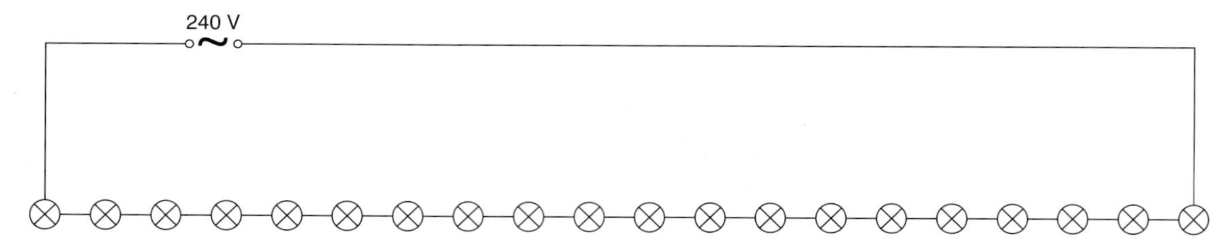

Δ Fig. 2.21 Many lamps connected in series

If the lamps are connected to a lower voltage, such as a 110 V supply, then the current will be approximately half of this through each lamp, and the lamps will glow very dimly.

If the components in the series circuit are not identical, then the current in each one will still be exactly the same, but those with a higher resistance will use up more of the supply voltage, and those with a lower resistance will use up less. The total voltage across all of the components together will, of course, be the supply voltage.

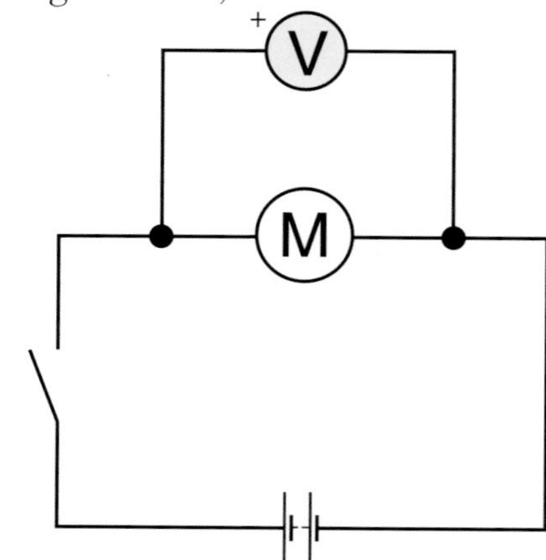

Δ Fig. 2.22 The voltmeter can be added after the circuit has been made. Testing with a voltmeter does not interfere with the circuit provided the voltmeter has a high resistance.

Measuring how current varies with voltage

The voltage across a component can be measured using a **voltmeter**, as shown in Figure 2.22.

To measure how current varies with voltage, you use the circuit in Figure 2.23. The component (which here is a resistor) is placed in a circuit with an ammeter to measure the current through the component, and with a voltmeter to measure the voltage across it. To power the circuit you could use a battery as shown, or you could use a power supply with a suitable output. To take readings, the circuit is switched on, and readings are made of the voltage and the current.

You can then plot a graph of voltage against current.

Note that the readings may change a little over the first few seconds. If so, this is probably because the component is heating up and its resistance is changing. If this happens, you will have to decide whether to take the readings before the component has heated up, and so measure the resistance at room temperature, or to wait until the readings have stopped changing. This would give you the steady-state resistance (the resistance of the component under normal operating conditions, once it has warmed up to room temperature) with the component at its usual running temperature.

You may wish to change the voltage of the battery by changing the number of cells (or you may adjust the output of the power supply).

Take care! If you try thick wire, the current will be extremely high for a very low voltage. The wire can get very hot very quickly and there is a risk of injury.

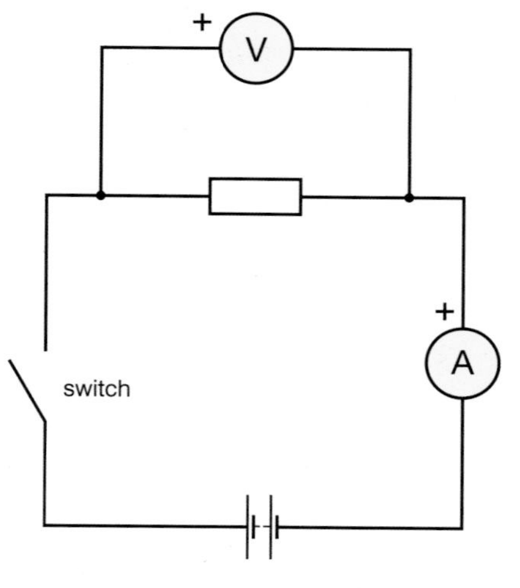

Δ Fig. 2.23 Circuit used to measure how current varies with voltage

QUESTIONS

1. A student connects two light bulbs, A and B, in parallel across a battery. Bulb A is brighter than bulb B. Which bulb has the higher resistance? Explain your answer.

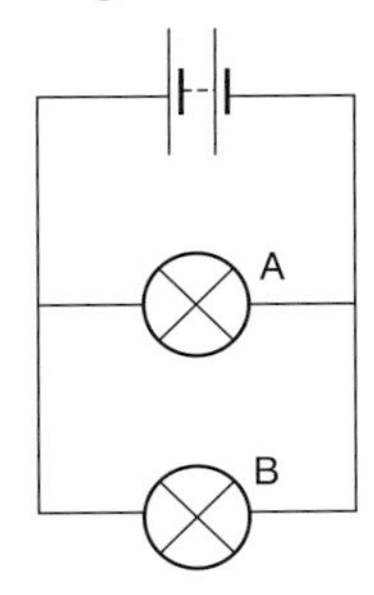

2. If the student connects the same bulbs in series with the same battery, which bulb will be brighter? Explain your answer.

3. Describe how to investigate how current varies with voltage across a resistor.

4. What is wrong with this circuit?

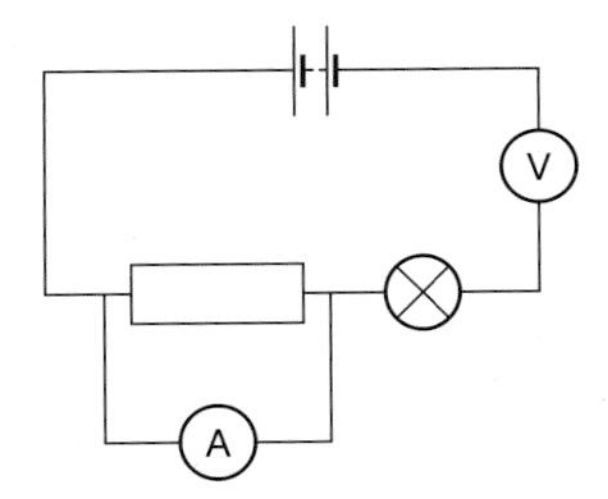

HOW CHANGING RESISTANCE AFFECTS CURRENT

Figure 2.24 shows the voltage/current graph for a component where the resistance remains constant, as shown by the constant gradient of the voltage/current graph. This would be the graph for components like **resistors** (a circuit component which is designed to resist the current by a set amount) and thick wires. The current in the resistor doubles if you double the voltage, triples if you triple the voltage, and so on. The resistance of the component does not change, and the extra current is caused solely by the increased pressure of the extra voltage. The current is **directly proportional** to the voltage, and the graph is a straight line. However, the resistance of most conductors becomes higher if the temperature of the conductor increases. As the temperature rises, the particles in the conductor vibrate more and provide greater resistance to the flow of electrons. For example, the resistance of a filament lamp becomes greater as the voltage is increased and the lamp gets hotter. The current is not directly proportional to the voltage because the heating of the lamp changes its resistance. Figure 2.24 shows the voltage/current graph for such a component. You can see that the gradient of the graph increases with increasing voltage.

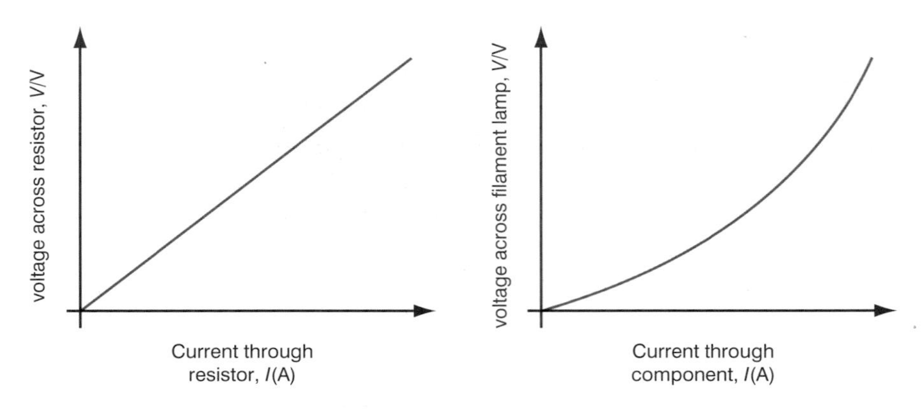

Δ Fig. 2.24 Left: Voltage/current graph for a component where resistance remains constant. Right: Voltage/current graph for a component where resistance increases as voltage increases.

It does not matter which way the current passes through a lamp, but a pocket calculator, say, could be destroyed if the battery is not inserted correctly. One way to prevent this is to add a **diode** to the circuit. A diode is a circuit component that only allows current to pass through it in one direction.

In this circuit the calculator is represented as a resistor. A calculator is far more complicated than that, but it does behave to the battery *as if* it were a resistor, drawing a small current *I* out of the battery.

The arrow on the diode shows the direction of the 'conventional current' (see page 106). When the battery is inserted the wrong way round, as shown in the right-hand part of Figure 2.25, there can be no current.

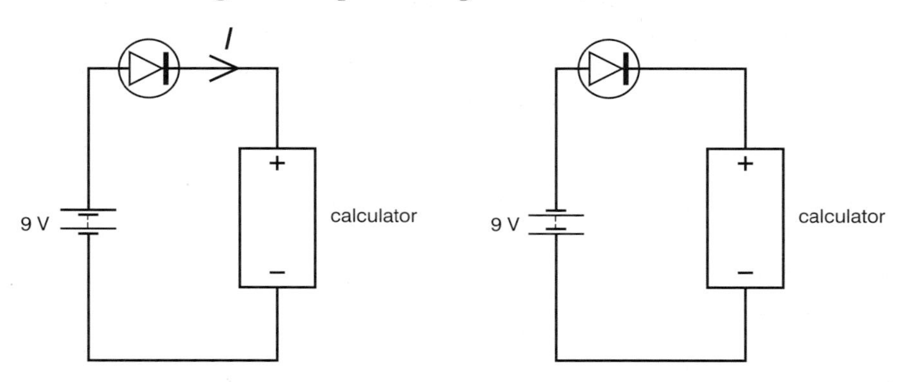

Δ Fig. 2.25 A diode only allows current to pass through it in one direction.

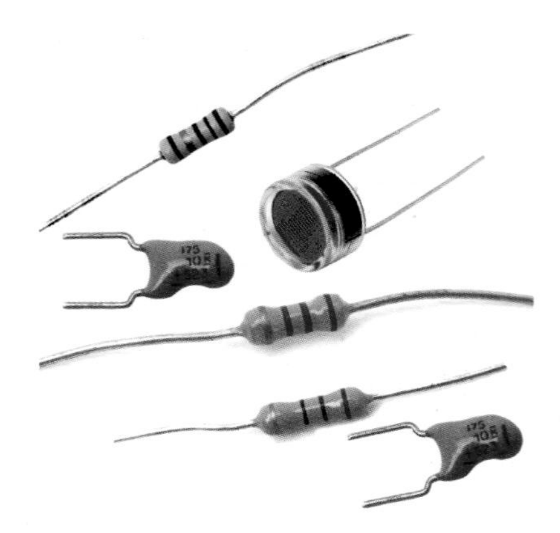

◁ Fig. 2.26 Some resistors and diodes

EFFECT ON CURRENT OF CHANGING RESISTANCE

As you have seen, using more voltage across a component increases the current in it. The amount of current in a circuit can also be controlled by changing the resistance of the circuit using a variable resistor or **rheostat**. Adjustment of the rheostat changes the length of the wire the current passes through. Since the wire resists the current, changing the length of the wire alters the resistance in the circuit. Variable resistors are often used, for example, to change the brightness of the lighting in a car.

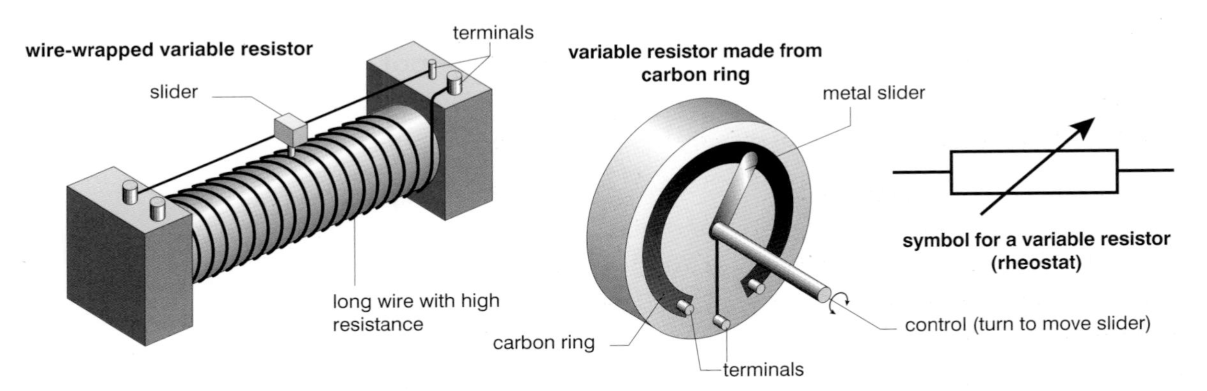

Δ Fig. 2.27 Variable resistors are commonly used in electrical equipment, for example in the speed controls of model racing cars or in volume controls on radios and audio systems.

THE THERMISTOR AND THE LDR

In some substances, increasing the temperature actually lowers the resistance. This is the case with **semiconductors** such as silicon. A semiconductor material is a material that does not conduct electricity as well as, for example, a metal, but conducts electricity better than an insulator, such as plastic. Silicon has very few free electrons to constitute an electric current and so behaves more like an insulator than a conductor. However, if silicon is heated, more electrons are removed from the outer electron shells of the atoms, which produces an increased electron cloud. The released electrons can move throughout the structure, allowing an electric current to pass more easily. This effect is large enough to outweigh the increase in resistance that might be expected from the increased vibration of the silicon ions in the structure as the temperature increases.

LDR
thermistor
diode
fixed resistor

Δ Fig. 2.28 Circuit symbols for components made using semiconductor materials.

Semiconducting silicon is used to make **thermistors**, which are used as temperature sensors, and **light-dependent resistors (LDRs)**, which are used as light sensors.

In LDRs it is light energy that removes electrons from the silicon atoms, increasing the electron cloud. So LDRs have a very high resistance in the dark, and a very low resistance in the light. LDRs are used in street lamps that switch on automatically at night, and in the type of burglar alarm that sets a light beam (usually an infra-red beam so that it is invisible) across the path of the burglar.

REMEMBER

In a thermistor, increasing temperature actually reduces the resistance. This is the opposite effect to that in a normal resistor. In a light-dependent resistor (LDR) an increase in brightness reduces the resistance.

QUESTIONS

1. A student connects a battery, an ammeter and an LDR to make a simple light-meter. Will the reading on the ammeter go up or down when the light gets brighter? Explain your answer.
2. Some descriptions of electric currents compare them to rivers flowing. Using this idea, explain why a diode can be thought of as a waterfall.
3. Describe how the resistance of a thermistor changes with temperature.

Developing investigative skills

A student investigates how the resistance of a thermistor varies with temperature. She uses a multimeter as an ohmmeter. To measure the temperature of the thermistor she immerses it in a water bath. At the start of the experiment she fills the beaker with water at 50 °C. She takes measurements of the temperature and the resistance at various temperatures as the water cools down. The student adds ice to help achieve lower temperatures and stirs the water regularly. The student's measurements are shown in the table.

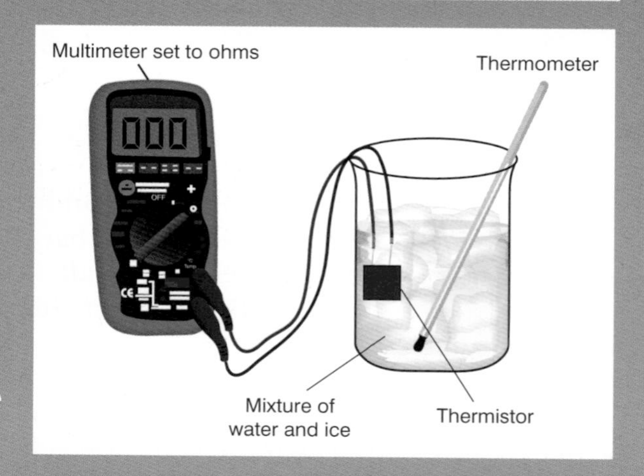

Δ Fig. 2.29 Apparatus for experiment.

Temperature (°C)	Resistance (kΩ)
10	12.62
15	8.47
20	6.61
25	5.45
30	4.25
35	3.54
40	2.79
45	2.11
50	1.12

Δ Table 2.1 Results of experiment.

Devise and plan investigations

1. Why is it important that the student stirs the water between readings?
2. The student uses a thermometer that measures from – 10 °C to + 50 °C and is accurate to the nearest 2 °C.
   a) What is the range of the thermometer?
   b) When the thermometer reads 40 °C, what values could the true temperature be between?

Analyse and interpret

3. Draw a graph of the student's results.
4. Describe the pattern (if any) shown by the graph.

Evaluate data and methods

5. The student allowed the water to cool down slowly during the experiment. How did this improve the accuracy of her results?
6. Use ideas about electron movement to suggest why the resistance of a thermistor changes in the way shown in this experiment.

VOLTAGE, CURRENT AND RESISTANCE

The relationship between voltage, current and resistance in electrical circuits is given by this equation.

voltage = current × resistance or $V = I \times R$

where V is the voltage in volts (V), I is the current in amps (A) and R is the resistance in ohms (Ω).

This equation is often called **Ohm's law**, but it is not a 'law', just the definition of the resistance of an object. The idea of resistance is useful because for a lot of objects their resistance does not change when you change the current through them. But there are many components, such as light bulbs, for which this is not true (see page 109).

It is important to be able to rearrange this equation when performing calculations. Use the triangle shown in Fig. 2.30 to help you.

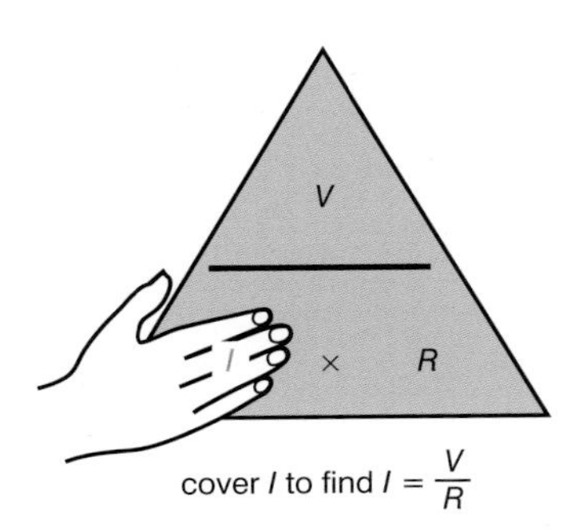

◁ Fig. 2.30 Voltage, current and resistance triangle to help with rearranging the equation.

Developing investigative skills

A student is asked to find the resistance of a piece of fuse wire. He sets up the circuit in the diagram and makes a note of the readings on the ammeter and the voltmeter for five different settings of the variable resistor. His measurements are shown in the table.

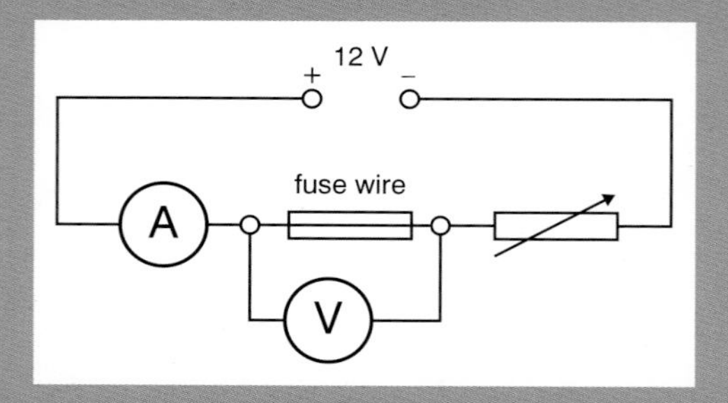

Δ Fig. 2.31 Circuit diagram for experiment.

Potential difference (V)	Current (mA)
0.0	0
1.0	88
2.0	177
3.0	275
4.0	363
5.0	451

Δ Table 2.2 Results of experiment.

Devise and plan investigations

❶ The student did not measure the voltage of the supply or the particular settings of the variable resistor. Explain why these measurements were not required.

❷ The student should check the ammeter and voltmeter for zero errors. What are these?

Analyse and interpret

❸ Draw a graph of the student's results.

❹ Use your graph to find the resistance of the wire.

Evaluate data and methods

❺ Another student suggests that drawing a graph is not necessary. They say that you could use the equation $R = V/I$ for each pair of measurements and then find a mean of these values. Explain why calculating the gradient of the graph is a better method.

❻ To get an accurate value for the resistance of the wire, the student needed to avoid any heating effects in the wire. Describe how the student could reduce heating effects when carrying out the experiment.

WORKED EXAMPLES

1. Calculate the resistance of a heater element if the current is 10 A when it is connected to a 230 V supply.

Write down the formula in terms of R: $R = \frac{V}{I}$

Substitute the values for V and I: $R = \frac{230}{10}$

Work out the answer and write down the unit: $R = 23\ \Omega$

2. A 6 V supply is applied to 1000 Ω resistor. What will the current be?

Write down the formula in terms of I: $I = \frac{V}{R}$

Substitute the values for V and R: $I = \frac{6}{1000}$

Work out the answer and write down the unit: $I = 0.006$ A

QUESTIONS

1. Calculate the potential difference across a 5 Ω resistor which has a current of 2 A though it.
2. A lamp has a potential difference. of 3.0 V across it and a current of 0.5 A through it. What is its resistance?
3. The resistance of a wire increases when the temperature increases. How would this cause the shape of the graph shown on the right to change? Explain why.

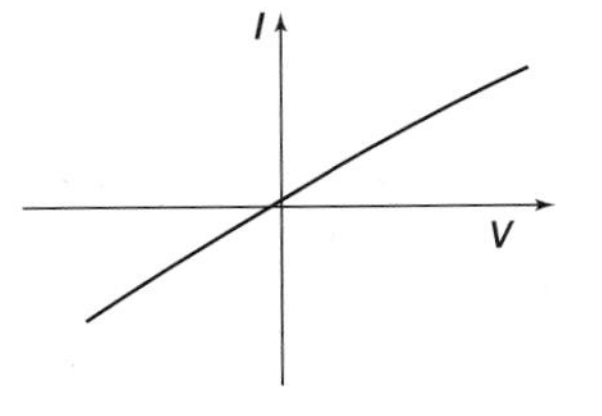

Calculations for resistors in series

Resistors can be combined in series. In Figure 2.31 three resistors are arranged in series in a circuit. The total resistance can be found by simply adding up the resistance values. Total resistance = 2 Ω + 3 Ω + 1 Ω = 6 Ω. We can also calculate the current through the resistors. As this is a series circuit, the current is the same everywhere in the circuit.

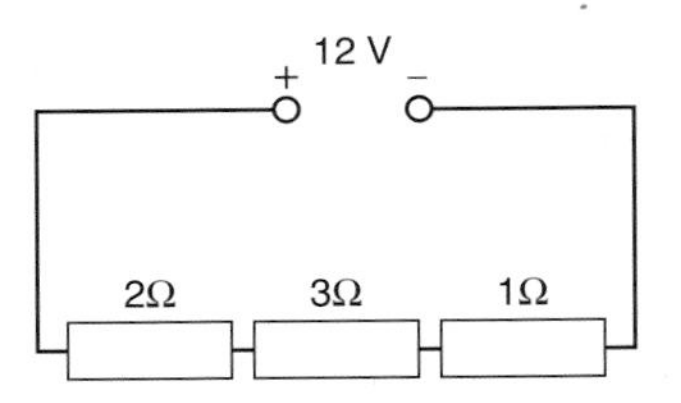

Δ Fig. 2.32 Three resistors in series.

WORKED EXAMPLES

1. Calculate the current through the resistors in Figure 2.31.

The voltage across all three resistors is 12 V.

Write down the formula: $V = IR$

Rearrange the formula: $I = V \div R$

Substitute the values for V and R: $I = 12 \div 6$

Work out the answer and write down the units: $I = 2$ A

2. A 12 V supply delivers a current of 3 A to three resistors in series. Calculate the resistance of resistor X in Figure 2.32.

Δ Fig. 2.33

First, calculate the total resistance of the three resistors:

Write down the formula: $V = IR$

Rearrange the formula: $R = V \div I$

Substitute the values for V and I: $R = 12 \div 3$

Work out the answer and write down the units: $R = 4\ \Omega$

The total resistance of all three resistors is 4 Ω.

Total resistance = 4 Ω = 1 + 1 + X

Resistor X must be 2 Ω.

Current is the rate of flow of charge

All materials contain **electrons**, but in many materials they are all 'locked' into the material's atoms and cannot move about. These materials cannot carry an electric current, and are called electrical insulators. Materials in which there are large numbers of electrons that are free to move around from atom to atom are called **conductors**.

When there is no current in a conductor, the free electrons move randomly between atoms, with no overall movement. When you connect it in an electrical circuit with a power source like a battery, there is a current in the conductor. Now the electrons drift in one direction, while still moving in a random way as well. The drift speed is very slow, often only a few millimetres each second. A current can pass through a conductor if it is connected in a complete circuit. If the circuit is broken, the current stops.

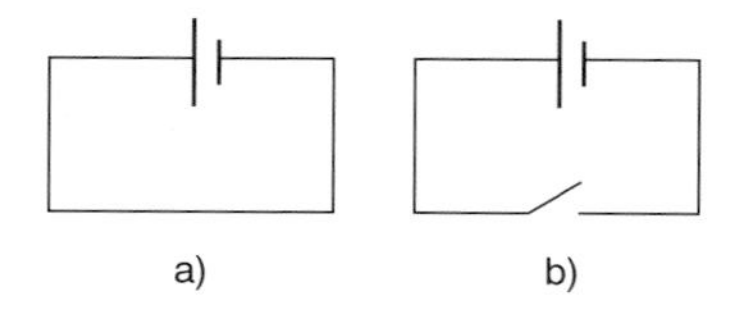

Δ Fig. 2.34 In circuit a), current can pass from the battery through the wires and back to the battery again. In circuit b), the open switch breaks the circuit so current cannot pass through the wires.

The size of an electric current depends on the number of electrons that are moving and how fast they are moving. But instead of measuring the actual number of electrons we use the total charge carried by the electrons round the circuit each second. The amount of charge on an object is measured in **coulombs** (C). A charge of 1 C is the charge on 6.24×10^{18} electrons.

Electric current is measured in **amperes**, or **amps** (A).

If one amp of current is passing down a wire, then one coulomb of charge is travelling past any point on the circuit each second. (1 A = 1 C/s.)

You use an ammeter to measure current in an electrical circuit. If the current is very small, you might use a milliammeter, which measures current in milliamps (1 mA = 0.001 A). Even smaller currents are measured with a microammeter. (1 μA = 0.000001 A)

If you want to measure the current in a particular component, such as a lamp or motor, the ammeter must be connected in series with the component.

The electric current is the amount of charge flowing every second – the number of coulombs per second:

charge = current × time or $Q = I \times t$

where I = current in amperes (A), Q = charge in coulombs (C) and t = time in seconds (s)

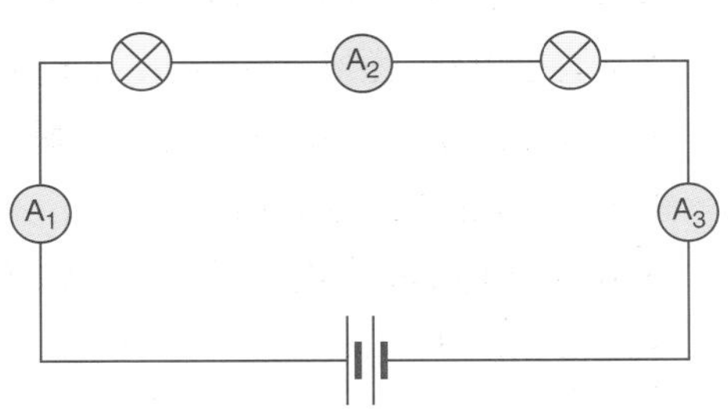

Δ Fig. 2.35 In this series circuit, the current will be the same throughout the circuit so $A_1 = A_2 = A_3$.

You can use this triangle to help you to rearrange the equation.

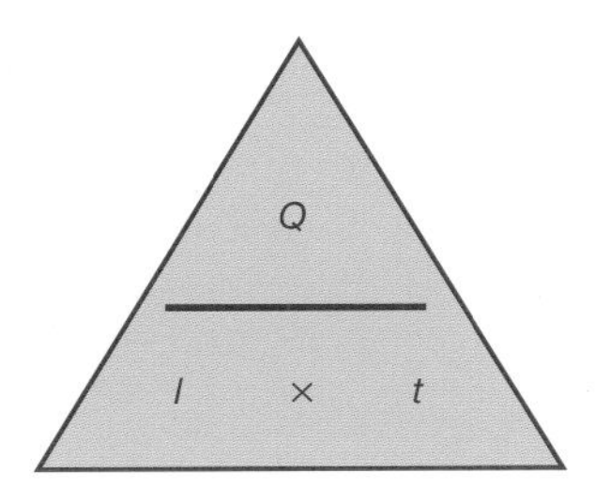

◁ Fig. 2.37 Equation triangle for charge, current and time.

Conservation of current at a junction in a circuit

In a series circuit, the current is the same no matter where the ammeter is put (Figure 2.35). This is not the case with a parallel circuit. The total current in Figure 2.36 is measured with ammeter A_1. When the current gets to the junction in the circuit the current splits so some goes through A_2 and the remainder goes through A_3. The current here is not lost or gained – it is **conserved**. Current is conserved because current is the rate of slow of charge and charge cannot be created or destroyed. If we add currents at A_2 and A_3 they will equal the current at A_1. In parallel circuits the total current remains the same and is conserved. So all the currents in the parallel parts of the circuit will add up to equal the current that leaves the power supply.

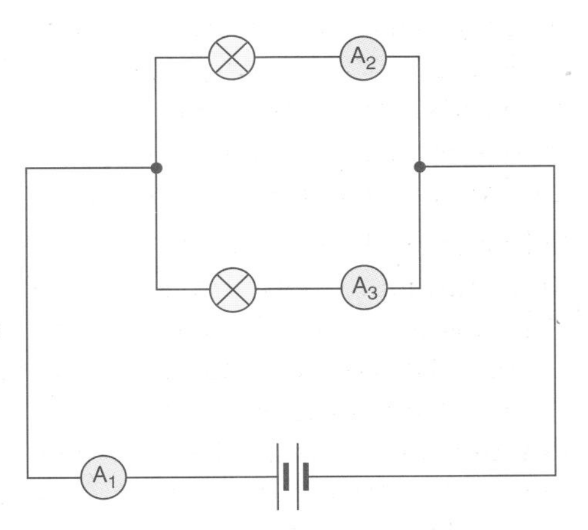

Δ Fig. 2.36 The flow of charge splits between the two branches of the parallel circuit so $A_1 = A_2 + A_3$.

REMEMBER

The current is the same in all parts of a series circuit but the potential difference across different components can be different.

QUESTIONS

1. Calculate the flow of charge in the following.

a) current 3 A for 5 s
b) current 2 A for 10 s
c) current 4 A for 23 s
d) current 1.5 A for 0.5 min

2. Current of 5 A passes for 30 minutes. How much charge has flowed past any given point in the circuit?

3. A charge of 60 coulombs produces a current of 0.5 A. How long does this take?

Current in metallic conductors

Scientists now know that electric current is really a flow of electrons around the circuit from negative to positive. Unfortunately, early scientists guessed the direction of flow incorrectly. Because of this, all diagrams were drawn showing the current going from positive to negative. This way of showing the current has not been changed and so the **conventional current** that everyone uses gives the direction in which positive charges would flow.

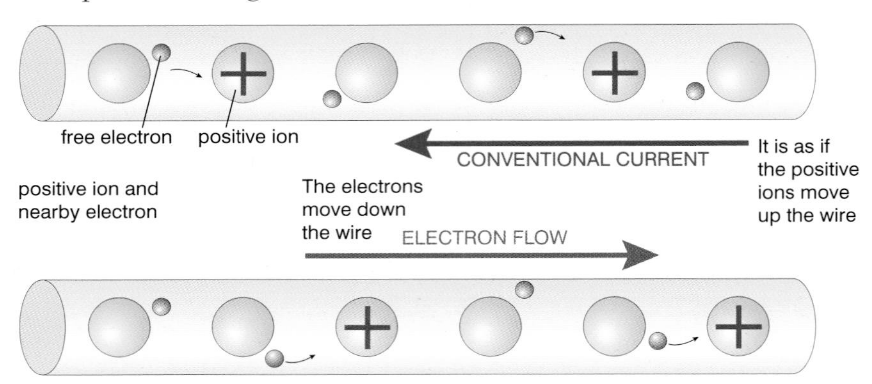

Δ Fig. 2.38 Conventional current is drawn in the opposite direction to electron flow.

QUESTION

1. How do the direction of conventional current and the direction in which electrons actually flow compare?

SUPERCONDUCTORS

In 1908, Heike Kamerlingh Onnes, a Dutch physicist, became the first person to produce liquid helium, which meant reaching temperatures lower than – 269 °C, the boiling point of helium. Having such a cold liquid meant that other low temperature experiments became possible as he could now cool down the apparatus sufficiently.

In particular, Kamerlingh Onnes looked at passing electric currents through extremely cold metals. In 1911 he was measuring the resistance of a sample of mercury. He found that below a particular temperature, called the critical temperature, the mercury behaved as if it had no electrical resistance at all – he had discovered superconductivity.

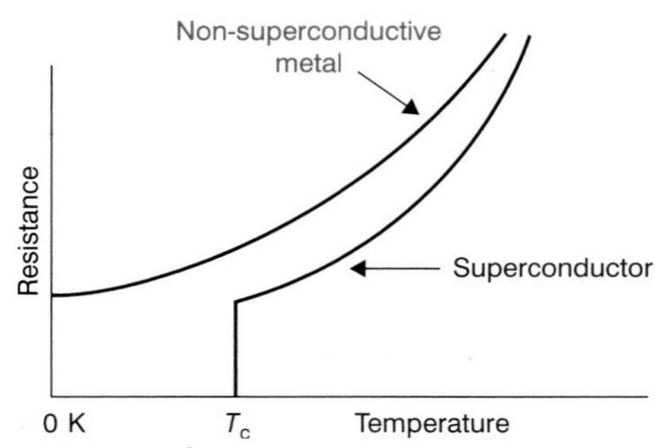

Δ Fig. 2.39 How resistance varies with temperature in a superconductor and a non-superconductor.

Following this discovery, many more metallic elements were found to have superconducting properties, but it wasn't until the 1950s that a theory to explain the behaviour was developed. It requires energy for the electrons to scatter as they move through the metal lattice (these scatterings are the 'collisions' that lead to heating in a resistance) and at such low temperatures this energy is not available, so the electrons move smoothly – with zero resistance. The two key features of this are that no energy is wasted through heating the conductor, which leads to being able to produce very large magnetic fields (see Section 7).

△ Fig. 2.40 A cross-section of a superconductor at CERN (Central European Organisation for Nuclear Research) in Geneva, Switzerland.

The search for superconductors has continued, with breakthroughs coming in the study of alloys rather than elements.

The discovery of materials that demonstrated superconductivity at temperatures up to −183 °C meant that liquid nitrogen could be used as the coolant – and liquid nitrogen is readily available commercially. The search for materials that superconduct at higher temperatures continues.

Superconductors are used in a variety of applications. They produce the strong magnetic fields required for MRI scanning in medicine and to confine beams of particles in accelerators such as the Large Hadron Collider. They even provide magnetic fields to support magnetic levitation (maglev) trains which 'float' above the track. On the small scale, superconductors are used in SQUID (superconducting quantum interference device) magnetometers which can measure the tiny magnetic fields associated with activity in the brain.

VOLTAGE, ENERGY AND CHARGE

The electrons flowing round a circuit have some **potential energy**, which can be referred to as electrical energy. As the electrons pass through the battery, or other power supply, they are given a particular quantity of potential energy, depending on the voltage of the power supply (a 12 V supply will give twice as much potential energy to the electrons as a 6 V supply). As they move around the circuit they transfer the energy that has been transferred to them by the power supply to the various components in the circuit. For example, when the electrons move through a lamp they transfer some of their energy to the lamp.

The amount of energy that a unit of charge (a coulomb) transfers between one point and another (the number of joules per coulomb) is called the **potential difference** (p.d.). Potential difference is the difference in energy of a coulomb of charge between two points in a circuit. Potential difference is measured in volts and so it is often referred to as voltage.

If the potential difference across a lamp, say, is 1 volt, then each coulomb of electricity that passes through the lamp will transfer 1 joule of energy to the lamp.

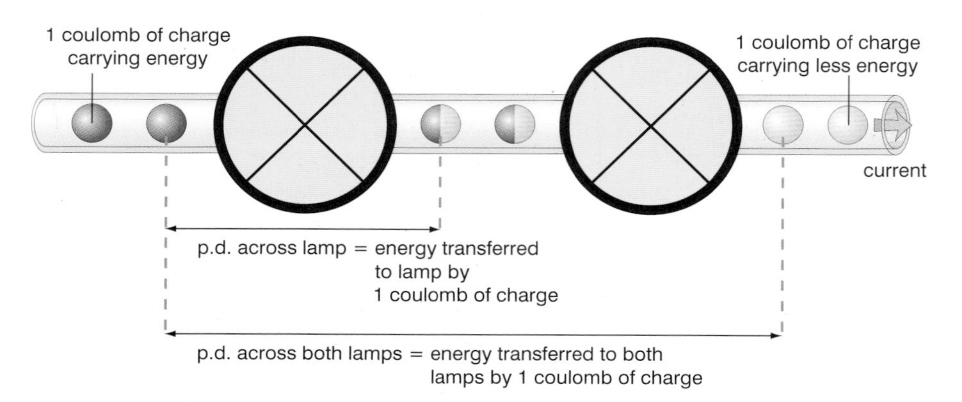

Δ Fig. 2.41 Potential difference (p.d.) is the difference in energy of a coulomb of charge between two parts of a circuit.

Energy transferred, charge and voltage

We have learnt about two relationships between energy transferred, current, voltage, charge and time:

energy transferred = current × voltage × time, $E = I \times V \times t$

and

charge = current × time, $Q = I \times t$

We can combine these two equations to give this useful relationship:

energy transferred = charge × voltage

or $E = Q \times V$

It can save time using a single equation rather than having to use two. The equation also shows that voltage is the energy transferred per unit charge passed.

WORKED EXAMPLE

A charge of 3600 C flows from a 12 V supply. Calculate the total energy transferred by the supply.

Write down the formula:	$E = Q \times V$
Substitute the values for Q and V:	$E = 3600 \times 12$
Work out the answer and write down the unit:	$E = 43\ 200$ J

REMEMBER

Potential difference is measured between two points in a circuit. It is like an electrical pressure difference and measures the energy transferred per unit of charge flowing.

Measuring electricity

Potential difference is measured using a voltmeter. If you want to measure the p.d. across a component then the voltmeter must be connected across that component. Testing with a voltmeter does not interfere with the circuit.

A voltmeter can show how the potential difference varies in different parts of a circuit. In a series circuit there are different values of the voltage depending on where you attach the voltmeter. You can assume that energy is only transferred when the current passes through electrical components such as lamps and motors – the energy transfer as the current passes through copper connecting wire is very small. It is only possible to measure a p.d. or voltage across a component. In a series circuit, the potential difference across the battery equals the sum of the potential differences across each lamp.

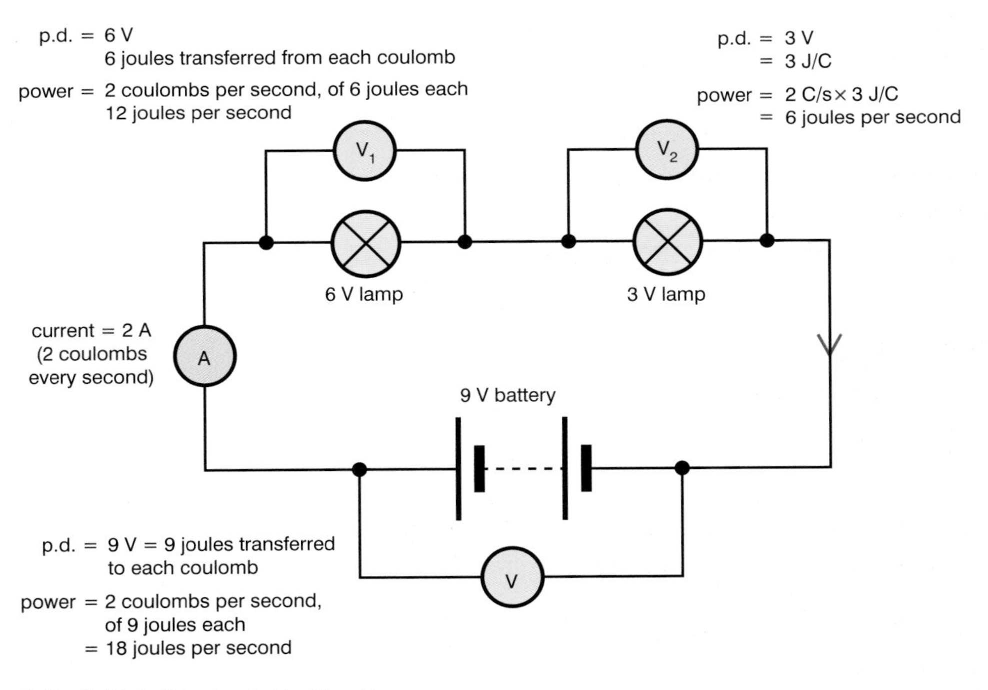

Δ Fig. 2.42 In this circuit, $V = V_1 + V_2$.

QUESTIONS

1. From where do the electrons in a circuit get their energy?
2. What do they do with this energy as they pass round a circuit?
3. What is 'potential difference' between two points in a circuit?
4. A voltage of 6 V means '6 joules of energy transferred by a charge of 1 coulomb'. Is this correct? Give a reason for your answer.
5. The potential difference between two points, A and B, in a circuit is 5 V. What does this mean about the energy of electrons at point A compared to point B?

EXTENSION

In science we use models to help us understand complex ideas and describe things we cannot see. Electrical circuits are sometimes modelled by comparing them with water moving downhill through pipes. A cell can be thought of as a pump, pushing water uphill, wires are like pipes that the water flows down through and a bulb can be thought of like a water wheel, transferring energy as the water moves past it.

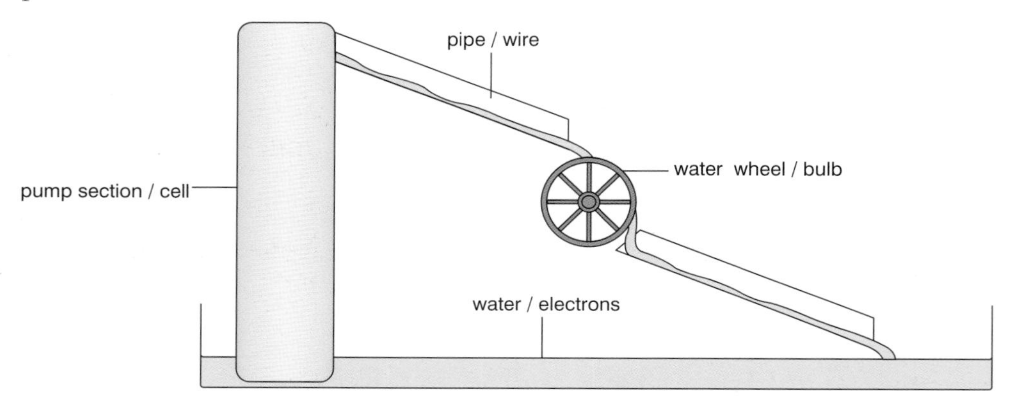

1. Using the same model, try to describe an ammeter and a voltmeter, current and potential difference and the difference between series and parallel circuits. You may use diagrams to help your explanations.
2. How is this a good model for electricity and where does the model have problems?

End of topic checklist

A **series circuit** is an electrical circuit with only one possible path for the current.

A **parallel circuit** is an electrical circuit with more than one possible path for the current.

Conventional current refers to the direction of electric current around a circuit, from positive to negative – the opposite direction to the flow of electrons.

A **coulomb** is the unit of charge.

A **diode** is a semiconductor device that allows an electric current through it in only one direction.

A **light dependent resistor** is a resistor with a resistance that decreases when light is shone on it.

Potential difference is the difference in electrical potential between two points in a circuit – also called the voltage between two points.

A rheostat is a resistor whose resistance can be varied.

A **semiconductor** is a material that does not conduct electricity as well as, for example, a metal, but conducts electricity better than an insulator, such as plastic.

A **thermistor** is a resistor made from semiconductor material: its resistance decreases as temperature increases.

The facts and ideas that you should understand by studying this topic:

- ❍ Know that lamps and LEDs can be used as indicators of current in a circuit.
- ❍ Know and use the relationship between voltage, current and resistance.
- ❍ Know and use the relationship between charge, current and time.
- ❍ Know that in metallic conductors it is negative-charged electrons that are flowing.
- ❍ Understand that current is the rate of flow of charge.

End of topic checklist continued

- ❍ Understand why a series circuit or a parallel circuit may be more appropriate in different situations.
- ❍ Understand why current is conserved at a junction in a circuit.
- ❍ Calculate current, voltage and resistance for components in series circuits.
- ❍ Know about voltages across components in parallel and series circuits.
- ❍ Know and use the relationship between energy transferred, charge and voltage.
- ❍ Understand that voltage is energy transferred per unit charge.
- ❍ Describe how the current varies with voltage, including how to test this experimentally, in
  - wires
  - metal filament lamps
  - resistors
  - diodes.
- ❍ Describe the variation of resistance with
  - light level in LDRs
  - temperature in thermistors.

End of topic questions

1. Look at the following circuit diagrams. They show a number of ammeters and in some cases the readings on these ammeters. All the lamps are identical.

a) For circuit X, what readings would you expect on ammeters A_1 and A_2? **(2 marks)**

b) For circuit Y, what readings would you expect on ammeters A_4 and A_5? **(2 marks)**

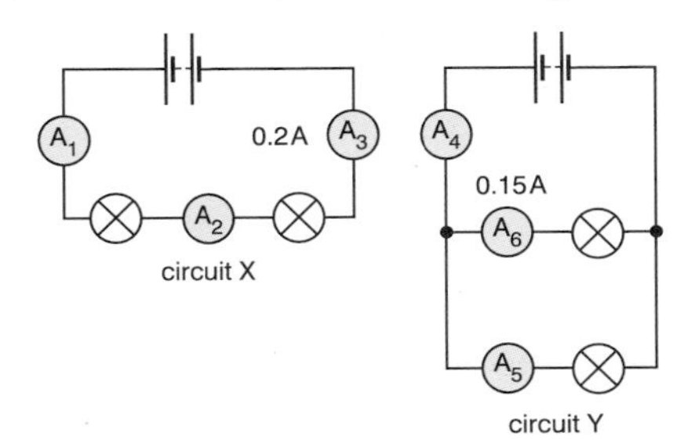

2. Look at the circuit diagram. It shows how three voltmeters have been added to the circuit. What reading would you expect on V_1? **(2 marks)**

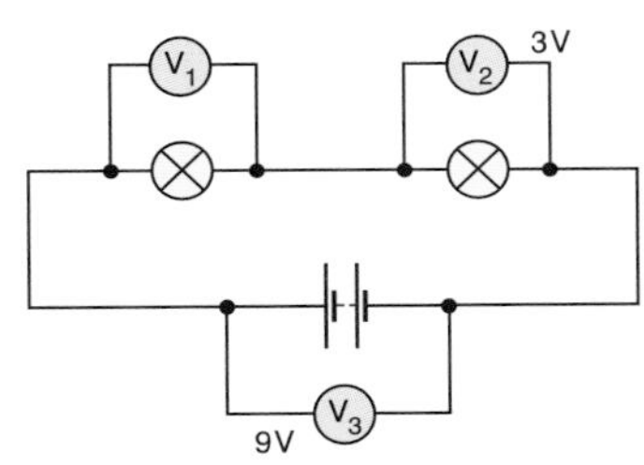

3. a) A charge of 10 C flows through a motor in 30 seconds. What is the current through the motor? **(2 marks)**

b) A heater uses a current of 10 A. How much charge flows through the lamp in:

i) 1 second **(2 marks)**

ii) 1 hour? **(2 marks)**

4. Use Ohm's law to calculate the following:

a) The voltage required to produce a current of 2 A in a 12 Ω resistor. **(2 marks)**

b) The voltage required to produce a current of 0.1 A in a 200 Ω resistor. **(2 marks)**

c) The current produced when a voltage of 12 V is applied to a 100 Ω resistor. **(2 marks)**

d) The current produced when a voltage of 230 V is applied to a 10 Ω resistor. **(2 marks)**

e) The resistance of a wire which under a potential difference of 6 V carries a current of 0.1 A. **(2 marks)**

f) The resistance of a heater which under a potential difference of 230 V carries a current of 10 A. **(2 marks)**

End of topic questions continued

5. Look at the circuit diagram on the right and answer the following questions:

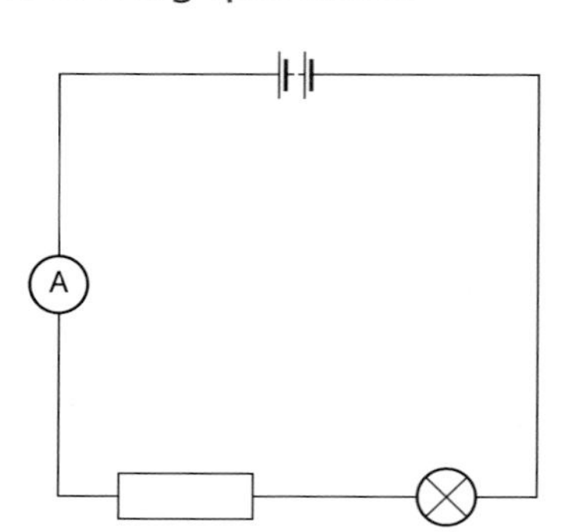

a) What supplies the voltage in this circuit? **(1 mark)**

b) What will happen to the current if another cell is added? **(1 mark)**

c) If the resistance of the circuit is increased, what will happen to the current in the circuit? **(2 marks)**

d) If the ammeter shows 0.3 A and the resistor is 5 Ω, what is the voltage of the cell? **(2 marks)**

6. An engineer was testing the resistance of a component and obtained the table of data shown below.

Voltage (V)	Current (A)
0.5	0.14
1	0.29
1.5	0.43
3	0.86
4	1.14
5	1.43
5.5	1.57
6.5	1.86

a) What type of graph should he draw and why? **(2 marks)**

b) Which was his independent variable? **(1 mark)**

c) Do you consider these results to be reliable? **(2 marks)**

d) How could he have improved the precision of his measurements? **(2 marks)**

e) Plot the graph and use the slope to calculate the resistance. **(3 marks)**

7. Copy and complete the following table, using the equation for Ohm's law. **(5 marks)**

Potential difference (V)	Current (A)	Resistance (Ω)
	0.15	2
6	0.2	
	0.5	12
12	3	
240		18.5

8. In the circuits below, what you would expect to read on each ammeter and voltmeter? All lamps are the same, and each cell produces 1.5 V. What *could* happen to these values if the bulbs had different resistances? **(12 marks)**

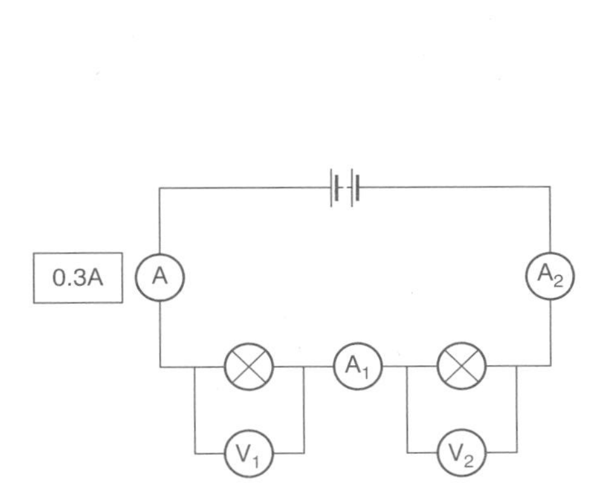

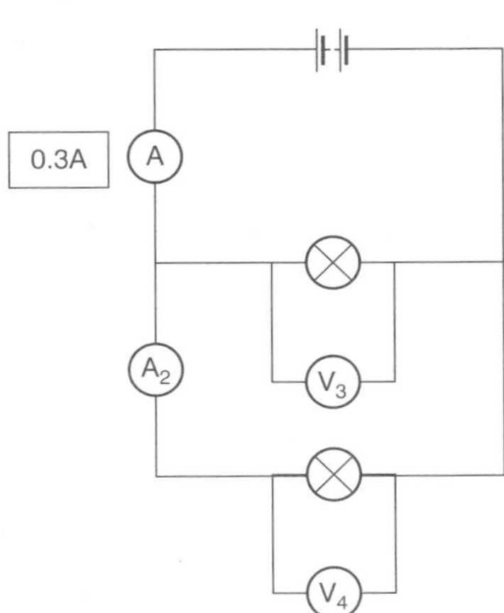

9. For the circuit below are the pairs of components connected in series or in parallel? **(1 mark)**

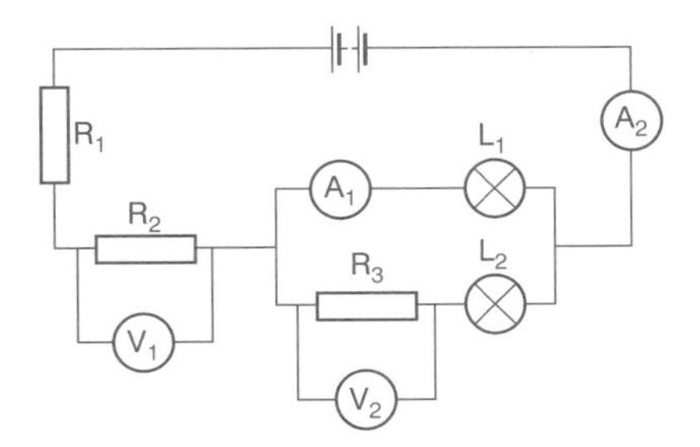

a) R_1, R_2 **(1 mark)**

b) L_1, L_2 **(1 mark)**

c) R_3, V_2 **(1 mark)**

d) R_3, L_2 **(1 mark)**

e) A_1, L_1 **(1 mark)**

f) R_2, V_1 **(1 mark)**

g) A_1, R_3 **(1 mark)**

h) R_1, A_2 **(1 mark)**

End of topic questions continued

10. A student has built the circuit shown on the right. She is trying to measure the current in each path of the circuit and voltage across each of the lamps as she changes the resistance of one path using the variable resistor.

a) Draw a diagram showing how the circuit should be built to make these measurements correctly. **(2 marks)**

b) Design the headings of a table of results you would use if doing this experiment. **(2 marks)**

11. A 24 V supply delivers a current of 6 A to three resistors in series. One of the resistors has a value of 1 Ω. The other two different resistors are identical. Calculate *X*, the value of the resistance. **(4 marks)**

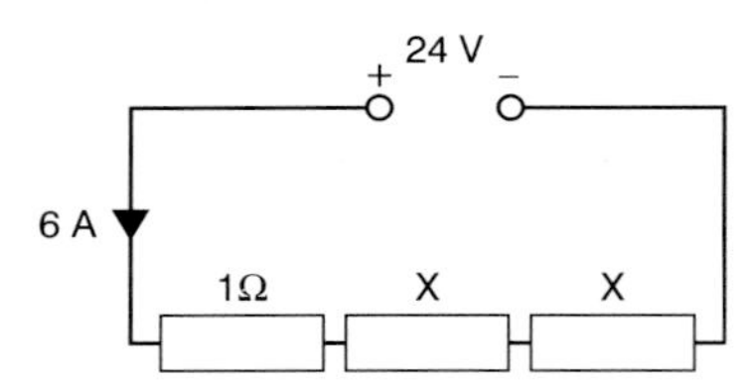

Electric charge

INTRODUCTION

Have you ever brushed your hair and seen individual hairs standing on end and wondered why this happens? Would you like to know how to stick a balloon to a surface without glue? This section explores how and why these things happen. They are both the result of movement of electric charge. You will learn about the forces between charges. You will also learn about the potential hazards of electrostatics, before looking at how electrostatic effects can be useful to us.

Δ Fig. 2.44 The blue light is caused by a flow of electric charge.

KNOWLEDGE CHECK

✓ Know that common electric effects are caused by imbalances in the number of electrons present.
✓ Know that forces can be attractive or repulsive.

LEARNING OBJECTIVES

✓ Be able to identify common materials which are electrical conductors or insulators, including metals and plastics.
✓ Be able to investigate how insulating materials can be charged by friction.
✓ Be able to investigate how insulating materials can be charged by friction.
✓ Explain how positive and negative electrostatic charges are produced on materials by the loss and gain of electrons.
✓ Explain common electrostatic effects in terms of the movement of electrons.
✓ Know that there are forces of attraction between unlike charges and forces of repulsion between like charges.
✓ Explain the potential dangers associated with electrostatic charges.
✓ Describe some uses of electrostatic charges.

CONDUCTORS AND INSULATORS

You have already seen that materials that allow an electric current to pass through them are called conductors; those that do not are called insulators.

Metals are conductors. In a metal structure, the metal atoms exist as **ions** surrounded by 'sea' or 'cloud' of negatively charged electrons. If a potential difference is applied to the metal, the electrons in this cloud are able to move, creating a current.

When the electrons are moving through the metal structure, they bump into the metal ions and experience resistance to the electron flow or current. Different conductors have different resistances according to how easily their electrons flow. For instance, copper is a better conductor than iron.

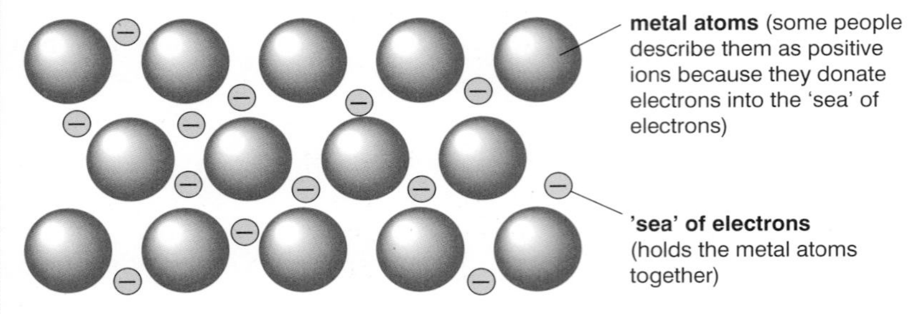

Δ Fig 2.45 In a metal structure metal ions are surrounded by a cloud or 'sea' of electrons.

The table below lists materials ranging from the best conductor to the best insulator. The range in the resistance of different materials is truly amazing. Silver is about 10^{27} times better at conducting charges than the plastic Teflon®. So to replace a 1 mm diameter wire of silver or copper in an electrical circuit, you would need a bar of Teflon® far larger in diameter than the Moon's orbit around the Earth.

Silver	metal	conductor (best)
Copper	metal	conductor
Aluminium	metal	conductor
Iron	metal	conductor
Graphite		conductor
Silicon		semiconductor
Most plastics		insulator
Oil		insulator
Glass		insulator
Teflon®		insulator (best)

Δ Table 2.3 Conductors and insulators.

CHARGING INSULATORS BY FRICTION

Materials such as glass, acetate and polythene can only become charged because they are insulators. Electrons do not move easily through insulating materials, so when extra electrons are added, they stay on the surface instead of flowing away, and the surface stays negatively charged. Similarly, if electrons are removed, electrons from other parts of the material do not flow in to replace them, so the surface stays positively charged. Conductors, such as metals, cannot be charged by rubbing.

Charge as the loss or gain of electrons

All atoms are made up of three main kinds of particles, called electrons, **protons** and **neutrons**. Electrons are the tiniest of these and have a negative charge. Protons and neutrons have about the same mass, but protons are positively charged, while neutrons have no charge. Protons and neutrons are found in the nucleus of the atom and electrons are found as a cloud surrounding (see section 7, Radioactivity and particles)

In most objects there are as many electrons as protons. So normally an object has no overall charge, because the size of the positive charge on the protons is equal to the size of the negative charge on the electron. If there are more electrons than protons, the object carries an overall negative charge.

If there are fewer electrons than protons, the object carries an overall positive charge.

When you charge an object, you are giving or taking away negatively charged electrons, so that the charge on the object overall is unbalanced. For example, when you rub a glass or acetate rod with a cloth, electrons from the rod get rubbed onto the cloth. So the cloth becomes negatively charged overall, and the rod is left with an overall positive charge. When an unbalanced charge collects on the surface of an object, the charge is called **static charge**. ('Static' means 'not moving'.)

When you rub a polythene rod with a cloth, electrons from the cloth get transferred to the rod, so the polythene carries a negative charge overall, and the cloth carries a positive charge.

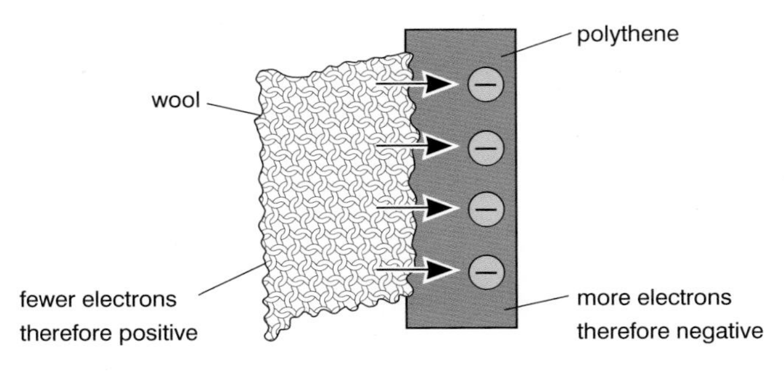

Δ Fig. 2.46 Transferring charge from a cloth to a rod.

REMEMBER

The amount of charge on an object is measured in coulombs. A charge of 1 C is the charge on 6.24×10^{18} electrons, so an object with a charge of +1.0 C has 6.24×10^{18} too few electrons. Typical electrostatic charges are less than 1 μC (1.0×10^{-6} C).

QUESTIONS

1. Have you ever 'got a shock' from a metal handrail? Explain why it was really the other way round – you *gave* a shock *to* the handrail.

ATTRACTION AND REPULSION

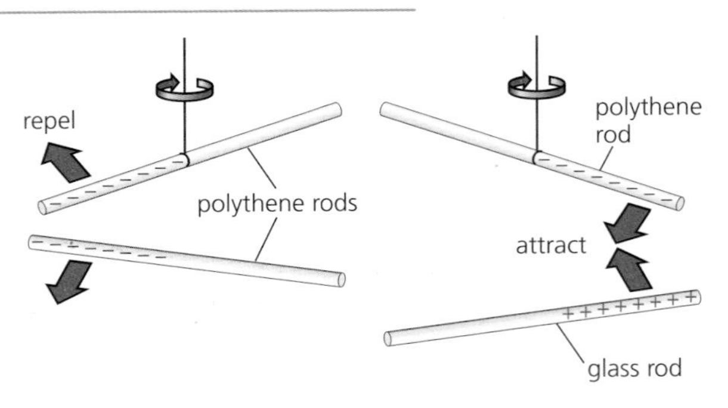

Δ Fig. 2.47 Like charges repel, unlike charges attract.

A charged object feels a force towards or away from another charged object. We say there are electrostatic forces between the charges. The strength of the force depends on:

- how close the particles are: the closer they are, the larger the force
- how much electrical charge they carry: the more charge, the larger the force.

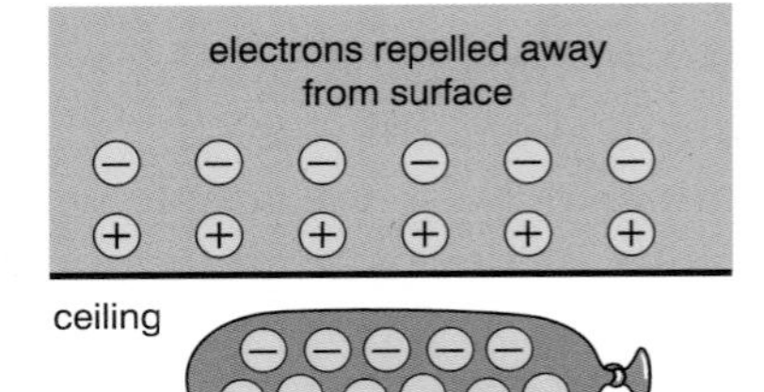

Δ Fig. 2.48 The balloon induces a charge on the ceiling's surface.

If you suspended charged polythene and glass rods so they could move freely, and brought the two close together, they would attract each other, because unlike charges attract.

Similarly, when a balloon is rubbed against clothing it will 'stick' to a wall or ceiling. This is because of the attraction between the negative charges on the balloon and the induced positive charges on the ceiling.

Developing investigative skills

A student investigates charging materials by friction (rubbing). He is asked to find out what sort of materials can be charged and what sort of forces they produce (attraction or repulsion). The student knows that electrostatic forces are small and small pieces of paper can be moved by them.

The following equipment is available:

Polythene rods, acetate rods, metal rods, dry cloths, small pieces of dry paper, nylon thread.

Devise and plan investigations

❶ How can the student find out which rods can be charged?

❷ How can he find out if the rods have the same charge or a different charge?

❸ Why is it important that the investigation is carried out in dry conditions?

Analyse and interpret

❹ One of the rods cannot be charged at all. Which rod cannot be charged? Suggest a reason why.

ELECTROSTATICS

As you can see, electrostatic forces look rather similar to magnetic forces. However, they are completely different. An electric field does not affect a magnet in any special way, and a magnetic field does not affect an electric charge (so long as it is not moving). You can even have a space that contains both types of field in different directions at the same time.

All the phenomena of electrostatics can be explained in terms of moving negative or moving positive charges. Early scientists did not know which it was, and it was only towards the end of the 19th century that they became sure that it was the negative charges that moved.

△ Fig. 2.49 Because the static charge on each hair is similar, the hairs repel each other and stick up in all directions.

QUESTIONS

1. The key discovery at the end of the 19th century was the electron. From your knowledge of atoms, explain why it is negative charges that move to create electrostatic effects.
2. Describe how electrostatic effects stop you from falling through the floor.
3. a) Copy and complete the diagrams below to show what happens when a cloth is rubbed on a material that gains electrons, and when a cloth is rubbed on a material that loses electrons.

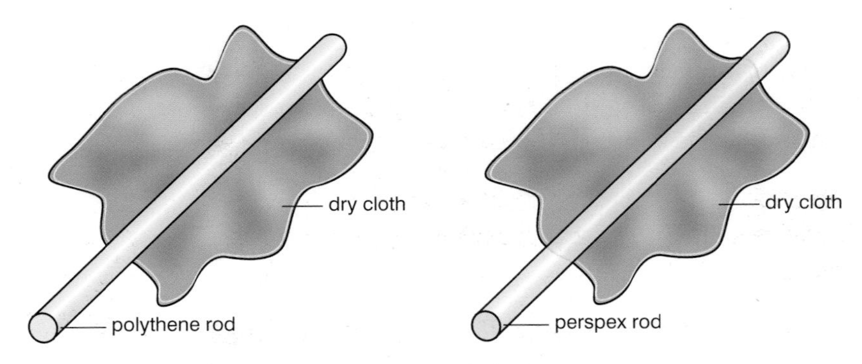

   b) What can you say about the sizes of the positive/negative charges between the cloth and the rod?

HAZARDS OF ELECTROSTATICS

The sudden discharge of electricity caused by friction between two insulators can cause shocks in everyday situations – for example:

- combing your hair
- pulling clothes over your head
- getting out of a car.

You may have noticed that you can get a nasty spark from your finger if you touch a metal object after rubbing your feet on a nylon carpet. This is similar to the effect you can sometimes feel if you touch a metal door handle. It is for this reason that workers who make sensitive electronic devices connect themselves to ground using devices such as antistatic wrist straps. These link to a grounding point (a point which is connected to 0 V) so that any static charge can discharge safely via the wrist strap and not damage the equipment before starting work. To protect against sparks of this type, aircraft are connected to the ground by a special wire before refuelling starts.

Δ Fig. 2.50 Using an antistatic wrist strap.

USES OF ELECTROSTATICS

Electrostatics can be useful: electrostatic scrubbers remove the dust from the smoke of coal-fired power stations, and photocopiers use electrostatics to move the ink powder to the right place on the paper. Electrostatic charges on the filter in an air cleaner or in a vacuum cleaner are responsible for much of the filter's effectiveness.

The electrostatic inkjet printer is used industrially to label the sell-by date onto the cans or bottles passing by on a conveyer belt. In this printer, a nozzle continuously produces a jet of droplets at a very high speed (50 m/s). The droplets are either collected up by the printer – or they are aimed at the cans – by using the fact that each droplet is given a small charge by friction as it passes through the nozzle. The printer can then steer the droplet by applying voltages to metal plates fixed near to the jet. The inkjet printer for the PC is much simpler, and uses small electrical heaters to squirt ink out of nozzles as the ink head passes over the places where ink is needed on the paper.)

The properties of static charges are also used in ink jet printers and photocopiers.

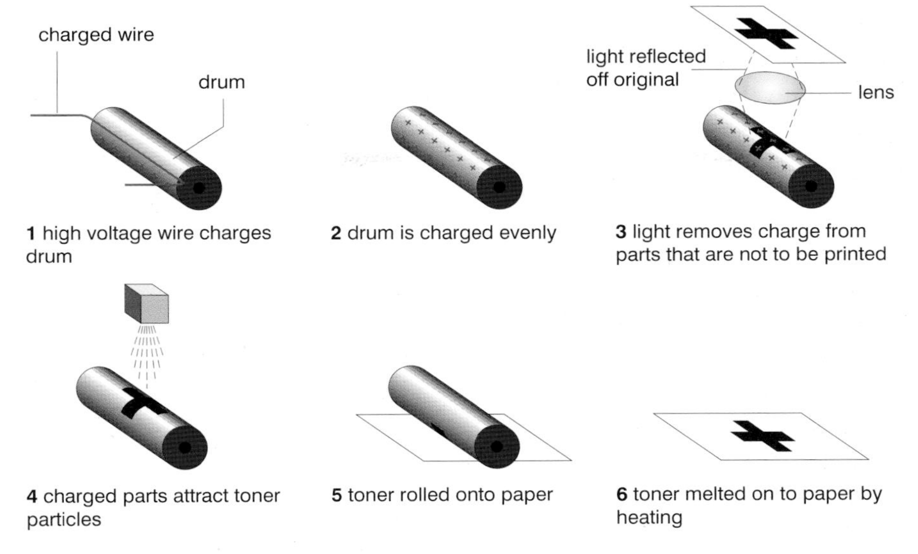

Δ Fig. 2.51 In a photocopier, charged particles attract the toner. Light is used to remove charge from parts that are not to be printed.

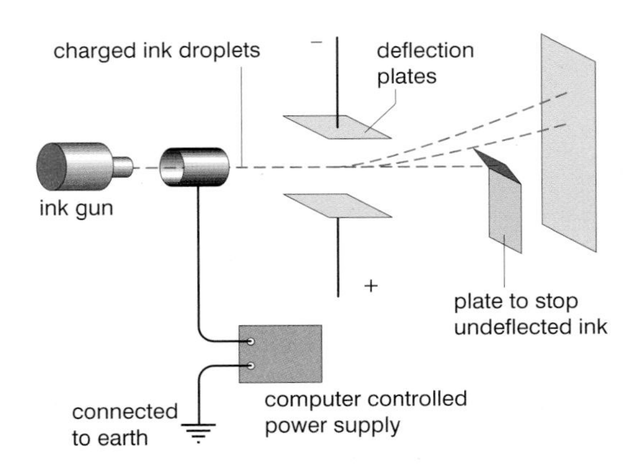

◁ Fig. 2.52 In ink jet printers, uncharged ink droplets do not reach the paper. This is how the spaces between words are made.

SPARKS

Normally, air does not conduct electricity. The electrons in each atom in the air remain bound to that atom and so there are no 'free' electrons available to move in an organised way (an electric current). However, if the nearby charge becomes strong enough – if there is a large enough electric field – then the force of attraction or repulsion on an electron may become so large that the attraction to the nucleus of the atom may not be enough to hold it. This then *ionises* the atom: that is, the electron is separated off from the 'main' part of the atom. As the atom is now missing an electron it has an overall positive charge and is called a positive *ion*.

Ionised air behaves very differently to 'normal' air. The presence of the free electrons and the positive ions means that an electric current can now pass through the air: the air becomes a conductor. As an electric current passes through it, heat is generated (since the air still has quite a high resistance) and the air emits light. This is the effect we would notice as a spark.

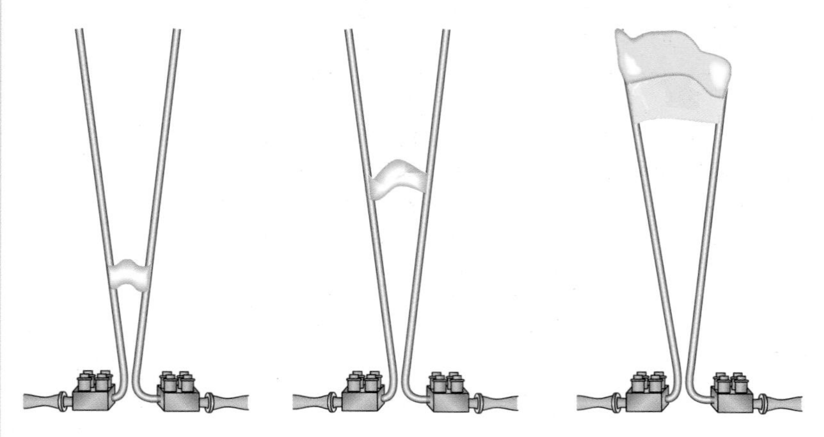

Δ Fig. 2.53 A spark between two electrical contacts.

As the spark (electric current passing through the air) heats up the air, it causes the air to expand rapidly. This expansion causes a vibration to pass through the air as the particles from the expanding air collide with nearby particles. This energy transfer is the sound wave we associate with a spark – it is the crackle we hear. If you have a piece of clothing that crackles when you take it off, then the crackles are caused by tiny sparks moving between fibres in the cloth. If you take off such a piece of clothing in the dark, you can see the sparks.

Sparks are dangerous in places such as petrol filling stations. It is the heating effect that poses the major risk. To ignite any fuel vapour in the air a source of heat is required and it the heat generated by the electric current in the air, the spark, that provides the heat source.

QUESTIONS

1. Give an everyday example of where you might receive a small electrostatic shock.
2. The table below lists some situations where static electricity can be dangerous to people or equipment in the workplace. Copy and complete the table by stating the danger and one way to prevent the danger. One has been done for you.

Situation	Danger	Prevention
Refuelling a car	Sparks could form as charge builds up, causing an explosion.	Tyres contain graphite – a conductor that transfers electrons to earth.
Making sensitive electronic equipment		
Refuelling an aeroplane full of passengers		

3. Give an example of where electrostatics are useful.
4. How are electrostatic forces similar to magnetic forces?

End of topic checklist

Common **electrostatic** effects are caused by having extra electrons or a shortage of **electrons**.

Electrostatic effects happen when **insulators** have gained or lost electrons.

Friction can cause insulators to become charged.

The facts and ideas that you should understand by studying this topic:

- ❍ Know that some materials are conductors and that others are insulators, giving examples of each.
- ❍ Be able to investigate how insulating materials can be charged by friction.
- ❍ Understand electrostatic effects in terms of movement of electrons.
- ❍ Understand that materials become negatively-charged overall when they gain electrons and positively-charged overall when they lose electrons.
- ❍ Understand that forces of attraction occur between unlike charges and forces of repulsion occur between like charges.
- ❍ Explain some uses of electrostatic charges.
- ❍ Explain the potential dangers of electrostatic charges.

End of topic questions

1. A plastic rod is rubbed with a cloth.

a) How does the plastic become positively charged? **(2 marks)**

b) The charged plastic rod attracts small pieces of paper. Explain why this attraction occurs. **(2 marks)**

2. a) A car stops and one of the passengers gets out. When she touches a metal post, she feels an electric shock. Explain why she feels this shock. **(2 marks)**

b) Write down two other situations where people might get this type of shock. **(2 marks)**

3. Give two examples of where electrostatic charges can be dangerous. **(3 marks)**

4. Your young cousin has noticed that when she combs her hair it sometimes crackles, and she asks you why this is. Write her a short, clear explanation of what is happening. **(3 marks)**

You then decide to show her the trick of picking up tiny pieces of paper with the charged comb. Again, explain to her what is happening to allow the comb to pick up the pieces of paper. **(3 marks)**

5. A factory-equipment manufacturer is attempting to build a new conveyor system using nylon rollers with a rubber belt around them. Using the table below, explain why these materials might cause a problem, and suggest what material pairing might be better to use. **(4 marks)**

		Suitable for use on:	
Material	**Charge collected**	**roller**	**belt**
nylon	positive	Y	Y
glass	positive	Y	N
leather	large positive	N	Y
wood	small negative	Y	N
polystyrene	negative	Y	N
polyurethane	negative	Y	Y
rubber	negative	N	Y

6. Which of the pairs of materials below are likely to build up a charge when rubbed together? Give a reason for your answer. **(8 marks)**

iron and steel rods	plastic comb and hair	duster and polythene spoon	duster and aluminium spoon
polyester clothes and hair	leather and rubber	aluminium and zinc rods	duster and glass rod

End of topic questions continued

7. Static can cause problems for people in a range of different situations. Here Jim and Rashida tell us about the ways they deal with static in their workplaces.

Jim

"I work in the electronics industry for a company that makes computer parts and repairs broken electrical equipment. If I get a build-up of static on my body, then it can end up discharging into the thing I'm working on. As I deal with a lot of sensitive electrical equipment, this kind of charge build-up can cause quite a bit of damage if it leaks. To combat this I wear an anti-static wrist strap, which links to a grounding point so the charge can be moved away safely."

Rashida

"I work at the airport on refuelling the aircraft. Obviously any kind of electrical spark would be a real danger, as it could ignite the fuel or the fumes that come off it. I wear a range of anti-static clothing to help keep me safe. This includes trousers and jackets made of materials that don't build up a charge easily when they are rubbed, and I have boots that have anti-static properties. The aircraft can build up a static charge on them as they fly through the air, so they have a grounding strap that discharges them as they land. Strangely, the job is more dangerous in the winter, as the air is less humid then so charge build-up can't dissipate as easily."

a) List the safety devices used by Jim and Rashida. **(4 marks)**

b) Why does Jim need to wear an anti-static wrist strap? **(2 marks)**

c) Can you think of any other jobs in which an anti-static wrist or ankle strap could be necessary? **(2 marks)**

d) What is the main hazard of Rashida's job? **(1 mark)**

e) What material might be good for Rashida's clothing to be made of? What would be a bad combination of materials for him to wear? **(2 marks)**

f) Do you think the anti-static boots are insulating or conductive? **(2 marks)**

g) Why is the dry air in winter more dangerous? **(2 marks)**

EXAMINER'S COMMENTS

a) The correct response has been given. The symbols '+' and '–' would also be acceptable.

b) 'He could rub it' scores one mark, although 'by friction' would have been a stronger phrase to use. There is a second mark for saying that the comb should be rubbed against an insulator (or you could give an example of an insulator, such as cloth). Always check the number of marks available.

c) One mark has been given for the correct response. The student indicates correctly that there will be an attraction between the comb and the pieces of paper.

d) This is a very vague answer. There are three marks available. The correct response needs to realise that Leon has become charged (perhaps by friction against a carpet), that these charges move when he touches the radiator, from Leon to the radiator. The third mark is for using correct scientific words. Relevant words here are: charging, electrons, earth, earthing.

Exam-style questions
Sample student answers

Question 1

This question is about electrostatics.

a) There are two kinds of electric charge.

Write down the names of both types of electric charge.

positive and negative ✓ (1) **(1)**

b) Leon wants to charge his plastic comb.

Write down one way he could charge his plastic comb.

He could rub it. ✓ (1) **(2)**

c) Leon holds his charged comb near some small pieces of paper.

Suggest what might happen to the papers.

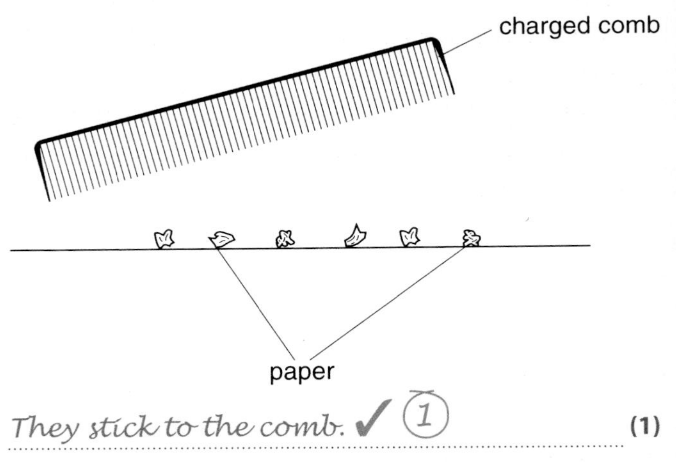

They stick to the comb. ✓ (1) **(1)**

d) Leon touches a metal radiator. He gets an electric shock.

Describe how Leon gets an electric shock.

(One mark is for the correct use of scientific words.)

The metal radiator is electric and gives Leon a shock. ✗ ✗ ✗ **(3)**

Exam-style questions continued

e) Leon paints cars.

Static electricity is useful in spraying paint.

i) Write down **one other** use of static electricity.

A photocopier ✓ (1) **(1)**

ii) Explain why static electricity is useful in spraying cars.

Use ideas about electric charge in your answer.

(One mark is for linking ideas.)

The paint is charged when it comes out of the sprayer. The car is also charged with the opposite charge. ✓ *This makes the paint stick to the car much better.* ✓ (2) **(4)**

(Total 12 marks)

e) i) One mark has been given. Alternative correct responses would include inkjet printers, dust precipitators or crop spraying.

ii) Two marks have been given. The student seems to have an idea of what is happening, but has failed to use the correct scientific terms accurately. One mark has been awarded for the idea that opposite charges attract, but saying the paint 'sticks' to the car is not accurate enough to gain a second mark – the student needed to say that the paint is attracted to the car. In a similar way, saying the paint covers 'much better' is too vague. At this level, the student should refer to the paint being attracted to the whole object, even parts not in a direct line, or that less paint is wasted. Another approach would be to state that like charges repel (one mark) and so the paint forms a fine spray (one mark), which produces an even coat (one mark).

Exam-style questions continued

Question 2

An electric iron has a power rating of 1100 W.

It is designed to run from a 230 V a.c. supply.

a) Explain why an electric current causes the iron to become hot. **(2)**

b) **i)** State the equation linking power, current and voltage. **(1)**

ii) Calculate the current in the iron when it is operating normally. **(2)**

c) The mains plug attached to the iron contains a fuse rated at 13 A.

Describe the purpose of the fuse. **(3)**

d) The mains connection for the iron also contains an earth wire.

i) Why is an earth wire needed for an electric iron? **(2)**

ii) Describe the operation of an earth wire. **(3)**

e) Explain the difference between an alternating current (a.c.) and a direct current (d.c.). **(2)**

(Total 15 marks)

Exam-style questions continued

Question 3

A student investigates how the current varies with voltage in a filament lamp.

This is the student's circuit diagram.

The lamp is fully bright when 12 V is applied.

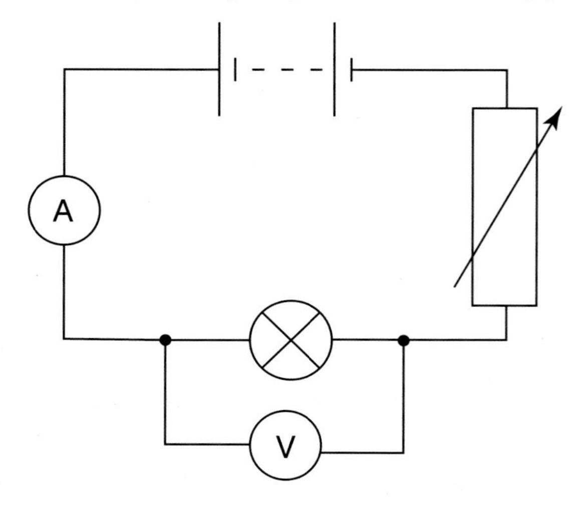

a) Describe how the student should use the circuit to carry out the investigation. **(2)**

b) **i)** Sketch the graph the student should obtain if the lamp operates normally. Put the voltage on the vertical axis and the current on the horizontal axis. **(3)**

ii) Explain the shape of the graph you have drawn. **(3)**

c) The student carries out the experiment again using a lamp that is designed to be fully bright when 24 V is applied.

Suggest how the graph obtained would be different if the student carried out the test using the same 12 V supply.

Explain your answer. **(3)**

d) The variable resistor used in the investigation has a resistance of 15 Ω.

i) State the equation linking voltage, current and resistance. **(1)**

ii) Calculate the current in the variable resistor if the voltage across it is 3 V. **(2)**

iii) State the equation linking charge, current and time. **(1)**

iv) Calculate the charge that passes through the variable resistor in 10 minutes when the voltage across it is 3 V.

State the correct unit. **(3)**

(Total 18 marks)

Exam-style questions continued

Question 4

A student investigates electrostatic charges.

She has some insulating rods made of different types of plastic.

She rubs each one with a cloth to create the electrostatic charge.

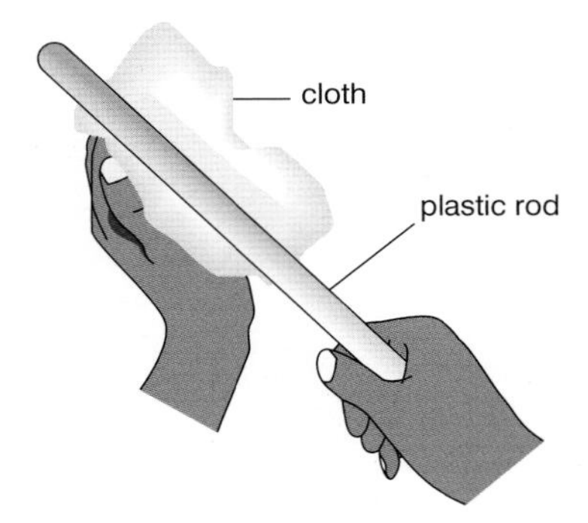

a) Using ideas about electrons, explain how an insulating rod can gain a positive or a negative electrostatic charge. **(2)**

b) A metal rod cannot gain an electrostatic charge in this way.

Explain why not. **(2)**

c) Describe an experiment the student could do to find out if two insulating rods had the same charge or opposite charges.

You should describe the equipment the student would need and how she should use it. **(4)**

d) Describe a situation where electrostatic charges can potentially be dangerous.

You should include:

what the hazard is

how the risk can be reduced. **(3)**

(Total 11 marks)

What is the connection between the waves you see on water and light? Light is a wave that behaves in the same way as water waves. Sound is another type of wave, as you will learn later in this section. Studying the behaviour of waves will help you to understand many of your everyday experiences, ranging from how you see objects to how you hear sounds.

You should already know that energy can be transferred as sound and light. You should also know that white light is made up of a range of different colours and that light can be reflected and refracted. You should also know how the frequency and amplitude of a sound wave are related to the pitch and loudness of the sound.

STARTING POINTS

1. Explain why a red object looks red.
2. Describe the pitch and loudness of the sound you hear if the sound wave has a large amplitude and the frequency is low.
3. Explain the meaning of the words translucent, transparent and opaque.
4. How does light travel through space?
5. Draw a diagram to show how light is reflected in a plane mirror.

CONTENTS

3 Waves

Δ Light is a wave and has many of the properties that a wave in the sea has.

Units

INTRODUCTION

For the topics included in waves, you will need to be familiar with:

Quantity	Unit	Symbol
angle	degree	°
time	second	s
frequency	hertz	Hz
length, wavelength	metre	m
speed	metre per second	m/s

Some of these you have met previously but the new units will be defined when you meet them for the first time.

LEARNING OBJECTIVES

✓ Use the following units: degree (°), hertz (Hz), metre (m), metre/second (m/s) and second (s).

Properties of waves

INTRODUCTION

The behaviour of waves affects us every second of our lives. Sound waves, light waves, infrared heat, television, mobile-phone and radio waves – the list goes on; waves are reaching us constantly. The study of waves is, perhaps, truly the central subject of physics. The woman in Figure 3.1 is surrounded by waves: she can feel the infrared waves from the Sun coming in through the window, she can hear the sound waves of her friend on the phone, the phone is using radio waves, and she can see around her with light waves.

Δ Fig. 3.1 There are many examples of waves in action.

KNOWLEDGE CHECK

✓ Know some simple examples of wave motion.
✓ Be able to measure lengths and times.
✓ Be familiar with everyday examples of reflection.

LEARNING OBJECTIVES

✓ Be able to explain the difference between longitudinal and transverse waves.
✓ Understand that waves transfer energy and information without transferring matter.
✓ Be able to describe different examples of wave motion using the terms amplitude, wavefront, frequency, wavelength and period.
✓ Be able to use the relationship between the speed, frequency and wavelength of a wave: $v = f \times \lambda$.
✓ Be able to use the relationship between frequency and time period: $f = 1/T$.
✓ Be able to explain why there is a change in the observed frequency and wavelength of a wave when its source is moving relative to an observer.
✓ Be able to explain how all waves can be reflected and refracted.

LONGITUDINAL AND TRANSVERSE WAVES

There are two types of waves: longitudinal and transverse.

Longitudinal waves. In a longitudinal wave, the vibrations are parallel to the direction of travel of the wave. This type of wave can be shown by pushing and pulling a spring. The spring stretches in places and squashes in others. The stretching produces regions of **rarefaction**, whilst the squashing produces regions of **compression**. Sound is an example of a longitudinal wave.

Transverse waves. In a transverse wave, the vibrations are at right angles to the direction of travel of the wave. Light, radio and other **electromagnetic waves** are transverse waves.

In the examples below, the waves are very narrow, and are confined to the spring or the string that they are travelling down. Most waves are not confined in this way. A wave on the sea, for example, can be hundreds of metres wide as it moves along.

Water waves are often used to demonstrate the properties of waves because their **wavefront** is easy to see. A wavefront is the moving line that joins all the points on the crest of a wave.

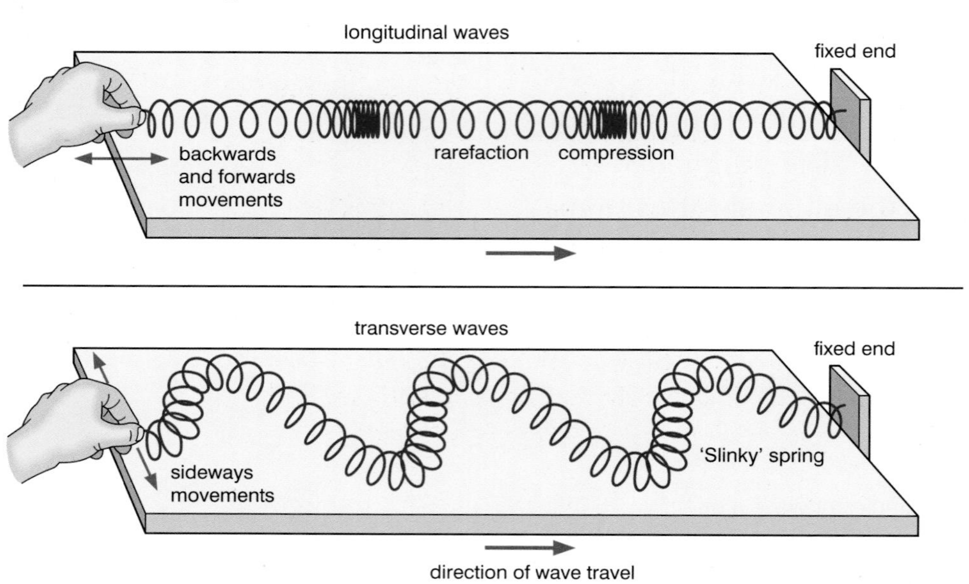

Δ Fig. 3.2 Longitudinal and transverse waves are made by vibrations. Both types of wave have a repeating shape or pattern.

AMPLITUDE, FREQUENCY, WAVELENGTH AND PERIOD

Waves have a wavelength, frequency, amplitude and time period.

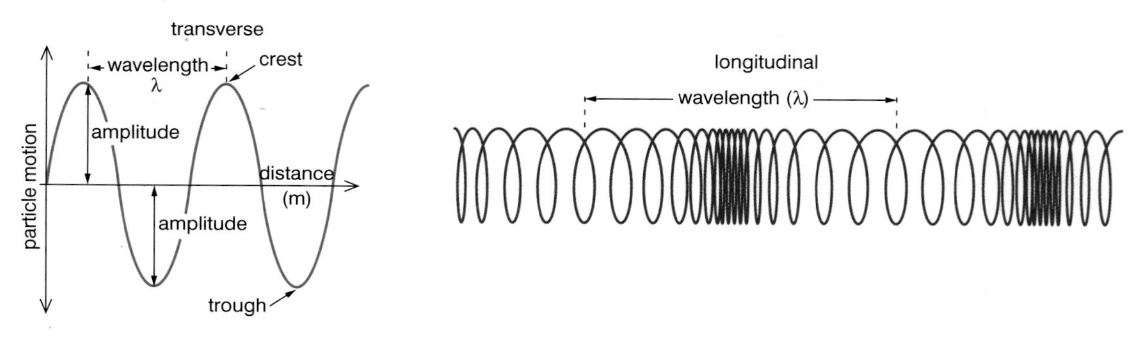

Δ Fig. 3.3 The wavelength and amplitude of a transverse wave and the wavelength of a longitudinal wave.

REMEMBER

Take care with graphs or diagrams like this. Make sure you notice if it is 'distance' or 'time' along the x-axis. Some labels only apply to one type of graph.

The **wavelength** is the distance in metres between two adjacent peaks or, if you prefer, the distance between two adjacent troughs of the wave. In the case of longitudinal waves, it is the distance between two consecutive points of maximum compression, or the distance between two consecutive points of minimum compression.

The **frequency** in **hertz** (Hz) is the number of peaks (or the number of troughs) that go past each second.

The **time period** is the time taken in seconds for each complete cycle of the wave motion.

The **amplitude** is the maximum particle displacement in metres of the medium's vibration from the undisturbed position. In transverse waves, this is half the crest-to-trough height.

The **wave speed** is the distance the wave travels in 1 s. The speed depends on the substance or medium the wave is passing through.

The largest ocean wave ever measured accurately had a wavelength of 340 m, a frequency of 0.067 Hz (one peak every 15 s), and a speed of 23 m/s. The amplitude of the wave was 17 m, so the ship that was measuring the wave was going 17 m above the level of a smooth sea and then 17 m below. (The waves went down 34 m from crest to trough.)

WAVES TRANSFER ENERGY AND INFORMATION

A wave carries energy and can carry information. You can feel the energy in infrared waves from the Sun as they strike your hands; you can see the energy contained in the ocean waves from a typhoon as they reach the coast after travelling hundreds of miles. And you can see the information contained in the light reaching your eyes from this page, or from a movie screen.

But note that in none of these cases has any object or matter travelled by vibrations from the source of the waves to the destination. Instead the wave is passed on from point to point along the route it takes. One good example is a piece of wood in the sea. It is jiggled up and down, and to and fro, by a wave, but after the wave has passed, the wood ends up where it started. Surfers can travel by catching a wave and 'riding' it, but they are outside the wave, not part of it.

QUESTIONS

1. Describe how the vibrations travel in **a)** a longitudinal wave **b)** a transverse wave.
2. Define the following terms for a wave: **a)** wavelength **b)** frequency **c)** amplitude.
3. How far does a wave with speed 5 m/s travel in 3 s?
4. What do waves transfer?

RELATIONSHIP BETWEEN SPEED, FREQUENCY AND WAVELENGTH

The speed of a wave in a given medium is constant. If you change the wavelength, the frequency must change as well. If you imagine that some waves are going past you on a spring or on a rope, they will be going at a constant speed. If the waves get closer together, then more waves must go past you each second, and that means that the frequency has gone up. The speed, frequency and wavelength of a wave are related by the equation:

wave speed = frequency × wavelength or $v = f \times \lambda$

where v = wave speed, usually measured in metres/second (m/s),
f = frequency, measured in cycles per second or hertz (Hz) and
λ = wavelength, usually measured in metres (m)

WORKED EXAMPLES

1. A loudspeaker makes sound waves with a frequency of 300 Hz. The waves have a wavelength of 1.13 m. Calculate the speed of the waves.

Write down the formula: $v = f \times \lambda$

Substitute the values for f and λ: $v = 300 \times 1.13$

Work out the answer and write down the unit: $v = 339$ m/s

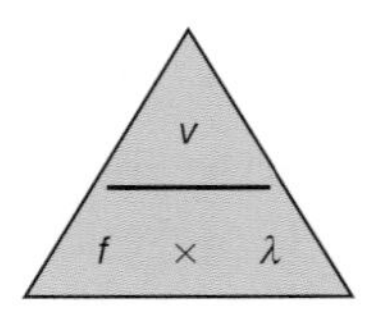

Δ Fig. 3.4 Use this triangle to help you to work out how to rearrange the formula.

2. A radio station broadcasts on a wavelength of 250 m. The speed of the radio waves is 3×10^8 m/s. Calculate the frequency.

Write down the formula with f as the subject: $f = \frac{v}{\lambda}$

Substitute the values for v and λ: $f = \frac{3 \times 10^8}{250}$

Work out the answer and write down the unit: f = 1 200 000 Hz or 1200 kHz

QUESTIONS

1. A water wave has a frequency of 5 Hz and travels at a speed of 6 m/s. What is its wavelength?

2. A radio wave has a wavelength of 12 m and a frequency of 25 MHz. What is its speed?

3. What is the frequency of a wave of speed 3×10^8 m/s and wavelength 0.2 m?

RELATIONSHIP BETWEEN FREQUENCY AND TIME PERIOD

The time period (T) is the time taken for each complete cycle of the wave motion. It is closely linked to the frequency (f) by this relationship:

$$\text{frequency (in hertz, Hz)} = \frac{1}{\text{time period (in seconds, s)}} \text{ or } f = \frac{1}{T}$$

WORKED EXAMPLE

1. A tuning fork is used to play a middle C, which has a frequency of 256 Hz. Calculate the time period of the vibration.

Write down the formula with T as the subject: $T = \frac{1}{f}$

Substitute the value of f: $T = \frac{1}{256}$

Work out the answer and write down the unit: $T = 0.0039$ s

QUESTIONS

1. A tuning fork has a frequency of 440 Hz. Calculate its time period.

2. What is the time period for a radio wave of frequency 1200 kHz?

3. What is the time period for a wave of frequency 25 MHz?

REFLECTION AND REFRACTION

When a wave is incident on (hits) a barrier, the wave is reflected. If it hits the barrier at an angle, then the angle of reflection will be equal to the angle of incidence. Figure 3.5 shows this.

Echoes happen because of the reflection of sound waves. The reflected wave takes longer to reach the listener's ear than the incident wave, and so the listener hears a repeat of the sound. Strictly speaking, an echo is a single reflection of the sound source. If there is more than one reflection, then the proper word to use is reverberation.

Waves are incident on a barrier at an **angle of incidence** i as shown in Figure 3.5. The waves reflect, with the **angle of reflection** r equal to the angle of incidence i. The reflected wave is the same shape as the incident wave.

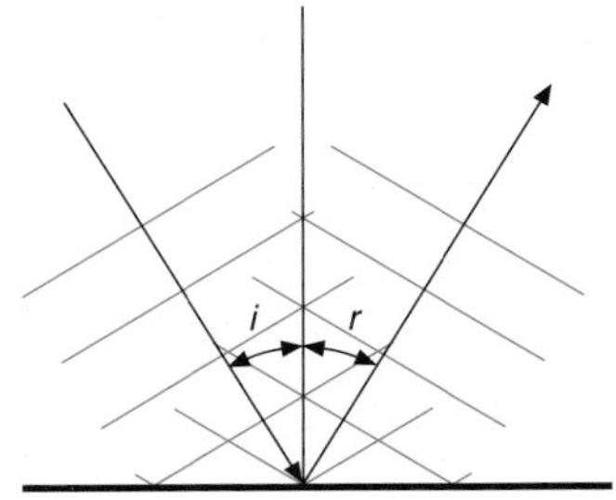

Δ Fig. 3.5 Waves being reflected from a barrier.

When a wave moves from one medium into another, it will either speed up or slow down. For example, a wave going along a rope will speed up if the rope becomes thinner. (This is why you can 'crack' a whip.) And sound going from cold air to hotter air speeds up. When a wave slows down, the wavefronts crowd together – the *wavelength gets smaller*. Figure 3.6 shows this happening as a wave moves from deep water to shallow water.

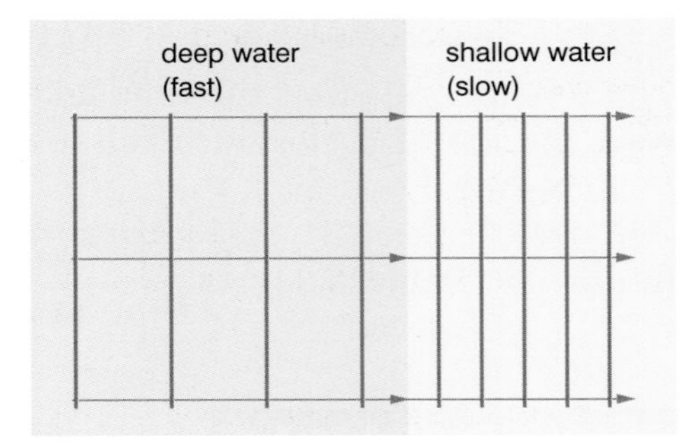

△ Fig. 3.6 When waves slow down, their wavelength gets shorter.

The wave slows down as it travels across the boundary to a more dense medium. The equation $v = f\lambda$ applies in both media. The frequency of the wave does not change as it crosses the boundary, so if the speed decreases and the frequency stays the same, the wavelength must also decrease. When a wave speeds up, the wavefronts spread out – the *wavelength gets larger*. If the speed increases but the frequency stays the same, then the wavelength must also increase. Note that in both cases the same number of waves will pass you per second; the wavelength may have changed, but the frequency has not.

△ Fig. 3.7 This surfer is successfully travelling along one wavefront. The next wavefront looks very close behind, but is probably still 50 m away. This is an illusion caused by telephoto camera lenses.

If a wave enters a new medium at an angle then the wavefronts also change direction. This is known as **refraction**. The amount that the wave is bent by depends on the change in speed. Water waves are slower in shallower water than in deep water, so water waves will refract when the depth changes as shown in Figure 3.8.

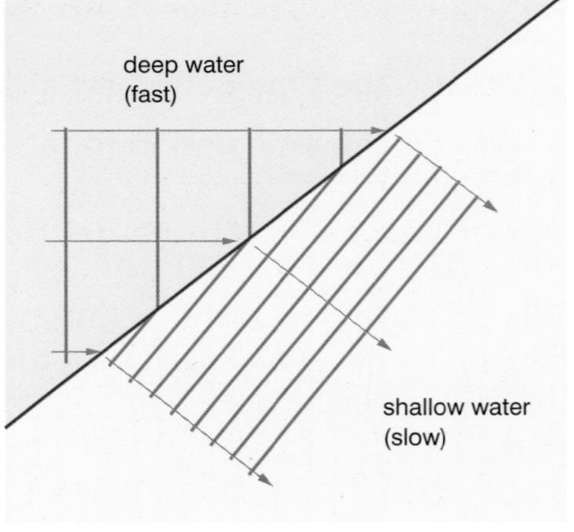

△ Fig. 3.8 If waves cross into a new medium at an angle, their wavelength and direction change.

THE DOPPLER EFFECT

When an ambulance approaches us with its siren on, the sound gets louder. It also gets quieter as it moves away.

But the frequency of the sound also changes.

The frequency of the siren increases as it approaches because the wavelengths are reduced. Look at the wavefronts coming from the siren in Figure 3.9. When the source is moving the sound wave changes shape so observers hear a higher pitch when the source is moving towards them and a lower pitch when the source is moving away.

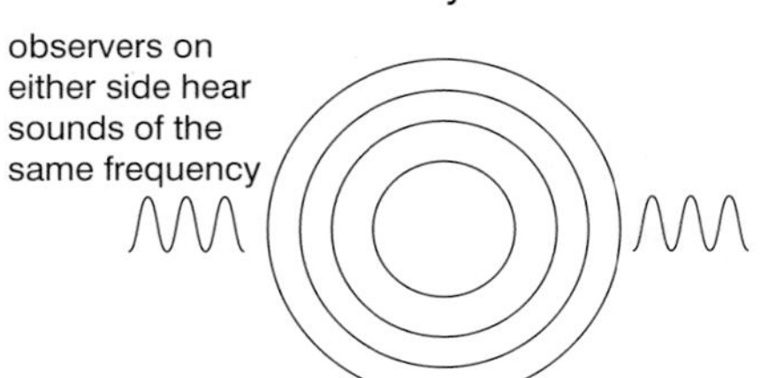

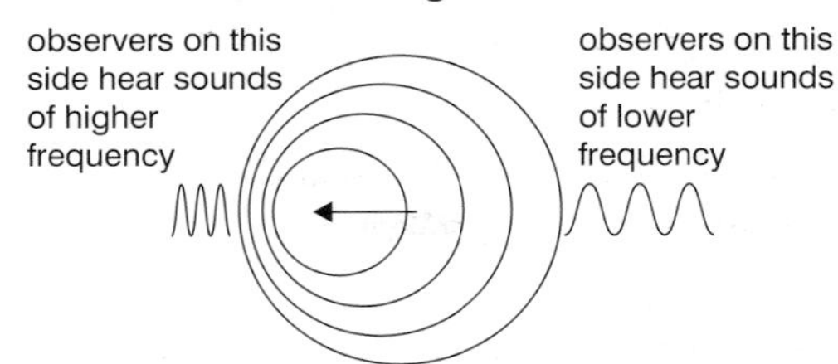

∆ Fig. 3.9 Waves from a stationary source and a moving source.

Many people notice this change in pitch with ambulances and other emergency vehicles. In Figure 3.10 the note coming from the siren is higher in front of the moving ambulance than behind it.

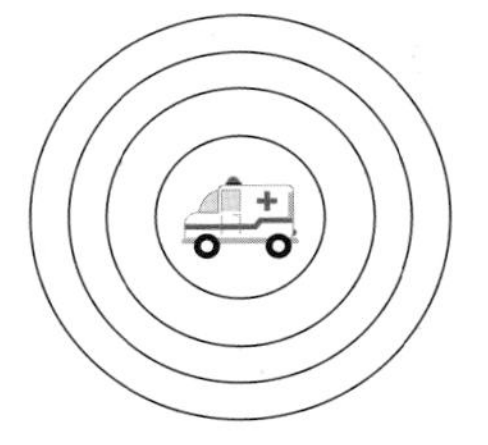

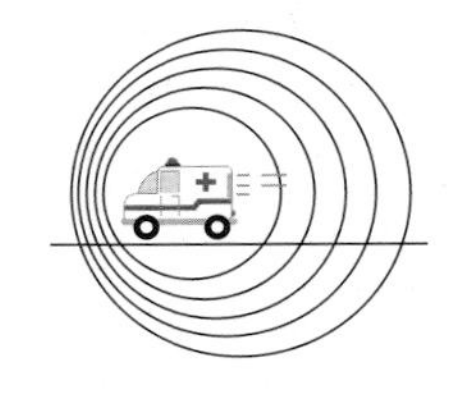

∆ Fig. 3.10 The frequency of the siren decreases as it moves away from an observer because the wavelengths are increased.

This change in pitch is called the **Doppler effect**. The Doppler effect also happens with other waves. Electromagnetic waves show a Doppler effect when the source is moving relative to an observer. The effect can be seen when objects such as galaxies move at very high speeds towards or away from the Earth. By looking at the colour (frequency) of the light emitted from the galaxy physicists can find out the direction and speed of the galaxy. Red-shifted (frequency-reduced) galaxies are moving away from us. Some galaxies are blue-shifted (increased frequency) which means they are moving towards us.

QUESTIONS

1. Explain why echoes occur.
2. Explain why wavefronts get closer together when a wave slows down.
3. On what does the amount that a wave is bent as it moves from one medium to another depend?
4. A man at a rail crossing hears a train approaching quickly at a steady speed. The frequency of the sound changes as the train approaches, passes and leaves the crossing.

   a) Which of the following is correct?

Answer	Frequency of sound from train in Hz		
	Approaching	Passing	Leaving
A	385	410	440
B	385	410	385
C	410	410	410
D	440	410	385
E	440	410	440

   b) Explain your answer.

Developing investigative skills

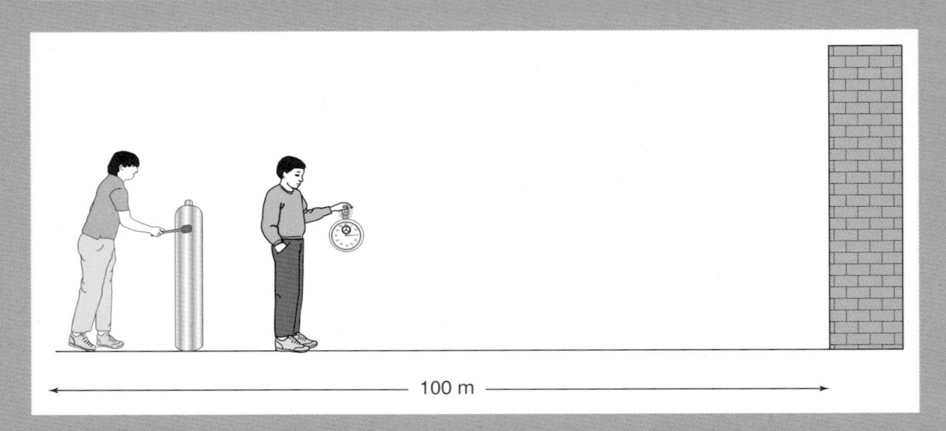

Δ Fig. 3.11 Two students measuring the speed of sound.

Two students are finding the speed of sound. They stand 100 m from a large wall. The first student strikes a piece of metal with a hammer. The sound this makes reflects back from the wall and the student hears an echo. The student hits the metal again in time with the echo and continues to do so, tapping out a steady rhythm.

The second student is in charge of timing. He knows that in between each sound that the first student makes the sound travels 200 m. By timing the interval for a number of strikes, the second student can record the data he needs.

Devise and plan investigation

❶ Explain whether the students should repeat their measurements for this experiment.

❷ The students stand 100 m away from the wall that provides the echo. Explain why this is a reasonable distance to use.

Analyse and interpret

❸ The students measure a time of 2.3 s from striking the metal to striking the metal after four echoes. Calculate the speed of sound given by this measurement.

❹ A better way to find the speed of sound would be through the use of a graph. Describe how the students could collect suitable data and how a graph could be used to find the speed of sound.

❺ Explain why using a graph to find the speed of sound would improve the accuracy of the experiment.

Evaluate data and methods

❻ What effect will reaction time have on the measurements made?

❼ Suggest how the timing in this experiment could be improved.

SUPERPOSITION AND HEARING

Listen carefully for a moment – what can you hear?

Perhaps you can hear people talking in the room where you are, perhaps there's some noise from outside, other people or even traffic. Maybe there's a TV or radio on, or perhaps the teacher is talking.

Now think about the sound waves. Each of the separate sounds you can hear was made as a pressure wave (vibrations) in the air. (Some of the sounds may be pressure waves moving through walls or windows, but we'll keep it simple for now and assume that all the sounds are coming to you through the air.) Now, any individual air particle can only be travelling in one direction at one time and the air particles are being pushed in different directions by the different sound waves. By the time the sounds reach your ear, shouldn't it all sound like one confusing mess?

The answer is yes and no.

If you have more than one wave source sending vibrations into the air, then you do indeed get a combined effect. However, it is an organised combination – the *principle of superposition* comes into play. Essentially, this says that the combined effect of a number of waves at any point is just the simple addition of each one separately. An example of the idea is shown in Figure 3.12.

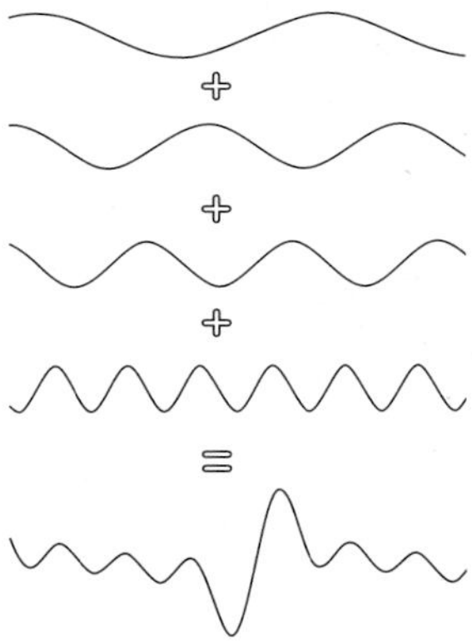

△ Fig. 3.12 Principle of superposition – adding waves together.

So a *single* waveform *does* arrive at your ears and it *is* a bit of a mess! However, hearing is about more than the sound wave that sets your eardrum vibrating. As you have grown, your brain has learned how to 'decode' the complicated waveform and recognise the separate sound waves that went into creating it. You can recognise which part of the sound is your friend talking and which part is from the traffic outside. If there are sounds you don't already know, then your brain can't help (perhaps it's someone speaking in a language you don't know).

Combining the sound waves is a straightforward piece of physics. The principle of superposition applies to any waveforms. *Hearing* is a little more complicated – it involves using your brain to 'make sense' of the combination.

End of topic checklist

The **amplitude** of a wave is the size of wave oscillations – for a mechanical wave, how far the particles vibrate around their central position.

The **wavelength** of a wave is the distance between two adjacent peaks or two adjacent troughs.

The **frequency** of a wave is the number of peaks or troughs that pass a point each second.

The **speed** of a wave is how far it travels in one second.

In **longitudinal waves** the particles move parallel to the direction of the wave.

A **compression** in a wave is an area where the particles are squashed.
A **rarefaction** is an area where the particles are more spread out.

In a **transverse** wave, the vibrations are at right angles to the direction of travel of the wave.

A **wavefront** is the moving line that joins all the points on the crest of a wave.

The **angle of reflection** is the angle between a wave reflected from a surface and the normal.

The **angle of incidence** is the angle between the wave hitting a surface and the normal.

Echoes are when waves are reflected off a surface.

When a wave enters a new medium and changes direction it is **refracted.**

The **Doppler effect** is a change in frequency of a wave when the observer or source move towards or away from each other.

The facts and ideas that you should understand by studying this topic:

- ❍ Be able to explain the difference between longitudinal and transverse waves.
- ❍ Know and use the definitions of amplitude, wavefront, frequency, wavelength and time period.
- ❍ Know and use the relationship between wave speed, frequency and wavelength: wave speed = frequency × wavelength
- ❍ Use the relationship between frequency and time period: $\text{frequency} = \dfrac{1}{\text{time period}}$.
- ❍ Understand that waves transfer information and energy without transferring matter.
- ❍ Be able to explain why there is a change in the observed frequency and wavelength of a wave when its source is moving relative to an observer, and know that this is known as the Doppler effect.
- ❍ Be able to explain that all waves can be reflected and refracted.

End of topic questions

1. A sound wave is observed on a cathode ray oscilloscope. It has a simple smooth repeating pattern.

a) Draw a trace of a sound wave and label it to indicate, the crest of the wave, the wavelength of the wave and the amplitude of the wave. **(3 marks)**

b) The frequency of the wave is 512 Hz. How many waves are produced each second? **(1 mark)**

2. a) Radio waves of frequency 900 MHz are used to send information to and from a mobile phone. The speed of the waves is 3×10^8 m/s. Calculate the wavelength of the waves. (1 MHz = 1 000 000 Hz, $3 \times 10^8 = 300\,000\,000$.) **(2 marks)**

b) What is the time period of the waves? **(2 marks)**

3. Calculate the wavelength of waves of speed 3×10^8 m/s with frequency of 400 MHz. **(2 marks)**

4. What is the frequency of waves of speed 3×10^8 m/s and wavelength 0.1 m? **(2 marks)**

5. What is the speed of a wave of 20 Hz with a wavelength of 4 m? **(2 marks)**

6. Calculate the missing quantities in the table. **(12 marks)**

Speed (m/s)	Frequency (Hz)	Wavelength (m)	Period (s)
	10	2	
20	5		
		6	0.2
340		34	
160	8		
		12	0.004

7. What are the most likely explanations of the following effects? Explain carefully.

a) The captain of an ocean-going ship is proceeding slowly into waves coming towards the ship. They suddenly notice that the waves change in two ways about 200 m ahead of where the ship is. The waves get further apart and change direction quite noticeably. **(3 marks)**

b) An observer is standing on the bank of a river. The wind is blowing from left to right, and waves are moving from left to right. However, he sees a small piece of wood that is moving slowly from right to left as it floats in the middle of the river. **(3 marks)**

8. A jet flies overhead past an observer at high speed. It makes a loud noise. The frequency of the noise is different when the jet approaches, passes and moves away from the observer.

Use diagrams with wavefronts to explain this change in frequency. **(3 marks)**

∆ Fig. 3.13 These satellite dishes are receiving radio waves which are part of the electromagnetic spectrum.

The electromagnetic spectrum

INTRODUCTION

As recently as 150 years ago, only the visible part of the electromagnetic spectrum was known. As the different sections of the spectrum have been discovered, created and studied, many technological applications have followed.

Familiarity with the electromagnetic spectrum is vital in the modern world. Whether it is for communication, cooking, medical imaging, medical treatment or any other of the many uses we make of them, electromagnetic waves continue to form an increasingly important part of everyday life.

KNOWLEDGE CHECK

✓ Be able to describe waves using key terms such as frequency, wavelength and amplitude.
✓ Understand the difference between transverse and longitudinal waves.
✓ Know that waves transfer information and energy without transferring matter.
✓ Know and be able to use the relationship between wavespeed, frequency and wavelength.

LEARNING OBJECTIVES

✓ Be able to identify the different regions of the electromagnetic spectrum and know that all these waves travel at the same speed in a vacuum.
✓ Be able to list the regions of the electromagnetic spectrum in order of increasing frequency and increasing wavelength.
✓ Be able to explain some of the uses of each region of the electromagnetic spectrum.
✓ Be able to describe some of the harmful effects on the human body of excessive exposure to regions of the electromagnetic spectrum.

A FAMILY OF WAVES

The **electromagnetic spectrum** is a 'family' of waves. Electromagnetic waves all travel at the same speed in a vacuum, that is, at the speed of light: 300 000 000 m/s. This can be written more conveniently as 3×10^8 m/s. This high speed explains how you can have a phone call between China and New Zealand with only a delay of 0.1 s before you hear the reply from the person at the other end. It takes the infrared signal this long to travel there and back through an optical fibre.

However, for astronomical distances the delays quickly become longer. Even when Mars is at its nearest to Earth, it takes 10 minutes to send a message to a robot on the surface and another 10 minutes for the robot's response to reach Earth. Getting a reply from the nearest star (apart from the Sun) would take 8½ years.

Note that all electromagnetic waves can travel through a vacuum, which is why we can see the light and feel the heat coming from the Sun. Other waves, such as sound waves, cannot travel through a vacuum.

Order of the electromagnetic spectrum

White light is a mixture of different colours and can be split by a **prism** (a glass or plastic block) into the **visible spectrum**. All the different colours of light travel at the same speed in a vacuum, but they have different frequencies and wavelengths. Red light has a wavelength almost twice as long as that of violet light. When the colours enter glass or perspex, they all slow down, but by different amounts, because their wavelengths are slightly different. The different colours are therefore refracted through different angles. Violet is refracted the most, red the least. Figure 3.14 shows the refraction of white light through a triangular prism, which splits it into the spectrum of visible light: red, orange, yellow, green, blue, indigo, violet (ROYGBIV).

Δ Fig. 3.14 The colours of visible light.

The visible spectrum is only a small part of the full electromagnetic spectrum. The electromagnetic spectrum has waves of wavelength of the order of 10^4 m right down to wavelengths of the order of 10^{-12} m. Visible light has wavelengths ranging from the order of 10^{-6} to 10^{-7} m. All electromagnetic waves travel at the same speed in a vacuum, which is 3×10^8 m/s. This means that the different wavelengths must have different frequencies, and that the longest wavelengths have the lowest frequencies. Figure 3.15 shows the full electromagnetic spectrum.

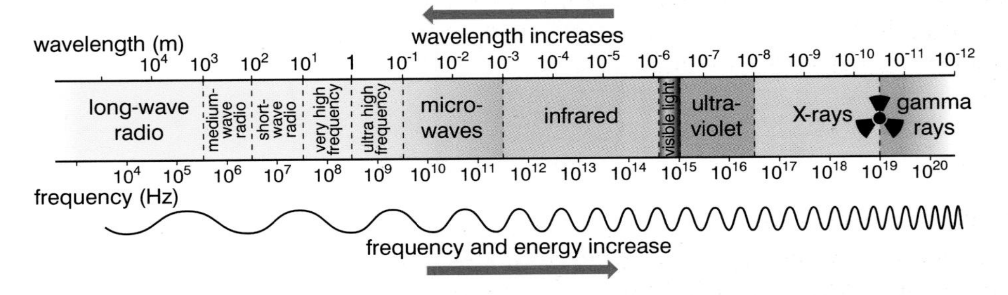

Δ Fig. 3.15 The electromagnetic spectrum.

REMEMBER

The energy associated with an electromagnetic wave depends on its frequency. The waves with the higher frequencies are potentially the more hazardous.

The visible light region of the spectrum contains the colours ranged from red, through orange, yellow, green and blue to violet. The red lies next to **infrared**, and violet is next to the **ultraviolet**.

Light of one wavelength – that is to say, of just one colour – is known as **monochromatic** light.

Δ Fig 3.16 A prism splits white light into the colourful spectrum of visible light.

QUESTIONS

1. Describe how white light can be split into a spectrum by a prism.
2. What is the speed of electromagnetic waves in free space?
3. Which has greater frequency: microwaves or X-rays?
4. Put these electromagnetic radiations into order of increasing wavelength.

   microwaves, gamma rays, infrared, visible light, ultraviolet

SCIENCE IN CONTEXT LASERS

Δ Fig. 3.17 A laser in a DVD player.

In 1917 Albert Einstein stated that it should be possible to stimulate an atom to release energy in the form of concentrated light. The laser, first demonstrated in 1960, is this effect in practice.

Laser stands for *light amplification by the stimulated emission of radiation*, but how does a laser work? In an atom, the electrons have particular energy states that they can be in ('energy levels'). If an atom 'absorbs' some energy, then what actually happens is that the energy allows an electron to move to a higher energy level. Some time later, the electron will 'fall' back to the lower energy level, releasing the extra energy it had as a pulse of light. Particular energy changes produce specific colours of light – for example, street lights often contain sodium vapour and the energy changes involved tend to produce the very familiar orange colour.

In a laser, the trick is to get a large number of atoms with electrons at higher energy levels. To do this, an energy supply, often another light source, is needed to stimulate the electrons to higher levels – this is called a population inversion.

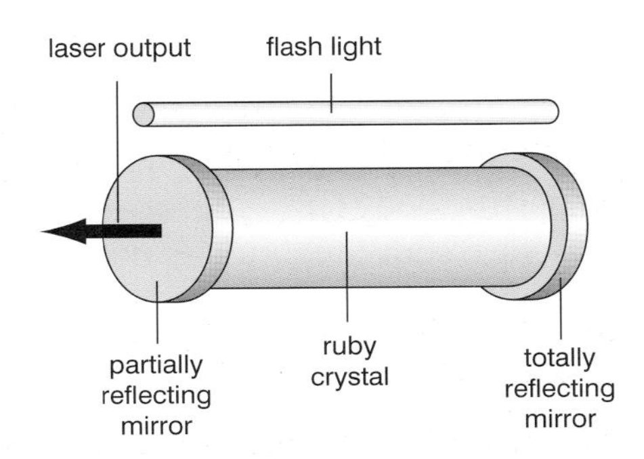

Δ Fig. 3.18 How a ruby laser works.

When an electron falls to a lower energy level it releases a pulse of light. The ends of the chamber are mirrors, so the light reflects backwards and forwards. As the light does this, it may collide with an atom, stimulating the emission of more pulses of light.

One of the mirrors is only partly reflecting, so the light can escape from that end. Because the pulses of light have all been produced from electrons falling between the same two energy levels, all the light is one single colour ('monochromatic') and because there are so many pulses of light it is a very concentrated light source.

Although lasers were first demonstrated in 1960, practical uses for them were slow to arrive. However, nowadays lasers are used in many applications, for example in military, industrial and medical situations. The bar code scanner at the supermarket is a laser, and so is the device 'reading' your CDs and DVDs although these work in a different way to the laser in Fig. 3.18 by using semiconductor materials.

Uses and hazards of electromagnetic radiation

From Figure 3.15, you will have seen that there are different types of electromagnetic radiation. Some of these types of radiation can have harmful effects on the human body if exposure to them is excessive.

Gamma rays, which as you see in Figure 3.15 are the electromagnetic waves with the shortest wavelength, and therefore the greatest frequency and energy, are produced by radioactive nuclei. They carry more energy than X-rays and can cause cancer or mutation in body cells. Gamma rays are frequently used in radiotherapy to kill cancer cells. The success of this treatment is greatly improved by the fact that cancer cells are easier to kill than ordinary cells.

Gamma rays are also used to sterilise medical equipment. The equipment is placed inside an air-tight bag. The radiation kills microorganisms on the equipment and the air-tight bag will keep the instrument sterile until the bag is opened. Food can also be treated with gamma rays to kill microorganisms, so that it will keep longer.

X-rays, which have the next highest energy radiation after gamma on the electromagnetic spectrum, are produced when high-energy electrons are fired at a metal target. Bones absorb more X-rays than other body tissue. If a person is placed between the X-ray source and a photographic plate, the bones appear to be white on the developed photographic plate compared with the rest of the body. X-ray images are also used in computerised tomography (CT) scans. X-rays have high energy, as shown in Figure 3.15, and can damage or destroy body cells: they may cause cancer. An X-ray may save your life, but it is not free from danger. A doctor will only arrange for an X-ray when it is clear that the benefits to you are far greater than the tiny risk that it will make you ill.

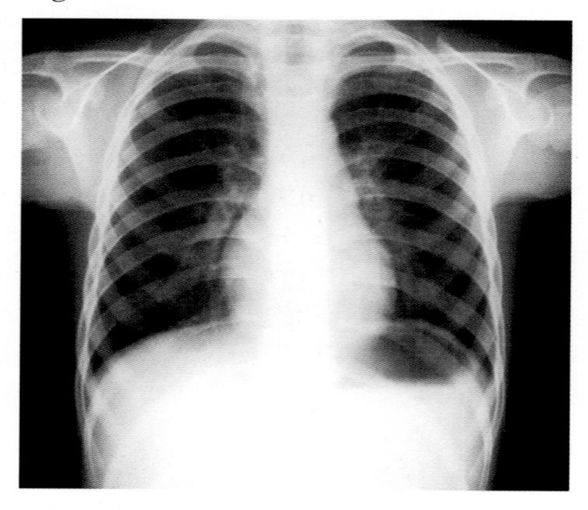

Δ Fig. 3.19 An X-ray of lungs.

However, X-rays (like gamma rays) are also used to treat cancer. They are targeted at the tumour with the aim of destroying the tumour cells while leaving healthy cells surrounding the tumour untouched.

Ultraviolet radiation (UV) is the component of the Sun's rays that gives you a suntan. UV is also created in fluorescent light tubes by exciting the atoms in a mercury vapour. The UV radiation is then absorbed by the coating on the inside of the fluorescent tube and re-emitted as visible light. Fluorescent tubes are more efficient than light bulbs because they do not depend on heating, so more energy is available to produce light. Ultraviolet can also damage the surface cells of the body, which can lead to skin cancer, and can damage the eyes, leading to blindness.

CELL DAMAGE CAUSED BY UV LIGHT

In 2007, scientists at the University of Virginia published a study about how cells protect themselves (or fail to protect themselves) from damage to their DNA caused by ultraviolet rays. The study showed that there is a simple switch mechanism inside cells, which is triggered by exposure to ultraviolet, and helps cells to survive and even thrive after exposure to ultraviolet rays.

After DNA damage caused by exposure to ultraviolet rays, normally cells stop moving and responding to stimuli until they are repaired. If the repair work is not carried out properly, the result can be cancer as the damaged cell keeps dividing.

All objects give out infrared radiation (IR). The hotter the object is, the more radiation it gives out. Thermograms are photographs taken to show the infrared radiation given out from objects. Infrared radiation grills and cooks our food in an ordinary oven and is used in remote controls to operate televisions and videos. Excessive exposure to infrared, such as when you touch a very hot object, can burn skin and other body tissue.

Microwaves are high-frequency **radio waves**. Microwaves are used to transmit mobile phone signals. The microwave signals are either sent directly between transmitter and receiver (along a line of sight), or to and from satellites in orbit. Microwaves are also used for cooking. Water particles in food absorb the energy carried by microwaves. They vibrate more making the food much hotter. Microwaves penetrate several centimetres into the food and so speed up the cooking process. Because of their ability to penetrate several centimetres, which is useful when cooking food, microwaves can heat body tissue internally, so care must be taken when using them.

REMEMBER

Infrared radiation is absorbed by the surface of the food and then the energy is spread through the rest of the food by conduction. In contrast, microwaves penetrate a few centimetres into food and then the energy is transferred throughout the food by conduction.

Radio waves have the longest wavelengths and lowest frequencies.

- UHF (ultra-high frequency) waves are used to transmit television programmes to homes.
- VHF (very high frequency) waves are used to transmit local radio programmes.
- Medium and long radio waves are used to transmit over longer distances because their wavelengths allow them to diffract around obstacles such as buildings and hills.
- Communication satellites above the Earth receive signals carried by high-frequency (short-wave) radio waves. These signals are amplified and re-transmitted to other parts of the world, and are used for television and radio.

QUESTIONS

1. Gamma rays and X-rays are both used in medical contexts. Describe how each is produced.
2. What is a thermogram?
3. How do microwaves cook food?

EXTENSION

There are many different departments in a hospital, for example:

- eye clinic
- X-ray
- nuclear radiation
- thermography
- physiotherapy.

Produce a table detailing the following:

- the name of the electromagnetic wave used in each department
- the wavelength of the wave used
- the application of the wave, detailing the specific medical use for the wave
- health and safety precautions that should be taken, where appropriate.

End of topic checklist

An **electromagnetic wave** is a wave that transfers energy – it can travel through a vacuum and travels at the speed of light.

The **electromagnetic spectrum** is electromagnetic waves ordered according to wavelength and frequency – ranging from radio waves to gamma rays.

The **visible spectrum** is the group of electromagnetic waves that we can see.

Infrared is the part of the electromagnetic spectrum that has a slightly longer wavelength than the visible spectrum.

Ultraviolet is the part of the electromagnetic spectrum that has a slightly shorter wavelength than the visible spectrum.

Radio waves have a long wavelength and are used for communications.

The facts and ideas that you should understand by studying this topic:

❍ Know that light is part of a continuous electromagnetic spectrum.

❍ Know that the electromagnetic spectrum consists of seven sections, which in order of increasing wavelength is: gamma, X-rays, ultraviolet, visible, infrared, microwaves, radio

❍ Know that this order is reversed for increasing frequency.

❍ Know all these waves travel at the same speed in free space.

❍ Know that visible light consists of a range of wavelengths, matching the spectrum of colours in the rainbow.

❍ Explain some uses of electromagnetic waves, including:

- radio waves – broadcasting and communications
- microwaves – cooking and satellite transmissions
- infrared – heating and night-vision equipment
- visible light – optical fibres and photography
- ultraviolet – fluorescent lamps
- X-rays – observing internal structures of objects and medical uses
- gamma – sterilising food and medical equipment.

❍ Understand the detrimental effects of excessive exposure to electromagnetic waves, including

- microwaves – internal heating of body tissue
- infrared – skin burns
- ultraviolet – damage to surface cells and blindness
- gamma – cancer, cell mutation.

End of topic questions

1. This is a list of types of wave:

gamma, infrared, microwaves, radio, ultraviolet, visible, X-rays

Choose from the list the type of wave that best fits each of these descriptions.

a) Stimulates the sensitive cells at the back of the eye. **(1 mark)**

b) Necessary for a suntan. **(1 mark)**

c) Used for rapid cooking in an oven. **(1 mark)**

d) Used to take a photograph of the bones in a broken arm. **(1 mark)**

e) Emitted by a video remote control unit. **(1 mark)**

2. Gamma rays are part of the electromagnetic spectrum. Gamma rays are useful to us but can also be very dangerous.

a) Explain how the properties of gamma rays make them useful to us. **(3 marks)**

b) Explain why gamma rays can cause damage to people. **(3 marks)**

c) Give one difference between microwaves and gamma rays. **(1 mark)**

d) Microwaves travel at 300 000 000 m/s. What speed do gamma rays travel at? **(1 mark)**

3. a) Write down the parts of the electromagnetic spectrum in order of increasing wavelength. **(4 marks)**

b) How would your list be different if you wrote it in order of increasing frequency? **(1 mark)**

4. Explain how heat energy from an electric heater travels to your hands if you put them close to the heater. **(3 marks)**

5. Copy the table. Fill in the gaps in the table. **(13 marks)**

Wave type	**Source of wave**	**Use**	**Property**
Radio	Radio transmitter and aerial		
X-rays	X-ray tubes	Security at airports	
	Lamps, Sun, flames		Can be dispersed into seven colours
Ultraviolet	Mercury vapour lamps	Security markings	
		Thermal imaging	Can cause burns
Gamma	Radioactive substances		Very penetrating
Sound		Hearing	
	Magnetron	Heating food quickly	

6. Explain how microwave ovens cook food. **(3 marks)**

7. Explain how radio waves are used to transmit radio and television programmes. **(3 marks)**

8. A house is situated behind a large hill. Short wave television waves and long wave radio waves both approach the house from the other side of the hill. Which will be better at the house, television or radio reception? Give a reason for your answer. **(3 marks)**

Δ Fig 3.20 A concert uses both light and sound.

Light and sound

INTRODUCTION

Visible light is just one part of the electromagnetic spectrum, but without it your life would be very different. Close your eyes for a moment and imagine a world of darkness. What would you miss the most? There are different processes involved in seeing the world around us, including reflection and refraction. You may already know that, as light enters your eyes, it is refracted by the lens in your eye and brought to a focus on your retina.

Now think about the importance of sound in your life. If there were no sound, which sounds would you miss most? Light and sound are both forms of wave, but light is a transverse wave and sound is a longitudinal wave. In this section you are going to find out how these two types of wave operate.

KNOWLEDGE CHECK

✓ Know how to describe waves using the terms wavelength, amplitude and frequency.
✓ Know the difference between longitudinal and transverse waves.
✓ Be able to describe the processes of reflection, refraction and diffraction.

LEARNING OBJECTIVES

✓ Know that light waves are transverse waves and sound waves are longitudinal waves.
✓ Know and use the law of reflection.
✓ Be able to draw ray diagrams to illustrate reflection and refraction.
✓ Be able to investigate the refraction of light using rectangular blocks, semi-circular blocks and triangular prisms.
✓ Be able to describe refraction of light, including ideas about refractive index.
✓ Be able to investigate the refractive index of glass.
✓ Describe how total internal reflection is used to transmit information along optical fibres and in prisms.
✓ Be able to explain the meaning of critical angle and how critical angle is related to refractive index.
✓ State the range of frequencies that humans can hear.
✓ Be able to investigate the speed of sound in air.
✓ Understand how an oscilloscope and microphone can be used to display waveforms of different sounds.
✓ Be able to investigate the frequency of a sound wave using an oscilloscope.
✓ Understand how the pitch and loudness of a sound relate to properties of the vibration that produces the sound.

LIGHT WAVES

Light waves have all of the properties of waves. You have already learned about their speed and wavelength. In addition, they are transverse waves.

POLARISED LIGHT

Transverse waves may have both a vertical and a horizontal vibration (think of the up-and-down or side-to-side motion of a slinky spring). It is sometimes possible for only one of these to travel through a particular medium – we call that medium a polarising medium. As a light wave approaches your eye, it could contain both the horizontal and vertical parts, or it could be vertically polarised or horizontally polarised. Light from an ordinary lamp is not polarised, but reflections from the surface of water or glass may be polarised – a fact that is important to anybody who wears polarising sunglasses.

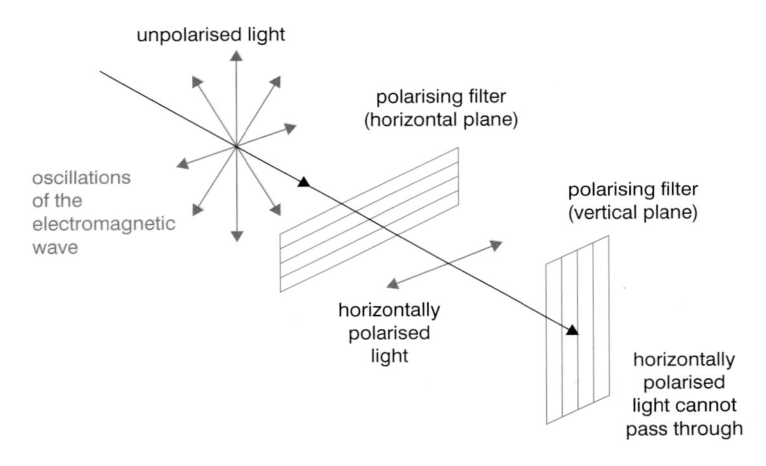

◁ Fig. 3.21 The difference between unpolarised and polarised light.

A good pair of sunglasses will change incoming light so that it is more comfortable for your eyes to deal with. Light intensity, or brightness is measured in lumens. Human eyes are comfortable with intensities of up to around 3500 lumens. However, on a sunny day intensities may be more than 6000 lumens on a large stretch of road. When the brightness of direct or reflected light gets to around 4000 lumens, your eyes have difficulty absorbing the light and you see these brighter areas as flashes of white light, which is called glare. At this intensity, your eyes try to reduce the discomfort by squinting.

Δ Fig. 3.22 This photo of tiny crystals was taken in polarised light.

Light from the Sun, or a light bulb, radiates in all directions. When light waves are incident on a surface, the reflected waves (which reach your eyes) are polarised to match the surface. Most of the glare that causes you to wear sunglasses comes from horizontal surfaces, such as water or roads. A horizontal surface that reflects a lot of light, like a lake, will produce a lot of horizontally polarised light. Polarising lenses in sunglasses are fixed so that only vertically polarised light can reach your eyes. So the glare from the lake (or other horizontal surface) will not reach your eyes and cause discomfort.

Like all other waves, light can be reflected and refracted. **Reflection** and refraction are easy to demonstrate with a mirror and a glass of water.

REFLECTION OF LIGHT AND RAY DIAGRAMS

A ray of light is a line drawn to show the path that the light waves take. What happens when an incident light ray (a light ray that is going to fall on a surface) reflects from a mirror?

Light rays are reflected from mirrors in such a way that **angle of incidence** (i) = **angle of reflection** (r).

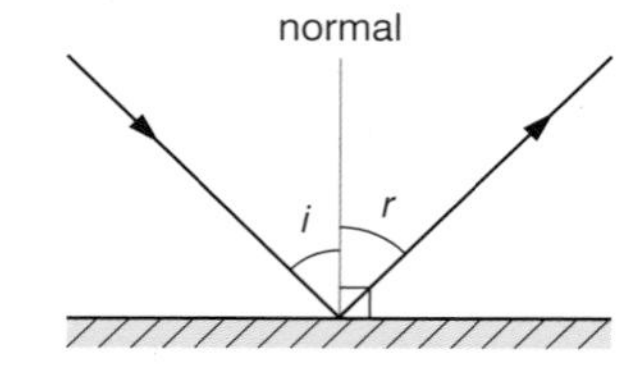

Δ Fig. 3.23 The angles of incidence and reflection are the same when a mirror reflects light.

The angles are measured to an imaginary line at 90° to the surface of the mirror. This line is called the **normal**. With a curved mirror it is more difficult to measure the angle between the ray and the mirror, but the same law still applies.

When you look in a **plane mirror** you see an **image** of yourself. The image is said to be laterally inverted because if you raise your right hand, your image raises what you would call its left hand. The image is formed as far behind the mirror as you are in front of it and is the same size as you. The image cannot be projected onto a screen. It is known as a **virtual image**.

Δ Fig. 3.24 The mirror image of the girl is laterally inverted.

In a plane mirror the image is always the same size as the object. Examples of plane mirrors include household 'dressing' mirrors, dental mirrors for examining teeth, security mirrors for checking under vehicles and periscopes.

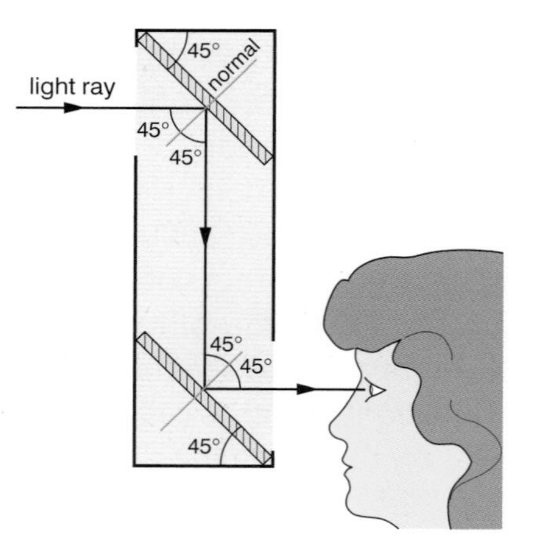

Δ Fig. 3.25 A periscope uses reflection to allow you to see above your normal line of vision – or even round corners.

CONCAVE AND CONVEX MIRRORS

There are other types of mirror that are curved. A concave mirror has the reflective surface on the inside. A convex mirror has the reflective surface on the outside.

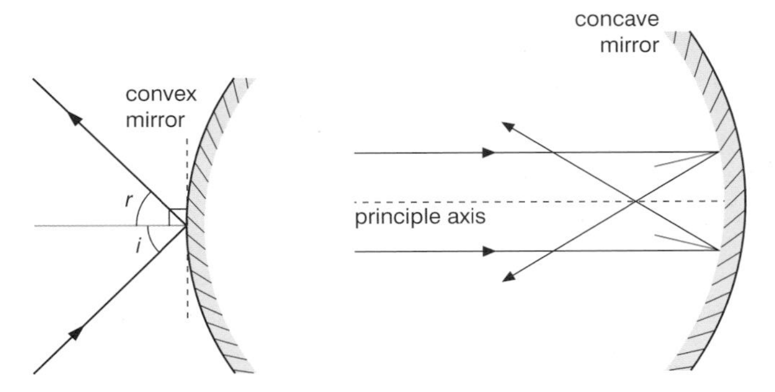

Δ Fig. 3.26 Reflection from a convex mirror and a concave mirror

Close to a concave mirror, the image is larger than the object: these mirrors magnify. They are used in make-up and shaving mirrors. They are also used in torches and car headlamps to produce a beam of light.

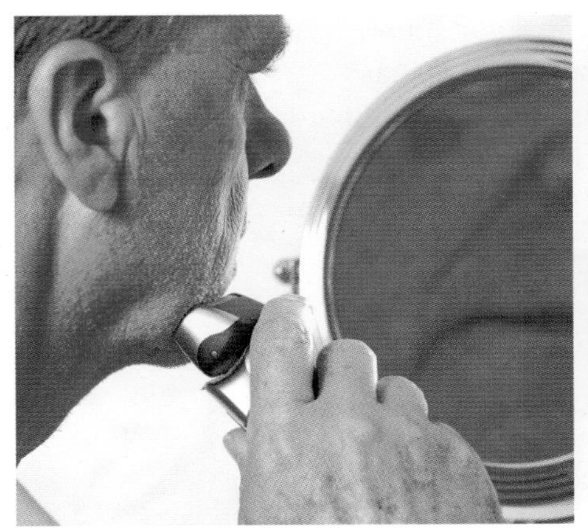

Δ Fig. 3.27 The reflection in a concave mirror is enlarged.

Δ Fig. 3.28 The size of the image in this convex mirror on a bus is reduced.

The light from a car headlamp bulb needs to be directed forward in a straight beam. If a bulb is placed at the focus of a parabolic mirror (a mirror that is curved like a parabola) then it will be reflected by the mirror into a parallel beam. Concave mirrors are also used to reflect light onto the cooking pot in the centre of a solar cooker, as shown in Figure 3.29.

The image in a convex mirror is always smaller than the object. Examples are a car driving mirror and a shop security mirror.

◁ Fig. 3.29 A solar cooker in Ladakh, India.

QUESTIONS

1. The angle of incidence on a plane mirror is 48°. What is the angle of reflection?

2. What is meant by the 'normal' of a mirror?

REFRACTION OF LIGHT

Light waves slow down when they travel from air into glass. If they are at an angle to the glass, they bend towards the normal as they enter the glass. When the light rays travel out of the glass into the air, their speed increases and they bend *away* from the normal. If the block of glass has parallel sides, the light resumes its original direction. This is why a sheet of window glass has so little effect on the view beyond. However, the view is shifted slightly sideways if you are looking through the glass at an angle.

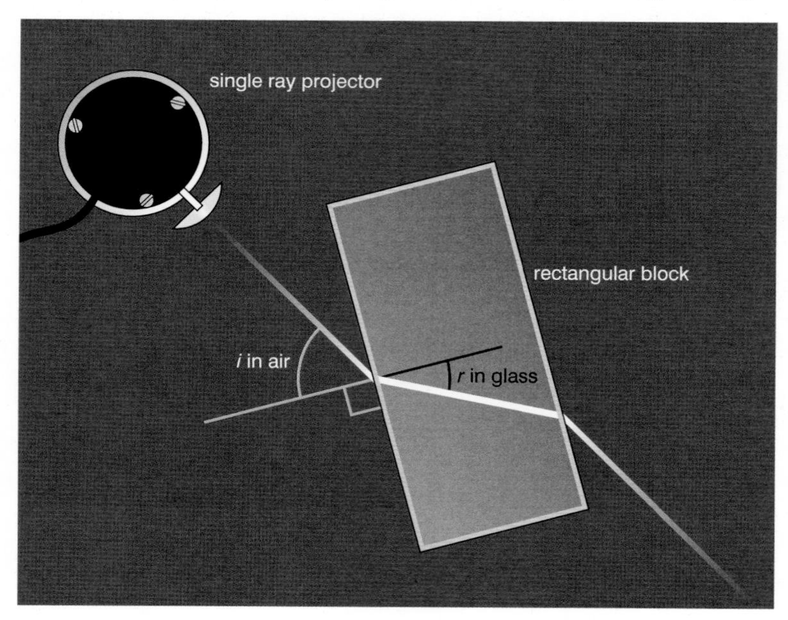

△ Fig. 3.30 Refraction of light in a rectangular glass block.

The angle of incidence, i, is the angle between the incident light ray and the normal to the surface. The **angle of refraction**, r, is the angle between the refracted light ray and the normal to the surface inside the material.

(Note that people tend to use the letter r both for the angle of reflection and the angle of refraction. But it will be clear from the context whether they are talking about reflection or refraction.)

Triangular prisms

Light also refracts when entering or leaving a **prism**. There is an overall change in the direction of the light (Fig. 3.31).

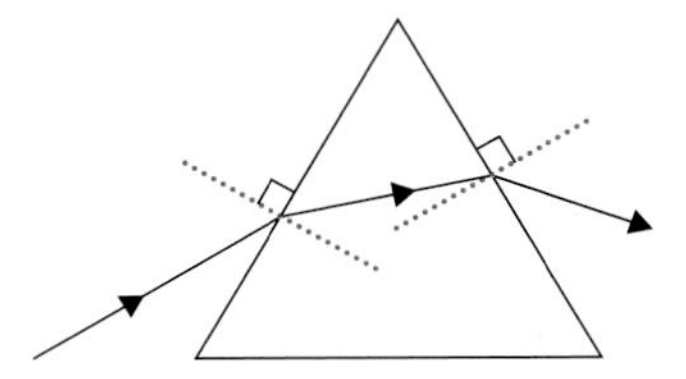

Δ Fig. 3.31 Refraction of light passing through a triangular prism.

Developing investigative skills

A ray box produces a narrow ray of light. A student is given a number of transparent blocks made from the same material.

Devise and plan investigations

❶ Describe how the student could use the ray box and transparent blocks to compare how light is refracted at the surfaces of different shaped blocks. Include a diagram of the arrangement.

❷ What is the independent variable in this investigation? What is the dependent variable? What variable is controlled?

Evaluate data and methods

❸ Suggest why it is difficult to obtain accurate measurements in this experiment.

Refractive index

The **refractive index** of a material indicates how strongly the material changes the direction of the light. It is calculated using the following formula:

$$\text{refractive index } n = \frac{\sin i}{\sin r}$$

where i = angle of incidence and r = angle of refraction

The refractive index of a vacuum is 1, and the refractive index of air is fractionally higher, but can be taken as 1. Other common refractive indexes are water 1.3; window glass 1.5; sapphire 1.75; diamond 2.4.

The refractive index n can also be defined as:

$$n = \frac{\text{speed of light in vacuum (or air)}}{\text{speed of light in the material}}$$

WORKED EXAMPLES

1. If the speed of light in a vacuum is 300 000 000 m/s, what is the speed of light in glass with a refractive index of 1.5?

Write down the formula: $n = \frac{\text{speed of light in vacuum (or air)}}{\text{speed of light in the material}}$

Rearrange the formula: $\text{speed in material} = \frac{\text{speed in vacuum}}{n}$

Substitute the values: $\text{speed in material} = \frac{300\ 000\ 000}{1.5}$

Work out the answer and write down the unit: speed of light in the material = 200 000 000 m/s $= 2 \times 10^8$ m/s

2. A light ray approaches a block of plastic at an angle of incidence of 60°. If the refractive index of the plastic is 1.4, what is the angle of refraction?

Write down the formula: $n = \frac{\sin i}{\sin r}$

If $i = 60°$, $\sin i = 0.866$.

Rearrange the formula: $\sin r = \frac{\sin i}{n}$

Substitute the values: $\sin r = \frac{0.866}{1.4}$

Work out the answer: $\sin r = 0.619$

From a calculator, if $\sin r = 0.619$, then $r = 38.2°$

The angle of refraction is 38.2°.

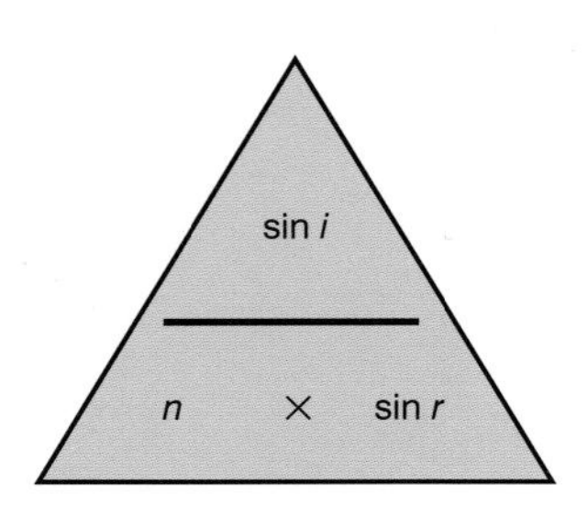

Δ Fig. 3.32 The equation triangle for refractive index.

QUESTIONS

1. Refraction of light at the surface of a pond can make the pond look shallower than it really is. Explain why.
2. Sound waves can be refracted when they travel through balloons filled with different gases. Suggest how the motion of a sound wave would be changed if it travelled through a balloon filled with carbon dioxide.
3. The refractive index of a particular glass is 1.5 and the angle of incidence is 50°. What is the angle of refraction?
4. The angle of refraction for a ray of light in a block of plastic with refractive index 1.4 is 25°. What was the angle of incidence?
5. The angle of incidence of a ray of light on a material is 55°. The angle of refraction in the material is 35°. What is the refractive index of the material?

EXTENSION

Light travels faster than anything else known to science, but it travels at different speeds in different materials. The refractive index of the material light passes through determines at what speed it travels.

Here is a table of refractive indices.

Material/substance	Refractive index
air	1.003
water	1.333
glass	1.52
diamond	2.417

Δ Table 3.1 Refractive indices.

1. Does light travel faster or more slowly in water than it does in air?
2. In which of these materials do you think light will travel slowest? Give a reason for your answer.
3. Calculate the speed of light in each of the materials in the table and confirm (or otherwise) your prediction from Question **2**.

Developing investigative skills

A student wants to find the refractive index of a rectangular block of glass. He draws around the block and marks the position of a ray of light that travels through the block. With the block removed, the student can draw in a normal line and then measure the angle of incidence and the angle of refraction. The student repeats this process for different angles of incidence. His measurements are shown in the table.

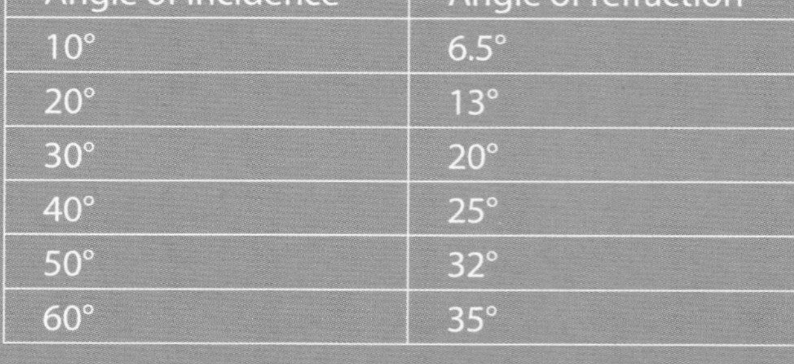

Angle of incidence	Angle of refraction
10°	6.5°
20°	13°
30°	20°
40°	25°
50°	32°
60°	35°

Δ Table 3.1 Results of experiment.

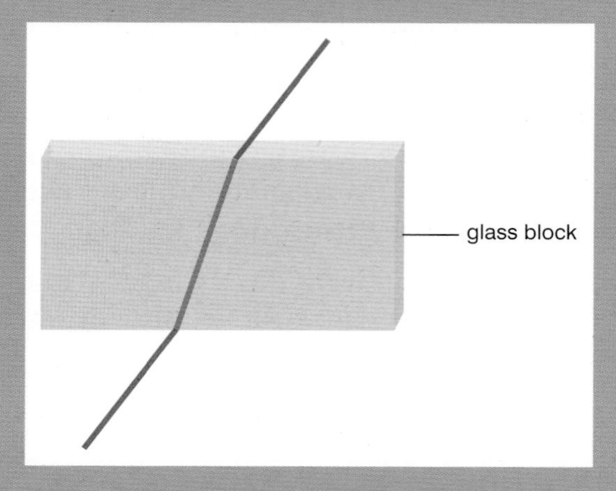

Δ Fig. 3.33 Ray of light being refracted by a glass block.

Devise and plan investigation

❶ Draw a diagram to show the measurements the student needs to make.

❷ How should the student mark the normal line during the experiment?

❸ What was the independent variable in this investigation? What was the dependent variable?

❹ What, if any, safety precautions should be taken if this experiment is carried out in the classroom?

Analyse and interpret results

❺ State the equation linking refractive index, sin *i* and sin *r*.

❻ Draw a graph of sin *i* (*y* axis) against sin *r* (*x* axis).

❼ Use your graph to find a value for the refractive index of the block.

Evaluate data and methods

❽ Describe two possible reasons the measurements may not be completely accurate.

❾ What difference would it make to the results if light of a different colour was used in the experiment?

TOTAL INTERNAL REFLECTION AND THE CRITICAL ANGLE

When rays of light pass from a dense medium to a less dense medium, they bend away from the normal.

As the angle of incidence increases, an angle is reached at which the emerging light rays would travel along the surface (the angle of refraction = 90°). At greater angles of incidence, all the rays are entirely reflected back inside the medium. This process is known as **total internal reflection**. Total internal reflection occurs when a ray of light tries to leave the glass. If the angle of incidence equals or is greater than the critical angle the ray will be totally internally reflected. In Figure 3.34 the angle is 2 or 3 less than the critical angle, and light is just managing to escape from the glass. The angle of incidence at which the angle of refraction becomes equal to 90° is known as the **critical angle** for the material.

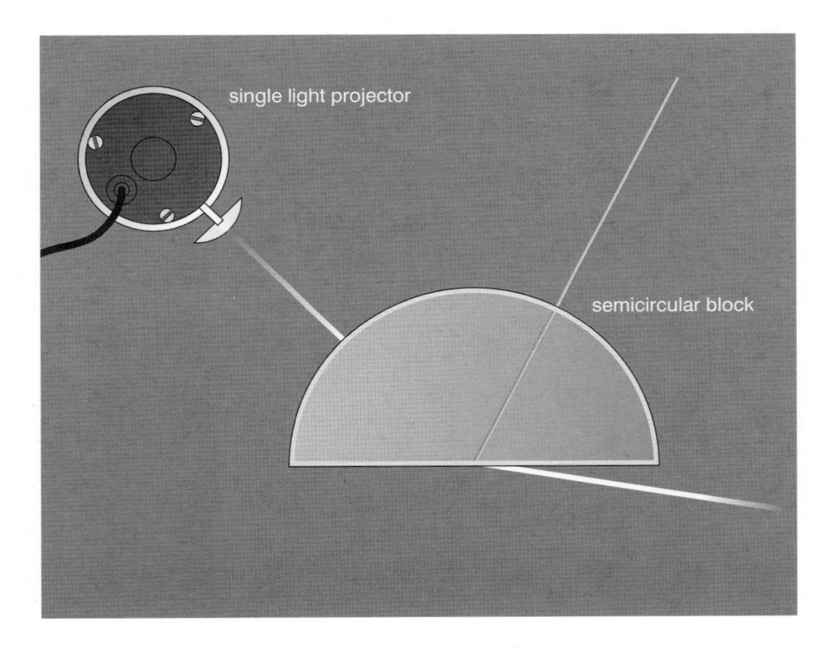

◁ Fig. 3.34 Total internal reflection.

The angle of incidence is just less than the critical angle here. The critical angle, c, is linked to the refractive index by this formula:

$$\sin c = \frac{1}{n}$$

So, for window glass, refractive index 1.5, we find that $\sin c = 1/1.5 = 0.67$. The critical angle for the window glass is therefore approximately 42° ($\sin^{-1} 0.67 = 42°$). Similarly, the critical angle for water is 49°.

Total internal reflection is used in fibre optic cables. A fibre optic cable can be made of a single glass fibre. The light continues along the fibre by being constantly internally reflected.

Δ Fig. 3.35 Fibre optic cable.

Light does not escape from the fibre because it always approaches the internal surface at an angle greater than the critical angle and is internally reflected.

Telephone and TV communications systems rely increasingly on fibre optics instead of the more traditional copper cables. Fibre optic cables do not use electricity – the signals are carried by infrared rays. The signals are very clear because they do not suffer from electrical interference. Other advantages are that they are cheaper than copper cables and can carry thousands of different signals down the same fibre at the same time.

Bundles of several thousand optical fibres are used in medical endoscopes for internal examination of the body. The bundle carries an image from one end of the bundle to the other, each fibre carrying one tiny part (one pixel) of the image.

OPTICAL FIBRES

Optical fibres transfer data in signals consisting of pulses of light (or infra-red radiation) following each other rapidly. The bandwidth of the fibre is the maximum number of pulses per second which can be transferred along the fibre and still be recognised as separate pulses at the receiving end. There is a limit to this bandwidth due to a process called pulse spreading. This is where the signal, originally a rectangular pulse, becomes weaker and spreads over a longer time interval as it travels along the fibre.

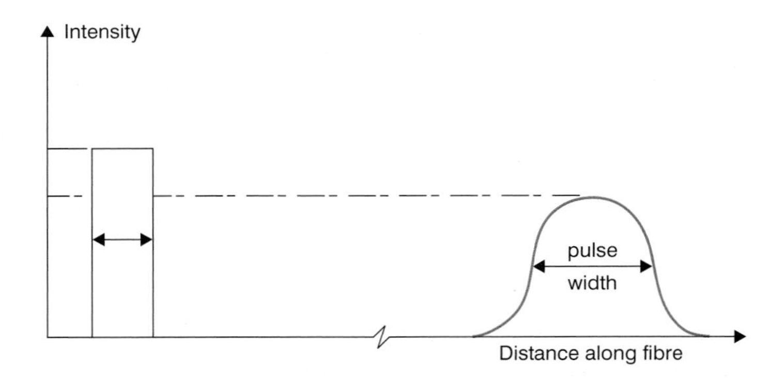

Δ Fig. 3.36 Signal transmitted down an optical fibre.

Spreading may cause pulses to become merged, meaning that the precise data transfer required for digital signals is not delivered with perfect accuracy.

The spreading of the pulse is due to two dispersion effects.

First, chromatic dispersion happens because the refractive index of the fibre is different for different frequencies (colours) of light. This effect also causes light to be 'split' into colours in a prism or in raindrops, producing a rainbow. In an optical fibre, chromatic dispersion means that the different frequencies of light will travel at different speeds in the fibre and lead to slightly different arrival times at the far end – the pulse will have 'spread'.

Secondly, if the width of the fibre is larger than the wavelength of the light there will be alternative paths (called 'modes') along the fibre.

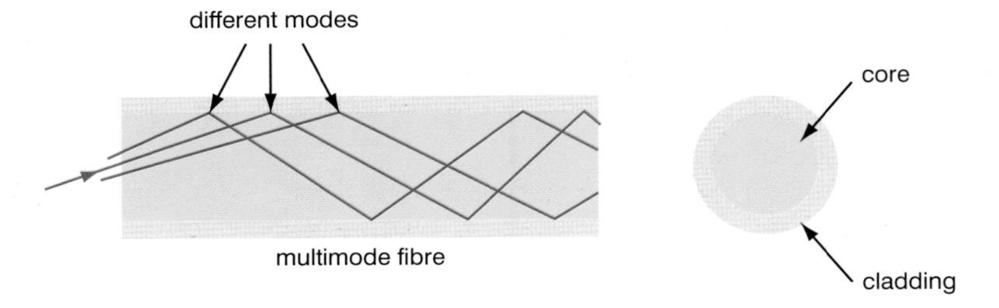

Δ Fig. 3.37 Different modes in an optical fibre.

Each path is a slightly different length so, again, the pulse arrives spread over a longer time. This multi-mode dispersion can be reduced either by using a narrower core or by using a graded-index material for the fibre core. Here, the refractive index of the core material reduces gradually from the centre to the edge.

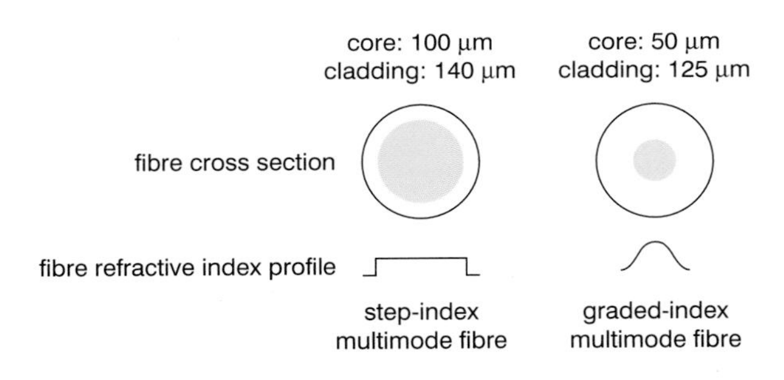

Δ Fig. 3.38 Multimode fibres.

The result of this is that the light travels more quickly towards the edge of the fibre, compensating for the extra distance it has to travel.

Glass prisms with internal angles of 45°, 45° and 90° are used as mirrors in periscopes and binoculars. In periscopes, light enters the prism through one of the smaller faces, is totally reflected off the inside of the larger face, and leaves through the other smaller face with its direction changed by 90°. Figure 3.39 shows how two such prisms are used in a periscope. Just the plain glass of the prism is used: no metal layer is needed to reflect the light.

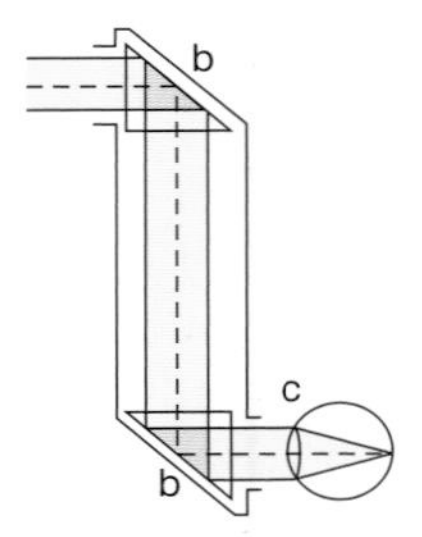

△ Fig. 3.39 In this periscope, two triangular prisms are used to redirect the light.

Binoculars consist of a pair of telescopes made shorter by 'folding up' the light path between the front and back lenses. Two 45°, 45°, 90° prisms are used, each changing the direction of the light by 180°. Light enters through one side of the larger face of the prism and is reflected off each of the two smaller faces in turn, being bent by 90° each time. It finally emerges from the other side of the prism's larger face, travelling in the opposite direction.

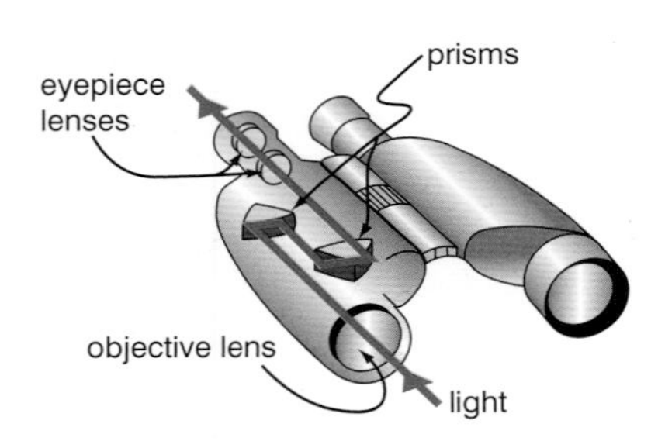

△ Fig. 3.40 How reflection is used in binoculars.

QUESTIONS

1. What is the critical angle?
2. What is the critical angle for a material of refractive index 1.4?
3. How does light travel along a fibre optic cable?
4. Why doesn't light escape from the cable?
5. How are television signals carried in fibre optic cables?
6. How are fibre optic cables used in medicine?

SOUND WAVES

Sound is caused by vibrations, of the front of a violin or a guitar, or of the column of air inside a trumpet. In the case of a loudspeaker it is particularly clear that the cone of the loudspeaker moves in and out and changes the pressure in the air in front of it. The sound travels as longitudinal waves. The compressions and rarefactions of sound waves result in small differences in air pressure.

Like other longitudinal waves, sound waves can be reflected and refracted.

QUESTIONS

1. What type of waves are sound waves?
2. What is the result of compressions and rarefactions in a sound wave?

Δ Fig. 3.41 This orchestra is creating a single longitudinal wave of very complicated shape. In ways that we barely understand, our brains can pick out the sounds of all the individual instruments that are playing together.

Sounds humans can hear

The human ear can detect sounds with pitches ranging from 20 Hz to 20 000 Hz. Sound with frequencies above this range is known as **ultrasound**. Ultrasound is used by bats for navigation. It can also be used to build up images of organs within the body, because different tissues reflect ultrasound waves in different ways. By combining the various reflections, an image is generated. A well-known example of the use of ultrasound scanning is for checking the development of a fetus during pregnancy.

SCIENCE IN CONTEXT

EAR DAMAGE

The ear is far more easily damaged than most people realise, and you should always take care both with the volume of sound and the length of time that the ear is exposed to it. The damage is cumulative, and so is not noticed at first. Many older rock stars have serious hearing problems, and many of the younger ones now wear ear plugs to prevent their own performances from damaging their hearing.

Earplugs worn by rock stars are designed so that the range of audio frequencies are reduced equally so that the wearer hears the upper and lower frequencies at the same relative levels as they would without the earplugs. (If this were not the case then the bass guitarist, for example, may feel that s/he was not playing loudly enough to balance the vocalist, simply because the lower frequencies were reduced in level too much relative to the higher vocal frequencies.)

Δ Fig. 3.42 A musician's earplug.

This type of earplug usually has a tiny diaphragm to reduce low frequencies and absorbent or damping material to reduce high frequencies. They are quite expensive and are intended to be used again and again. They reduce noise levels by about 20 dB and are not intended to protect the wearer from noise levels above 105 dB.

MEASURING THE SPEED OF SOUND IN AIR

The simplest method to measure the speed of sound in air uses two microphones and a fast recording device such as a digital storage scope.

1. A sound source and the two microphones are arranged in a straight line, with the sound source at one end.
2. The distance between the microphones (x), called microphone basis, is measured.
3. The time of arrival between the signals (delay) reaching the different microphones (t) is measured.
4. Then speed of sound = x/t.

Measuring the speed of sound by an echo method

Hard surfaces reflect sound waves. An **echo** is a sound that has been reflected before you hear it. For an echo to be clearly heard, the obstacle needs to be large compared with the wavelength of the sound. So, you will hear an echo when you make a loud noise when you are several hundred metres from a brick wall or a cliff, for example. You will not hear an echo when you are just a couple of metres from the wall or cliff. There will still be an echo, even when you are much closer to the wall, but because sound travels very quickly, the echo will return in such a short time that you will probably not be able to distinguish it from the sound that caused it.

The following worked example illustrates how echoes may be used to measure the speed of sound.

WORKED EXAMPLE

Two students stand side by side at a distance of 480 m from the school wall. Student A has two flat pieces of wood, which make a loud sound when clapped together. Student B has a stopwatch.

As student A claps the boards together, student B starts the stopwatch. When student B hears the echo, he stops the stopwatch. The time recorded on the stopwatch is 2.9 s.

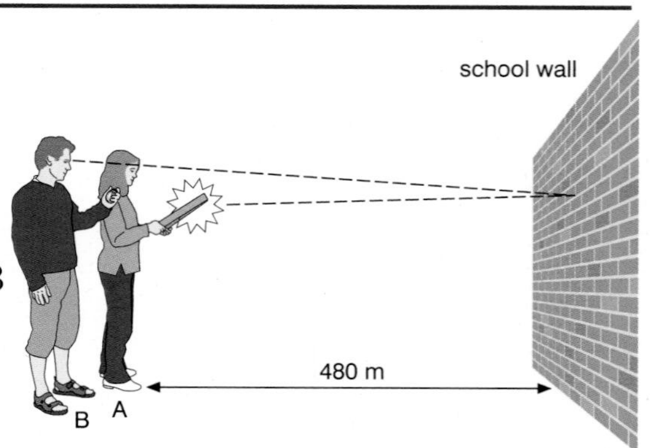

Δ Fig. 3.43 Two students carrying out investigations into the speed of sound.

Calculate the speed of sound.

Write down the formula: speed of sound = distance travelled / time taken

Work out the distance:	distance to wall and back	= 2 × 480
		= 960 m

Record the time the sound took to travel there and back:

	time	= 2.9 s
Substitute in the formula:	speed of sound	= 960 / 2.9
		= 331 m/s

See if you can think of some things that might be done to improve the accuracy of this experiment.

Developing investigative skills

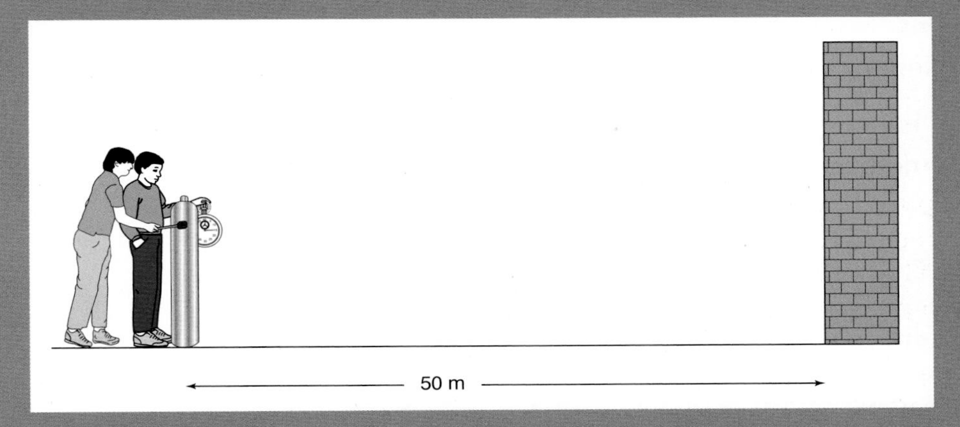

Δ Fig. 3.44 Two students measuring the speed of sound.

Two students are finding the speed of sound. They stand 50 m from a large wall. The first student strikes a piece of metal with a hammer. The sound this makes reflects back from the wall and the student hears an echo. The student hits the metal again in time with the echo and continues to do so, tapping out a steady rhythm.

The second student is in charge of timing. He knows that in between each sound that the first student makes the sound travels 100 m (to the wall and back again). By timing the interval for a number of strikes, the second student can record the data he needs.

Observing, measuring and recording

1. Explain whether the students should repeat their measurements for this experiment.
2. The students stand 50 m away from the wall that provides the echo. Explain why this is a suitable distance to use.

❸ The students measure a time of 2.3 s from striking the metal to striking the metal at the eighth following echo. Calculate the speed of sound given by this measurement.

❹ A better way to find the speed of sound would be through the use of a graph. Describe how the students could collect suitable data and how a graph could be used to find the speed of sound.

❺ Explain why using a graph to find the speed of sound would improve the accuracy of the experiment.

Handling experimental observations and data

❻ What effect will reaction time have on the measurements made?

❼ Suggest how the timing in this experiment could be improved.

QUESTIONS

1. What is the frequency range for human hearing?
2. What is ultrasound?
3. Describe a simple method for measuring the speed of sound in air.

USING AN OSCILLOSCOPE AND MICROPHONE TO DISPLAY SOUND WAVES

Sound waves can be displayed on an oscilloscope by using a microphone and a loudspeaker as shown in Figure 3.45. This produces a voltage against time graph for the sound wave on the screen of the oscilloscope. From the voltage-time graph, you can find the frequency of the sound wave since you will know the time taken for one cycle of the wave (which is the period).

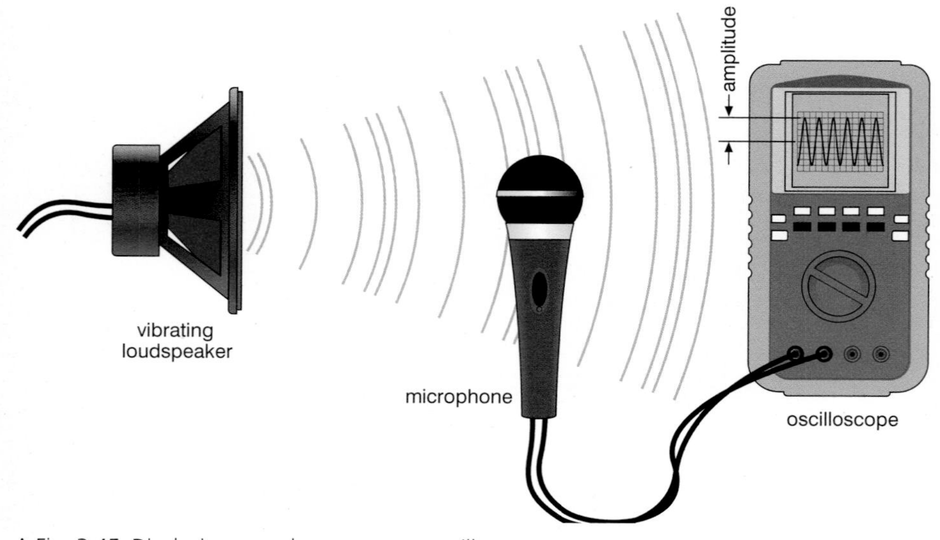

Δ Fig. 3.45 Displaying sound waves on an oscilloscope screen.

Developing investigative skills

A student uses a microphone, different tuning forks and an oscilloscope to investigate the frequency of different musical notes. He is trying to identify the frequency and note for each tuning fork.

He adjusts the oscilloscope settings so that the trace of any sound will move horizontally across the whole screen in 0.1 s. He takes screen shots of the wave patterns for each note and counts the number of wave cycles that each tuning fork produces in this time. From this he can calculate the frequency of the sound wave.

Look at his results.

Tuning fork	Number of wave cycles	Time (s)	Frequency (Hz)	Note of tuning fork
P	39	0.1		
Q	22	0.1		
R	25	0.1		
S	33	0.1		
T	24	0.1		
U	29 ½	0.1		
V	35	0.1		

Devise and plan investigations

❶ How should the student make sure that the microphone detects only the sound from the tuning forks?

Analyse and interpret results

❷ Calculate the frequency, in Hz, for each tuning fork.

❸ The student finds some data on the internet. Look at this information.

Musical note	Frequency (Hz)
A	220.00
B	246.94
C	261.63
D	293.67
E	329.63
F	349.23
G	391.99

Use this information to identify the note for each tuning fork.

Evaluate data and methods

❹ The student is unsure about the note from tuning fork 'R'. Explain why.

❺ Explain ways to improve the investigation to improve results and increase the confidence in his findings.

PITCH, FREQUENCY, AMPLITUDE AND LOUDNESS

Sounds with a high pitch have a high frequency. Examples of high-pitched sounds include bird-song, and all the sounds that you hear from someone else's personal stereo when they have set the volume too high. Low-pitched sounds have a low frequency. Examples of low-pitched sounds include the horn of a large ship and the bass guitar.

Loud sounds have high amplitude, whereas quiet sounds have low amplitude. The loudness of sounds can be compared using decibels.

Typical sound wave patterns are shown in Figure 3.46.

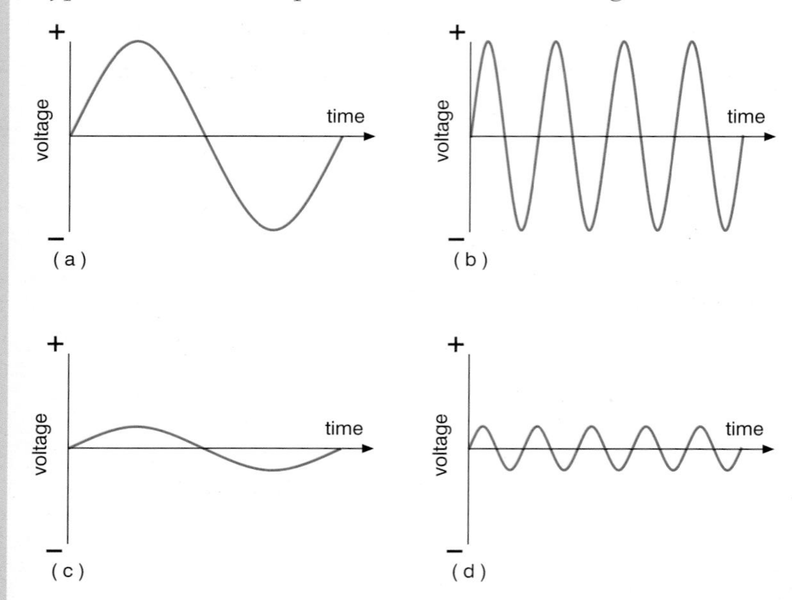

Δ Fig. 3.46 Typical sound wave patterns: a) a loud sound of low frequency; b) a loud sound of high frequency; c) a quiet sound of low frequency; d) a quiet sound of high frequency.

End of topic checklist

The **angle of incidence**, i, is the angle between the incident light ray and the normal to the surface. The **angle of refraction**, r, is the angle between the refracted light ray and the normal to the surface inside the material.

The **refractive index** of a material indicates how strongly the material changes the direction of the light.

The **critical angle** (c), is the angle of incidence that produces the maximum refraction. If the angle of incidence is increased more than the critical angle, then **total internal reflection** (TIR) will occur.

The facts and ideas that you should understand by studying this topic:

- ❍ Know that light and sound waves can be reflected and refracted.
- ❍ Know that in a reflection, angle of incidence = angle of reflection.
- ❍ Know how to draw ray diagrams to illustrate reflection and refraction.
- ❍ Be able to investigate refraction in rectangular, semi-circular and triangular prisms.
- ❍ Know and use the relationship refractive index, $n = \sin i/\sin r$.
- ❍ Know and use the relationship refractive index, $n = 1/\sin c$.
- ❍ Be able to determine the refractive index of a glass block.
- ❍ Explain the meaning of the term critical angle.
- ❍ Describe the role of total internal reflection in transmitting signals in optical fibres and prisms.
- ❍ Understand that sound waves are longitudinal waves, whilst light waves are transverse.
- ❍ Understand that the range of human hearing is 20 Hz–20 000 Hz.
- ❍ Be able to investigate the speed of sound in air.
- ❍ Understand how an oscilloscope and a microphone can be used to display sound waves.

End of topic checklist continued

- ❍ Be able to use an oscilloscope and a microphone to determine the frequency of a sound.
- ❍ Know how the pitch of a sound relates to the frequency of vibration of the source.
- ❍ Know how the loudness of a sound relates to the amplitude of vibration of the source.

End of topic questions

1. a) Rays of light can be reflected and refracted. State one difference between reflection and refraction. **(1 mark)**

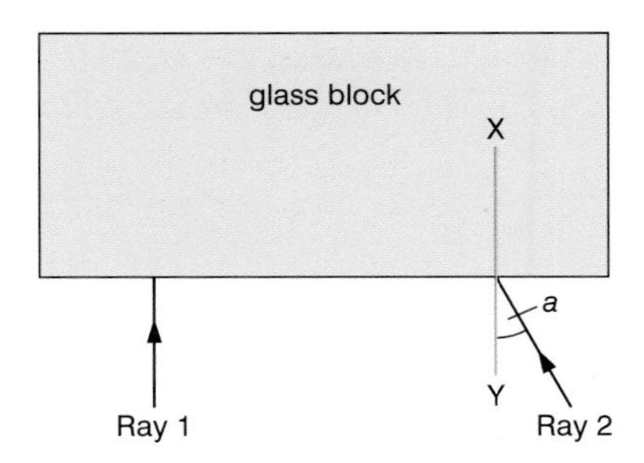

b) The diagram shows a glass block and two rays of light.

i) Complete the paths of the two rays as they pass into and then out of the glass block.

ii) What name is given to the angle marked *a*?

iii) What name is given to the line marked XY? **(3 marks)**

2. The diagram shows light entering a prism. Total internal reflection takes place at the inner surfaces of the prism.

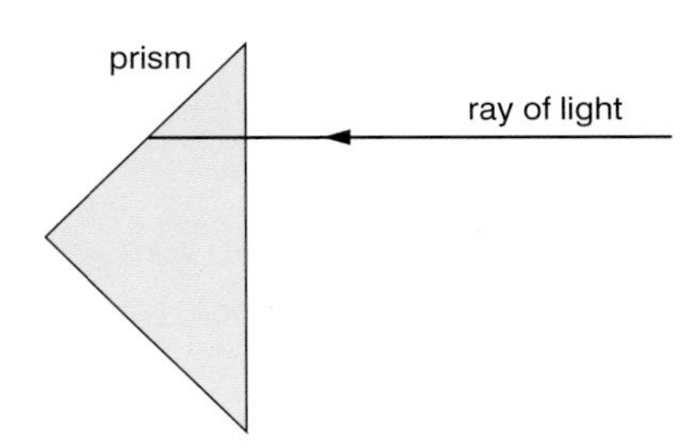

a) Copy the diagram and complete the path of the ray. **(1 mark)**

b) Suggest one use for a prism like this. **(1 mark)**

c) Explain what happens at the second face when the angle of incidence is:

i) 36°

ii) 42° (critical angle)

iii) 46° **(3 marks)**

End of topic questions continued

3. A student traces the path of a red light beam through a rectangular block of plastic, and finds that the angle of incidence is 50°, and the angle of refraction is 21.7°.

a) What is the refractive index of the block? **(2 marks)**

b) What will be the critical angle for this material? **(2 marks)**

4. a) What causes a sound? **(1 mark)**

b) Explain how sound travels through the air. **(3 marks)**

c) Astronauts in space cannot talk directly to each other. They have to speak to each other by radio. Explain why this is so. **(3 marks)**

5. Ayesha and Salma are doing an experiment to measure the speed of sound. They stand 150 m apart. Ayesha starts the stopwatch when she sees Salma make a sound and she stops it when she hears the sound herself. She measures the time as 0.44 seconds. Calculate the speed of sound in air from this data. **(3 marks)**

6. The speed of sound in air is approximately 340 m/s.

A student hears two echoes when she claps her hands. One echo is 0.5 s after the clap, and one echo is 1.0 s after the clap. She decides that the two echoes are from two buildings directly in front of her. What is the distance between the two buildings? **(2 marks)**

7. Refraction and total internal reflection lead to the 'heat haze' that you can see over hot surfaces, particularly roads in the summer. Use ideas about how heating affects the density of air to explain how this happens. **(3 marks)**

8. Optical fibres must not be too wide. By thinking about the different paths the light could take in the optical fibre, explain why optical fibres must not be too wide. **(3 marks)**

9. Suggest a reason why the signals, sent along a fibre optic cable, are laser signals that contain a single wavelength of light or infrared radiation. **(2 marks)**

10. Draw an oscilloscope trace for each of the following sounds.

a) low frequency quiet sound **(1 mark)**

b) high frequency loud sound **(1 mark)**

c) high frequency quiet sound **(1 mark)**

d) low frequency loud sound **(1 mark)**

11. A polystyrene ball is suspended touching the prongs of a vibrating tuning fork. The ball kicks away from the tuning fork and then moves back to it.

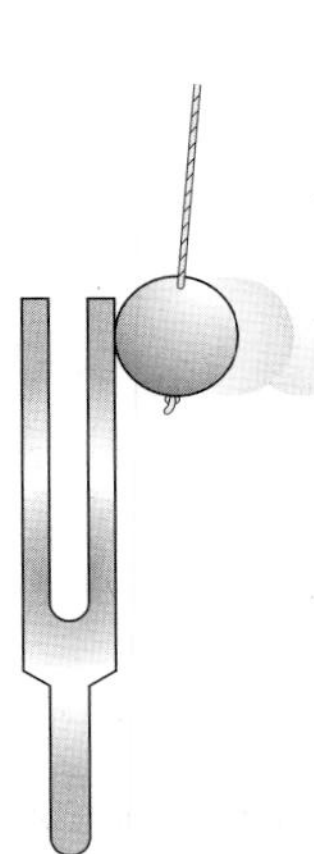

a) Explain the behaviour of the ball by:

i) describing how the prongs of the tuning fork move

ii) describing how a sound wave is created by the tuning fork

iii) making a drawing of the sound wave, labelling the key features. **(3 marks)**

b) Explain what you would see if the tips of the tuning fork were dipped into a beaker of water. **(2 marks)**

End of topic questions continued

12. Copy the diagram below to show what happens to light when it travels through air, then glass, followed by air again. The incident ray has been drawn for you.

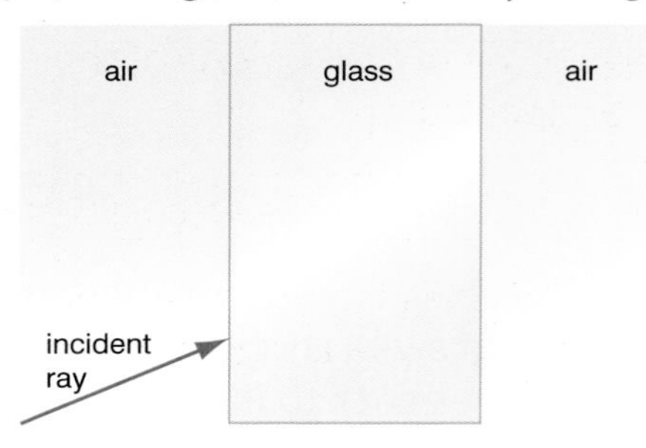

Draw and label:

a) the refracted ray **(1 mark)**

b) the emergent ray **(1 mark)**

c) the normal. **(1 mark)**

d) Use your protractor to measure and record the different angles of the light rays when entering and leaving the glass block.

e) Finally, explain why the ray of light is refracted when it enters the glass block. **(1 mark)**

13. Red light, of wavelength 650 nm, passes into a container of paraffin oil of refractive index 1.44. (Hint: 1 nm = 10^{-9} m)

a) What is the wavelength of the light in the paraffin? **(2 marks)**

b) If this wavelength calculated in part **a)** were in air, what colour would it be? **(2 marks)**

c) What is the frequency of the wave in paraffin? **(1 mark)**

Exam-style questions

Sample student answers

Question 1

The chart shows some of the regions of the electromagnetic spectrum.

gamma	X-rays	A	visible	infrared	B	radio

a) i) Complete the chart be writing the names of the missing regions (A and B) of the spectrum.

A: UV ✗

B: microwaves ✓ (1) **(2)**

ii) The table above lists the spectrum in what order?

A. increasing amplitude

B. increasing density

C. increasing frequency

D. increasing wavelength

D ✓ (1) **(1)**

b) i) Describe one situation in which X-rays are useful.

in a hospital ✗ **(1)**

ii) Explain why X-rays are useful in the example you have given.

X-rays can show broken bones because they will pass through skin and flesh ✓ (1)

but there will be a shadow where the bones are. ✓ (1) **(2)**

c) i) Describe one situation in which X-rays can be harmful.

when they get into your body ✗ **(1)**

EXAMINER'S COMMENTS

a) i) The question clearly asks for the name of the missing regions. Writing 'UV' is not a name, only an abbreviation, so it loses a mark. Make sure you read the question and answer in exactly the way it asks you to.

ii) You should know the different ways to describe the order of the spectrum. Here the answer is given in terms of increasing wavelength, but you should also be able to describe the sequence in order of frequency and possibly in order of energy.

b) i) Stating 'in a hospital' is too vague to be a 'situation' as the question asks.

ii) Clearly, the student was aware of the correct situation – imaging – so should have written this in their answer.

c) i) Again, the student has not described a situation, so cannot gain the first mark.

Exam-style questions continued

ii) Describe how X-rays can be harmful and how the risks can be reduced.

They can damage body cells ✓ (1)

and cause some cells to become cancer cells. ✗ **(2)**

d) A remote control for a television uses infrared signals.

The human body detects infrared radiation as heat.

Explain why the infrared signal from a television remote control does not make your skin feel hot.

The signal from the remote is not strong enough to make you feel hot. ✗

There is not enough energy to burn you. ✓ (1) **(2)**

(Total 11 marks)

ii) The student has two valid points, but there is only one mark available. The second was for describing how the risks can be reduced and the student has failed to give any response to this. Always read the question carefully and answer each point required.

d) 'Not strong enough' is too vague. The correct term 'amplitude' is required here. The second part of the answer correctly links to energy and the student might have thought a little further about this, possibly then getting the link from 'strong' to 'amplitude'.

Exam-style questions continued

Question 2

The diagram shows some waves on the surface of water in a ripple tank.

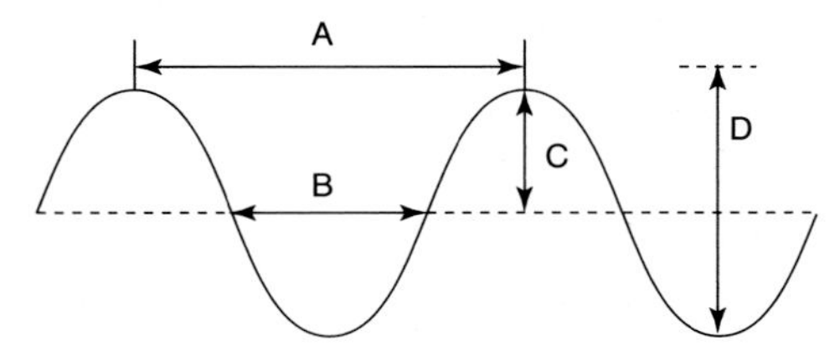

a) i) Which letter represents the wavelength of the wave? **(1)**

ii) Which letter represents the amplitude of the waves? **(1)**

b) The waves in the water are transverse waves.

i) Give another example of a transverse wave. **(1)**

ii) Describe the difference between transverse waves and longitudinal waves. **(2)**

c) A student makes some observations about the waves in the ripple tank.

number of waves passing by in 6 seconds - 10

distance between wave crests - 5 cm

i) Show that the time period for these waves is 0.6 s. **(1)**

ii) Use the equation frequency $= \frac{1}{\text{time period}}$ to calculate the frequency of the waves. **(1)**

iii) State the equation linking wave speed, frequency and wavelength. **(1)**

iv) Calculate the wave speed of these waves in m/s. **(2)**

(Total 10 marks)

Question 3

a) Describe an experiment to find the refractive index of glass, using a glass block.

You should include:

- the equipment needed
- a diagram of how the equipment will be used
- what measurements should be made
- how the measurements will be used to find the refractive index. **(6)**

b) Use ideas about reflection and refraction to explain how information can be sent along optical fibres.

You may use diagrams to help your explanation. **(5)**

(Total 11 marks)

Energy is essential to everyday life. When you ride a bicycle, the energy store in your body decreases while the energy store of the bicycle increases. When a bus engine burns petrol or diesel, the energy store in the fuel decreases and the energy store of the bus increases. Since the engine is hot, it transfers energy to the surroundings. In this section you will learn about different types of energy transfer. You will then look at how work and power are defined in physics, which is not quite the same as in everyday usage. Finally, you will look at the important topic of renewable and non-renewable energy resources and how electricity is generated.

STARTING POINTS

1. What is the source of the energy store in your body?
2. Why are energy transfers never 100% efficient?
3. Name some energy stores.
4. What do you understand by the terms conduction, convection and radiation in relation to energy?
5. What is the principle of conservation of energy?
6. Name some ways in which electricity is generated. Which of these use renewable sources and which use non-renewable sources?

CONTENTS

4 Energy resources and energy transfers

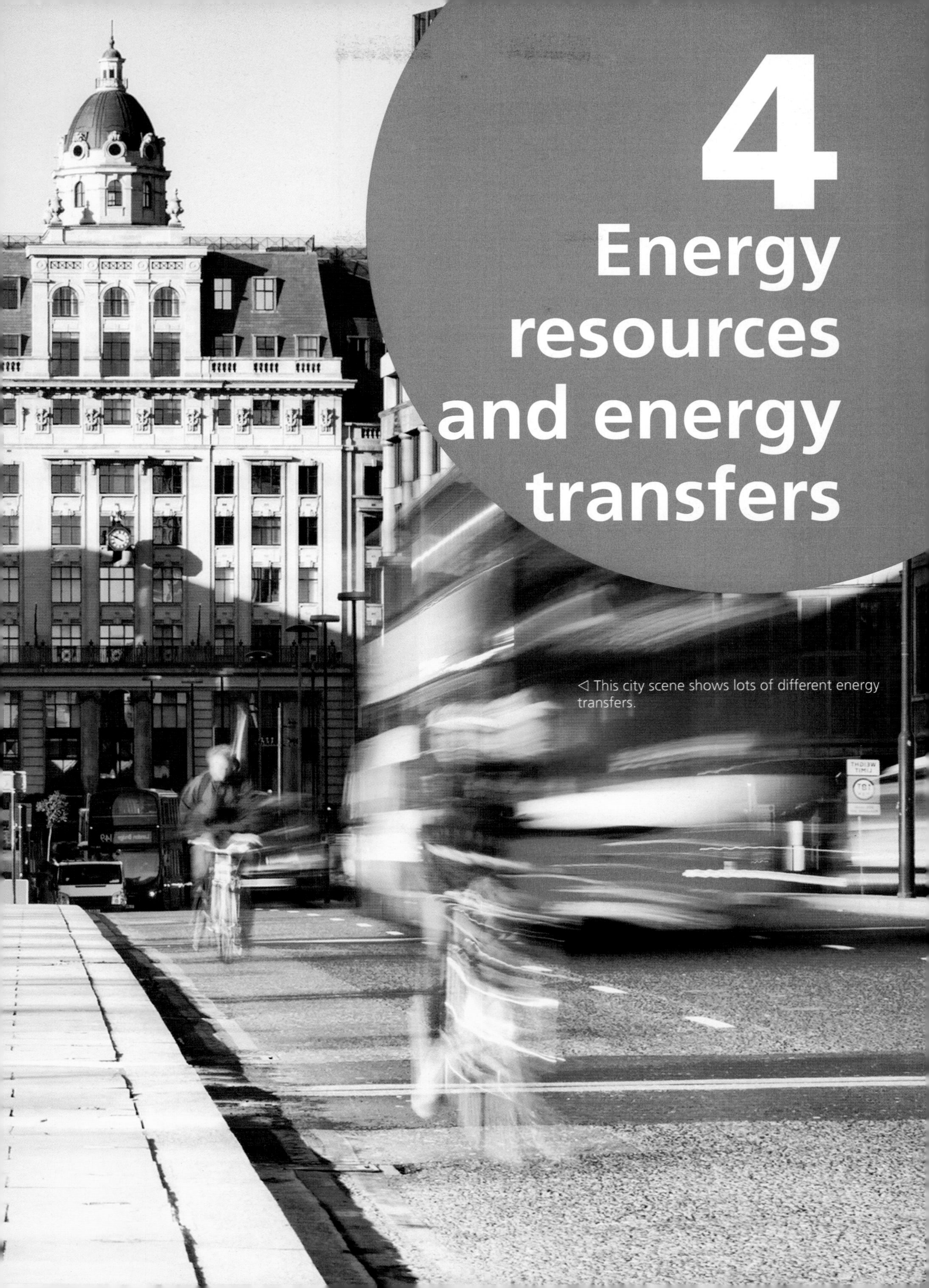

◁ This city scene shows lots of different energy transfers.

Units

INTRODUCTION

For the topics included in energy resources and energy transfer, you will need to be familiar with:

Quantity	Unit	Symbol
mass	kilogram	kg
length	metre	m
time	second	s
speed	metre per second	m/s
acceleration	metre per second squared	m/s^2
force	newton	N
energy	joule	J
power	watt	W

Some of these you have met previously but the new units will be defined when you meet them for the first time.

LEARNING OBJECTIVES

✓ Use the following units: kilogram (kg), joule (J), metre (m), metre/second (m/s), metre/second^2 (m/s^2), newton (N), second (s) and watt (W).

Energy

INTRODUCTION

The study of energy, how it it is stored and what it does when it is transferred, is at the heart of physics. But what actually is energy? It can be surprisingly hard to pin down. We have an intuitive feeling that when we have lots of energy we can get things done. When we are feeling 'drained', then it is much harder. Being able to track where the energy is stored, and how it is transferred, is a key skill that will help to explain many aspects of physics.

Δ Fig. 4.1 As a skier skis down a mountain, the store of gravitational potential energy decreases and the store of kinetic energy increases.

KNOWLEDGE CHECK

✓ Know some everyday uses of energy.
✓ Be able to describe devices that transfer energy from one store to another.

LEARNING OBJECTIVES

✓ Be able to describe energy transfers involving energy stores.
✓ Be able to describe and recognise the energy stores: chemical, kinetic, gravitational, elastic, thermal, magnetic, electrostatic, nuclear.
✓ Be able to describe and recognise energy transfers to include: mechanically, electrically, by heating, by radiation (light and sound).
✓ Be able to explain how emission and absorption of radiation are related to surface and temperature.
✓ Be able to investigate thermal energy transfer by conduction, convection and radiation.
✓ Be able to explain ways of reducing unwanted energy transfer, such as insulation.

EXAMPLES OF ENERGY STORES

All vehicles need fuel to move. Steam trains use coal as a fuel, and many cars now use diesel, petrol or even alcohol that has been produced from sugar. Some vehicles use chemical energy stored in a battery. In the future, cars may use hydrogen as a fuel. But whatever fuel you use, you are buying something with a store of **chemical energy**.

Δ Fig. 4.2 Fuel, whatever form it comes in, gives a car the ability to move.

A clock needs energy to make the hands move, and this energy can be stored in a spring that you wind up with a key, in an electrical battery, or in weights that are raised up.

If a spring is stretched or compressed, the spring will store **elastic energy** as shown in Fig. 4.4.

Δ Fig. 4.3 The weights attached to this clock store gravitational energy when they are raised. As the weights slowly fall, energy is transfered to the mechanism that turns the hands.

Δ Fig. 4.4 The spring in this shock absorber stores elastic energy when it is compressed.

If a load is raised above the ground, it has stored **gravitational potential energy** as shown in the top part of Fig. 4.5. If the spring is released or the load moves back to the ground (bottom of Fig. 4.5), the stored energy is transferred to a store of movement energy, which is called **kinetic energy**.

In all of the examples above, the potential energy can be used to make an object move, and hence give it kinetic energy. Kinetic energy can also be transferred to other energy stores, which can be seen most clearly in a pendulum. At each end of its swing, the pendulum has a maximum amount of gravitational energy. At the middle of its swing some of the stored gravitational energy has been transferred to increase the store of kinetic energy (the pendulum is moving fastest), as shown in Fig. 4.6.

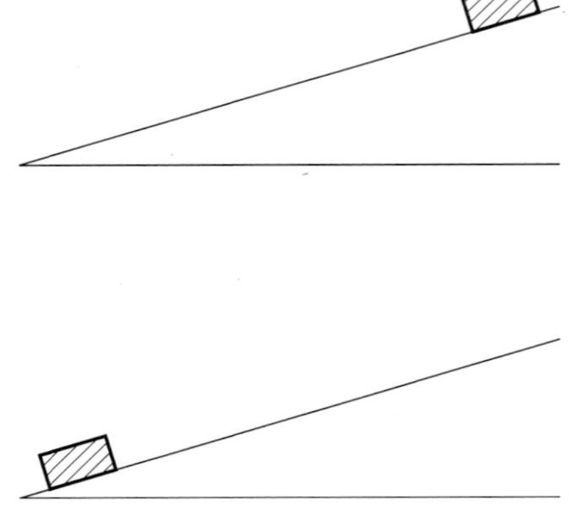

Δ Fig. 4.5 Block on a ramp.

ENERGY STORES AND TRANFERS

We have already met some ways of storing energy. The following are examples of **energy stores**.

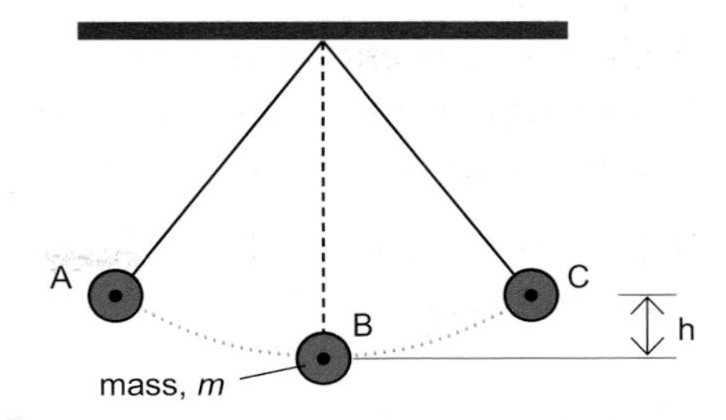

Δ Fig. 4.6 Energy changes in the swing of a pendulum. The pendulum has maximum stored gravitational energy at A and C, and maximum kinetic energy at B.

- Petrol in a car stores **chemical** energy, which can be released when the fuel combines with oxygen.
- A trampoline changes shape when a person lands on it. For a short amount of time a trampoline stores **elastic** energy.
- A moving object has a store of **kinetic** energy.
- A dam that is used to trap water increases the water's store of **gravitational** energy.
- Hot water stores **thermal** energy. A hot object contains atoms that are vibrating rapidly.
- The forces between particles in the nucleus of an atom are very large indeed. These force fields store **nuclear** energy which is released in radioactive decay, fission or fusion.
- Two separated electric charges that are attracting, or repelling, store **electrostatic** energy.
- Two magnets that are attracting, or repelling, store **magnetic** energy.

Energy can be transferred from one store to another. The following are examples of **energy transfers**.

- A stretched trampoline transfers energy **mechanically**, applying a force to increase the person's kinetic and gravitational stores.
- In a battery-operated fan, an electric current transfers energy **electrically** from the energy store in the battery to the movement of the motor.
- A flame transfers energy to the surroundings by **heating**.
- An LED transfers energy by **radiation** (light, an electromagnetic wave) to illuminate a traffic light.
- A vibrating guitar string transfers energy to the room as **sound**. A sound wave carries a very small amount of energy from the vibrating string, setting air particles around it into vibrations. These vibrations are passed through the air as a longitudinal wave. When the wave reaches the ear it causes the ear drum to vibrate.

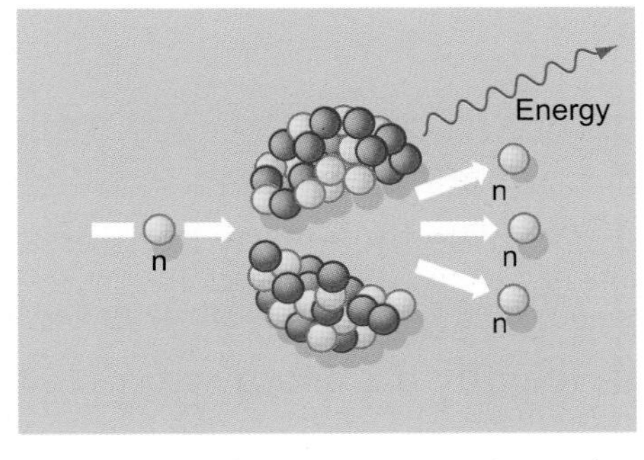

Δ Fig. 4.7 Splitting an atom can release a lot of energy.

In some cases this transfer can be done between just two stores, such as between kinetic energy and gravitational energy in a pendulum. In other cases, energy is transferred to more than one type of store. For example, if you brake when you are riding your bicycle, work done by friction transfers energy to the brake pads, which heat up. A small amount of energy is also transferred to the surroundings by sound – the brakes produce a noise.

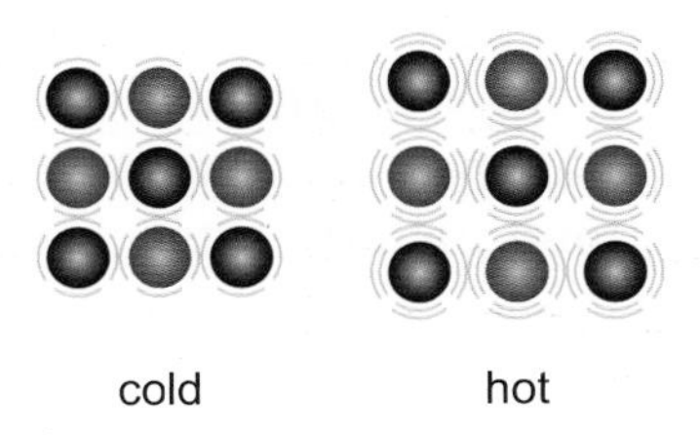

Δ Fig. 4.8 The atoms in a hot object vibrate more because they have more energy.

ENERGY AND THE EARTH

Energy and matter are constantly interacting on our planet. Part of this interaction produces volcanoes, glaciers, mountain ranges, ocean currents and the weather. The energy required for all Earth processes comes from two sources: energy from the Sun, which keeps the oceans and the atmospheric cycles (such as the water cycle) going; and nuclear energy, which comes from radioactive decay in the Earth's interior, and is transferred to the Earth's outer layers by conduction and convection. This thermal energy from the Earth's interior drives geological processes.

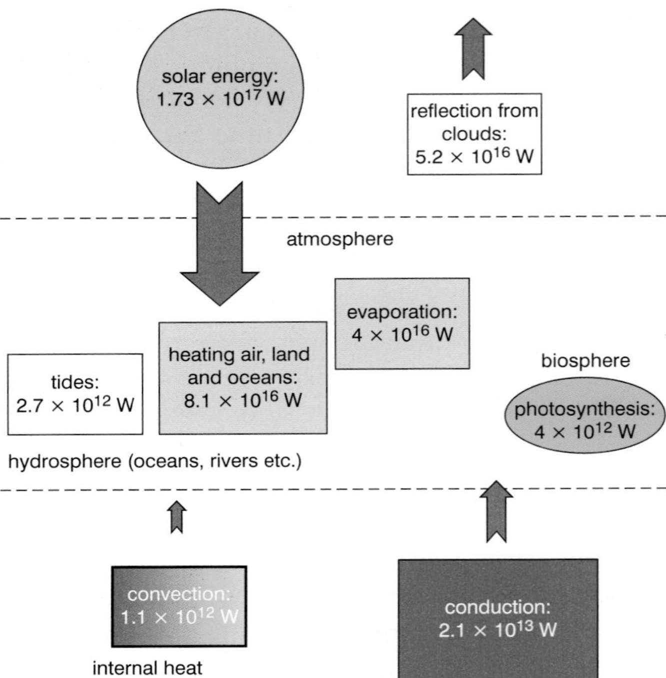

Δ Fig. 4.9 Energy transfers in the Earth's system.

The amount of energy that moves through the system is huge: it is of the order of 1.74×10^{17} W. Most of this comes from the Sun. The diagram shows the energy transfers that take place in the Earth's system.

QUESTIONS

1. Describe the energy transfers that take place as a pendulum swings.

2. Where can elastic energy be found?

3. Identify the energy stores in the following examples:

a) Uranium fuel used in a nuclear power station
b) Diesel oil
c) A moving wind turbine
d) A bow drawn back before the arrow is released
e) A hot water bottle
f) A ball at the top of a slope
g) A rubbed balloon attracting hair.

4. Identify how energy is being transferred in the following examples:

a) A siren on a police car
b) A battery-powered torch is switched on
c) An oven cooks a potato.

THE PRINCIPLE OF CONSERVATION OF ENERGY

In every energy transfer, not all the input energy is *usefully* transferred. For example, as a pendulum swings, it transfers energy between gravitational and kinetic stores but as the pendulum moves through the air it increases the movement of the air particles and so heats the air. This energy transfer increases the thermal energy store of the surroundings. The pendulum eventually stops swinging (has a zero store of kinetic energy and does not increase its store of gravitational energy).

REMEMBER

You may need to describe how energy is transferred in different situations, but remember that total energy is always conserved: the energy at the start and at the end must have the same total value.

The **law of conservation of energy** says that energy cannot be created or destroyed, simply transferred. So you must account for the *total* amount of energy output, and that includes the energy that will have been transferred to the surroundings as thermal energy by heating, as well as perhaps by light or sound.

In another example, the motor in a tram transfers energy electrically to kinetic energy but also gets hot. Only some of the energy output is useful. The rest of the output energy is transferred to a thermal energy store, which is not useful. But energy is always conserved – there is no change to the total amount of energy.

Δ Fig. 4.10 The current in the overhead wires transfers energy electrically to the motor. Some of the energy transferred turns the wheels and makes the tram move. The rest of the energy transferred increases the thermal energy store of the surroundings.

QUESTIONS

1. State the law of conservation of energy.
2. Describe how energy is being transferred between stores when a ball rolls down a hill, and how the total energy changes
3. Consider a moving car.

   a) Identify the energy store that allows the car to move.

   b) What is the useful energy output?

   c) What is the 'wasted' energy output?

Efficiency of energy transfer

Energy is always conserved – the total amount of energy after the transfer must be the same as the total amount of energy before the transfer. Unfortunately, in nearly all energy transfers some of the energy will end up as 'useless' thermal energy, by heating the surroundings.

Energy **efficiency** can be calculated from the following formula:

$$\text{efficiency} = \frac{\text{useful energy output}}{\text{total energy input}} \times 100\%$$

For example, the electric motor that is used to power a train may take in 10 kJ of energy electrically and give out 9.5 kJ of kinetic energy. The useful energy output is therefore 9.5 kJ and the energy input is 10 kJ. Using the equation

we find that $\frac{9.5}{10} \times 100\% = 95\%$

so the motor is 95% efficient. The other 5% of energy ends up increasing the temperature of the motor and passes to the surrounding air. This energy is wasted, and in fact the motor will need cooling fans to prevent it from overheating.

Developing investigative skills

A student investigates the efficiency of a small electric motor. He uses a motor to lift a mass through a constant distance of 1 m. He times how long it takes to lift the mass and makes a record of the potential difference across and the current through the motor. His data are shown in the table.

Mass lifted (kg)	Distance lifted (m)	Useful work done (J) (= *mgh*)	p.d. across motor (V)	Current through motor (A)	Time to lift the mass (s)	Electrical energy supplied (J)
0.01	1.0		2.4	0.20	22.0	
0.03	1.0		2.4	0.22	24.4	
0.05	1.0		2.4	0.25	26.5	
0.07	1.0		2.3	0.28	27.6	
0.09	1.0		2.3	0.29	28.7	

Devise and plan investigations

1. What measuring instruments would the student need for this investigation?
2. The timer measured to 0.01 s, but the student decided to record values to a precision of 0.1 s. Suggest why he did this. Was it the correct thing to do?

Analyse and interpret results

❸ Complete the table to show the useful energy output and the total energy input. (Take $g = 10$ N/kg, and use the equation $E = IVt$ to calculate the electrical energy E transferred to the motor.)

❹ Use your data to plot a graph of useful energy out (y-axis) against total energy in (x-axis)

❺ Describe the pattern (if any) shown by the graph.

Evaluate data and methods

❻ How does the pattern on the graph relate to the efficiency of the motor?

❼ If you were to take more measurements with this equipment, what value(s) of mass would you choose? Explain your answer.

Sankey diagrams

Energy transfers can be summarised using simple energy transfer diagrams or **Sankey diagrams**. The key point of Sankey diagrams is that they allow you to show visually the flow of different quantities, with the width of the bars related to the amounts following each path. To put it another way, the thickness of each arrow is drawn to scale to show the amount of energy. This can be extended to a wide variety of situations.

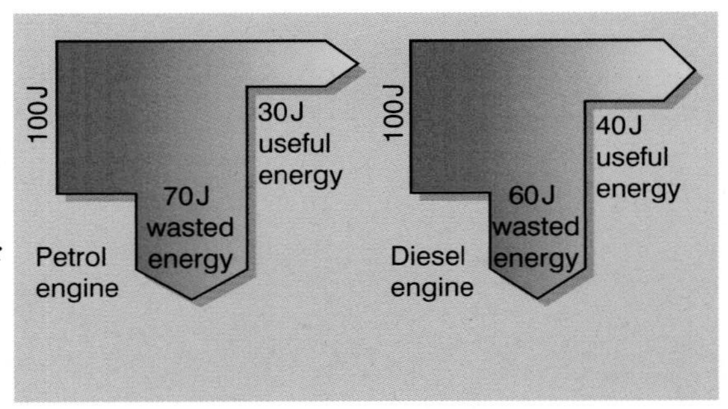

△ Fig. 4.11 Sankey diagrams for petrol and diesel engines.

A diesel engine is more efficient than a petrol engine. Figure 4.11 shows Sankey diagrams for the two types of engine. You can see that the diesel engine can give out 40 J of useful energy for every 100 J input. This gives an efficiency of 40%, which is very good efficiency for any process that transfers energy by heating. You will note that the engine has to lose 60 J of thermal energy, much of it down the exhaust pipe, but it still needs a large radiator as well.

In a power station as much as 70% of the energy transfers do not produce useful energy. The power station is only 30% efficient.

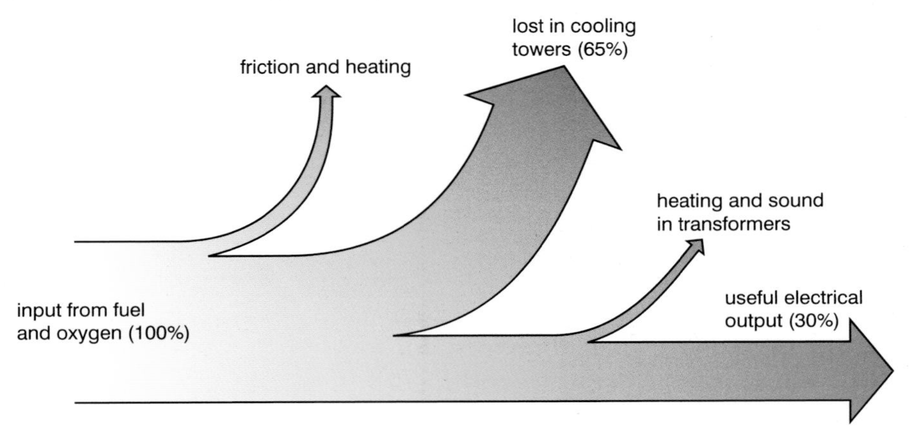

△ Fig. 4.12 Sankey diagram for a power station.

Scientists are working hard to increase the efficiency of power stations. Many power stations are now trying to make use of the large amounts of energy 'lost' in the hot water. In some cities, the houses of whole regions of the city are heated by hot water from the power station. Some of the most modern fossil-fuelled power stations have had efficiencies nearer to 40%. This may not seem much, but if all power stations in the world could use 25% less fuel, it would save millions of tonnes of coal or gas per year. Combined cycle gas-fired power stations can have efficiencies of almost 50%, while combined heat and power installations may be over 70% efficient.

SCIENCE IN CONTEXT

EXAMPLES OF SANKEY DIAGRAMS

Sankey diagrams are named for Matthew H. Sankey, who published the first diagram of this type in 1898. In it, he compared the energy efficiency of a steam engine to a perfect heat engine that would have no energy losses. This is very similar to the way you are using Sankey diagrams in International GCSE Physics. Here are a few examples, some related to physics, some not. There are many more examples available if you search online.

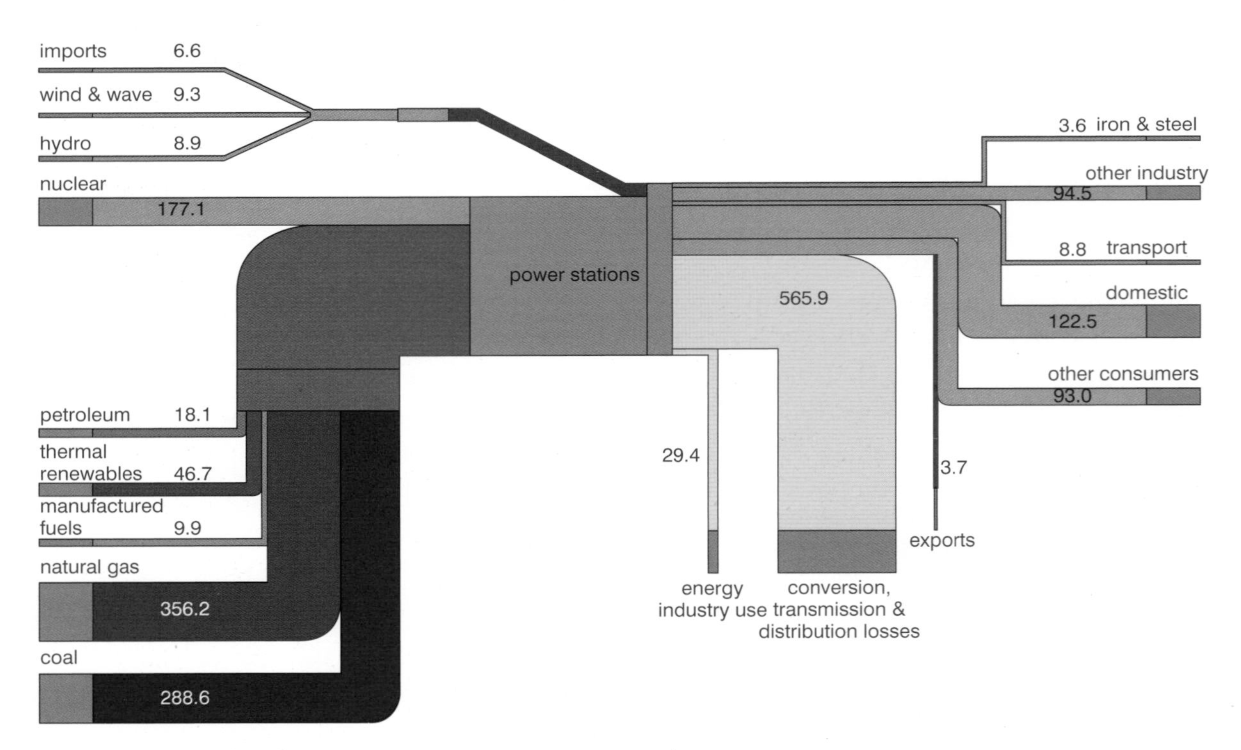

Δ Fig. 4.13 Energy flow for the UK 2009. Figures are in terawatt hours.

This chart shows the energy resources used in the UK in 2009. It shows clearly how the UK is still heavily reliant on fossil fuels for energy production. It also indicates how inefficient the whole system is – look at the proportion on the right hand side labelled 'conversion, transmission and distribution losses'.

Figure 4.14 shows what happens to the water used in a family home. The width of arrows indicates the relative amounts following each path.

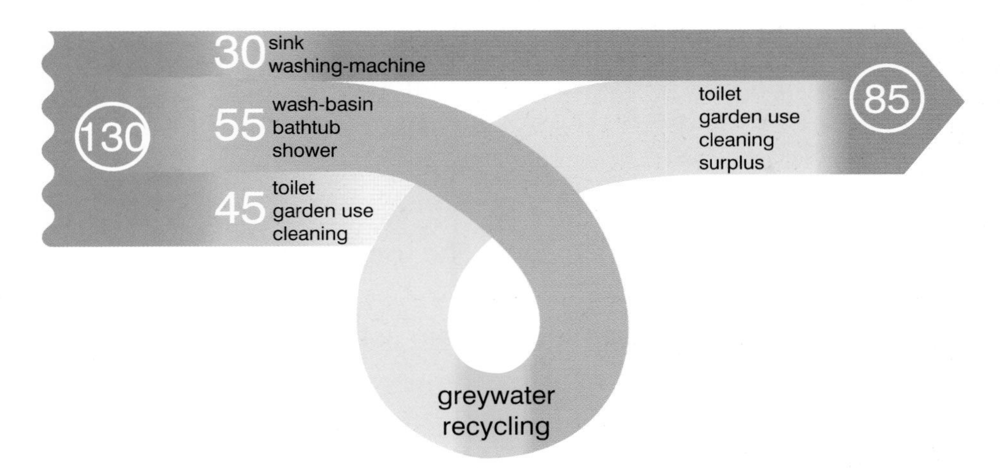

Δ Fig. 4.14 Sankey diagram for water use in a family home.

Figure 4.15 shows the materials used in Hawaii in 2005–6. This chart describes the volumes of materials and products. You can clearly link the inputs and outputs. Notice that, like the water chart above, this also has a 'recycling' element to it.

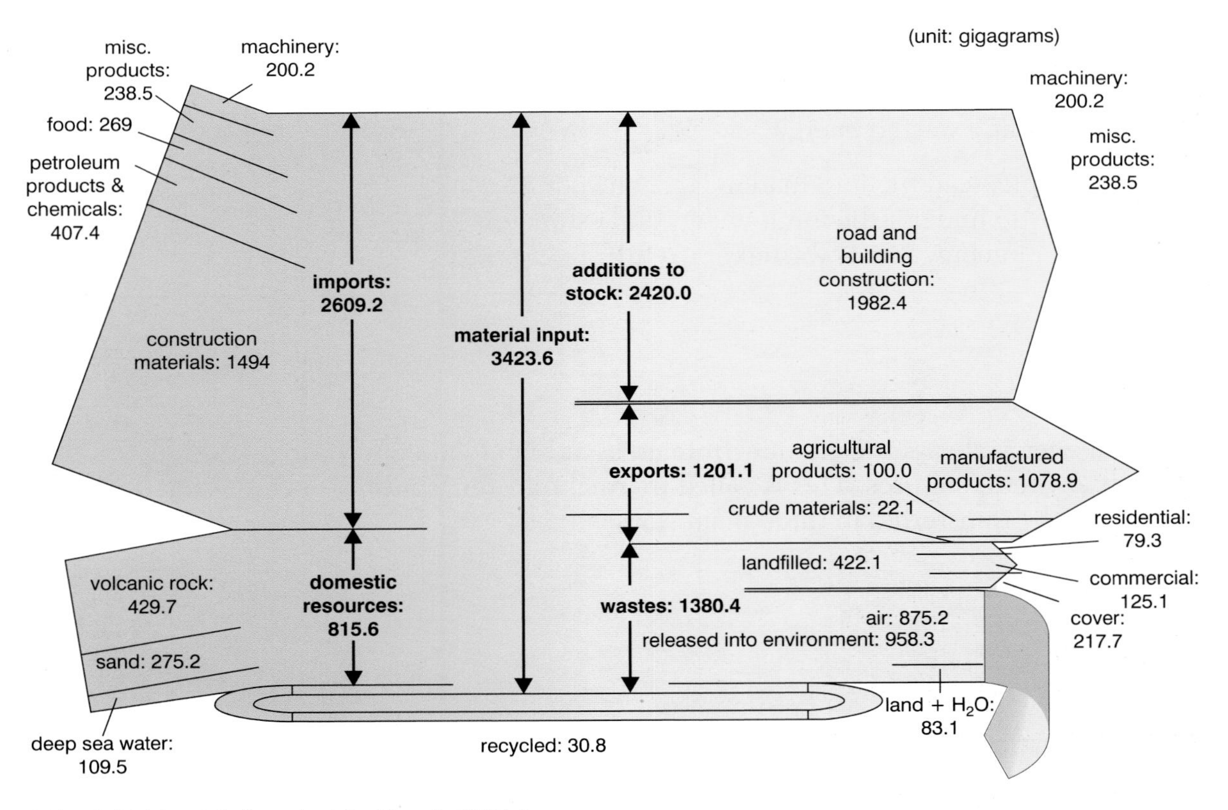

Δ Fig. 4.15 Materials flow chart for Hawaii, 2005–6.

QUESTIONS

1. Explain what the width of an arrow in a Sankey diagram shows.
2. A Bunsen burner heats water in a beaker. 1000 J of energy are transferred into 400 J of useful energy, to warm the water, and 600 J of wasted energy to heat the room. Draw a Sankey diagram to scale to show this.
3. What is the efficiency of the Bunsen burner in question 2?
4. An athlete gains 40 000 J of energy from her food. She uses 800 J to train.
   a) How much energy is not used by the athlete to train (wasted energy)?
   b) What is her efficiency?

EXTENSION

Sankey diagrams are a useful method for showing all of the information that is known when energy transfers from one store to another, for example input energy, useful output energy, wasted output energy and the values of all of these energies.

Using three objects found in your house, produce a Sankey diagram for each one to show input energy, useful energy and wasted energy.

Where possible, read the information label on your chosen items to find out the input energy and estimate the amount of energy that is wasted or useful.

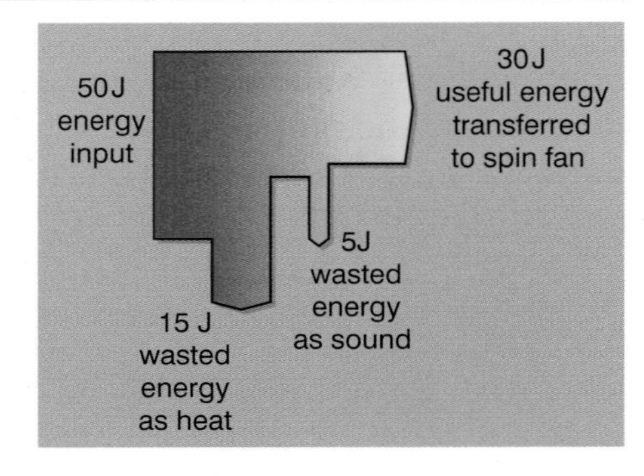

Δ Fig. 4.16 A Sankey diagram for an electric fan.

CONDUCTION, CONVECTION AND RADIATION

Thermal energy always tries to flow from areas at high temperatures to areas at low temperatures. This is called thermal transfer. Thermal energy can be transferred in three main ways:

- conduction
- convection
- radiation.

Conduction

Materials that allow thermal energy to transfer through them quickly are called **thermal conductors**. Those that do not are called **thermal insulators**. **Conduction** cannot occur when there are no particles present, so a **vacuum** is a perfect insulator.

REMEMBER

If someone talks about an 'insulator', you will have to work out from the context if it refers to a thermal insulator or to an electrical insulator. If the context is energy, then it could be a thermal insulator. If the context is electricity, it is likely to be an electrical insulator.

If one end of a conductor is heated, the atoms that make up its structure start to vibrate more vigorously. Because the atoms in a solid are linked together by chemical bonds, the increased vibration can be passed on to other atoms. The energy of movement (kinetic energy) passes through the whole material.

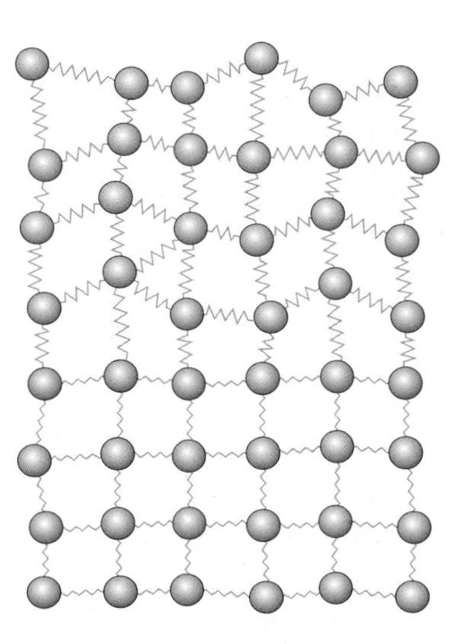

Δ Fig. 4.17 Conduction in a solid.

Metals are particularly good thermal conductors because they contain freely moving electrons. These free electrons move faster when the metal is heated. As the electrons travel all over the piece of metal, they transfer some of their kinetic energy to other atoms and electrons in the metal. This transfer of energy takes place very rapidly. This is in addition to the thermal energy that is transferred by vibrations of the atoms making up the structure of the metal. Figure 4.17 shows conduction in a solid. Particles in a hot part of a solid (top) vibrate further and faster than particles in a cold part (bottom). The vibrations are passed on through the bonds from particle to particle.

Conduction can be demonstrated using the equipment shown in Figure 4.18. The rods are made of different metals, so the heat conducts along them at different rates. The better the conductor, the quicker the wax at the end of the rod melts.

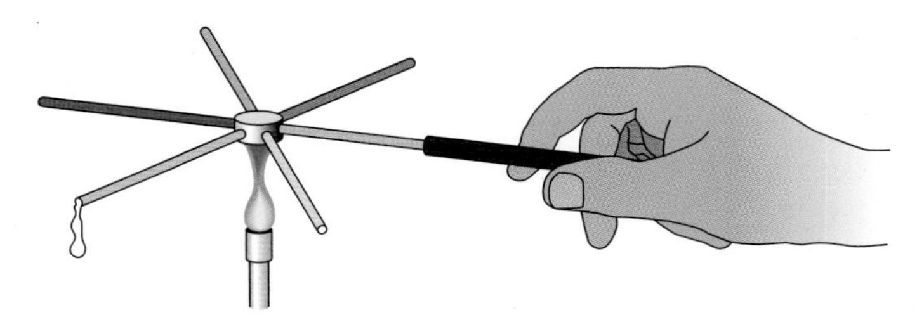

◁ Fig. 4.18 Experiment to show conduction.

QUESTIONS

1. Why are metals particularly good conductors?
2. Why is outer space a perfect insulator?
3. How is energy transferred in a thermal conductor?
4. Describe an experiment to demonstrate conduction.

EXTENSION

Devise an experiment to find out whether or not the rate of energy transfer varies along a strip of copper. Explain how you could tell if any change is linear.

Convection

Convection occurs in liquids and gases because these materials flow (they are fluids). The particles in a fluid move all the time. When a fluid is heated, energy is transferred to the particles, causing them to move faster and further apart. This makes the heated fluid less dense than the unheated fluid. The less dense warm fluid rises above the more dense colder fluid, causing the fluid to circulate as shown in Figure 4.19. This convection current is how the thermal energy is transferred.

Δ Fig. 4.19 Demonstrating convection with potassium manganate(VII) crystals.

Potassium manganate(VII) crystals in water demonstrate convection as shown in Figure 4.19. The warmer water expands, becomes less dense and rises, making a trail as some of the dissolved potassium manganate(VII) is carried along as well.

If a fluid's movement is restricted, energy cannot be transferred. That is why many insulators, such as ceiling tiles, contain trapped air pockets. Wall cavities in houses are filled with fibre to prevent air from circulating and transferring thermal energy by convection.

QUESTIONS

1. Why does convection occur in liquids and gases?
2. Explain why warm air rises.
3. How could you demonstrate convection in a laboratory?
4. Describe how cavity wall insulation in houses reduces thermal energy transfer by convection.

Radiation

Radiation, unlike conduction and convection, does not need particles at all. It can travel through a vacuum. This is clearly shown by the radiation that arrives from the Sun. Radiated thermal energy is carried mainly by infrared radiation, which is part of the electromagnetic spectrum.

All objects take in and give out infrared radiation all the time. Hot objects radiate more infrared than cold objects. The amount of radiation given out or absorbed by an object depends on its temperature and also on the colour and texture of its surface (see Table 4.1).

Thermograms (Figure 4.20) give a visual representation of the amount of infrared radiation that is given out by an object at a particular moment.

△ Fig. 4.20 A thermogram of a house.

Type of surface	As an emitter of radiation	As an absorber of radiation	Examples
dull black	good	good	Emitter: cooling fins on the back of a refrigerator are dull black to radiate away more energy. Absorber: The surface of a black bitumen road gets far hotter on a sunny day than the surface of a white concrete one.
bright shiny	poor	poor	Emitter: a polished, shiny steel teapot is a poor radiator so it will stay hotter for longer than a black one. Absorber: Fuel storage tanks are sprayed with shiny silver or white paint to reflect radiation from the Sun.

△ Table 4.1 Comparison of different surfaces as emitters or absorbers of infrared radiation.

A Leslie's cube has sides with different surfaces – shiny, dull, dark, light – and can be used to show how they emit and absorb thermal radiation at different rates. Because all sides of the cube are heated by the same water inside the cube, any differences in the way they radiate energy can only be due to the differences in their surfaces.

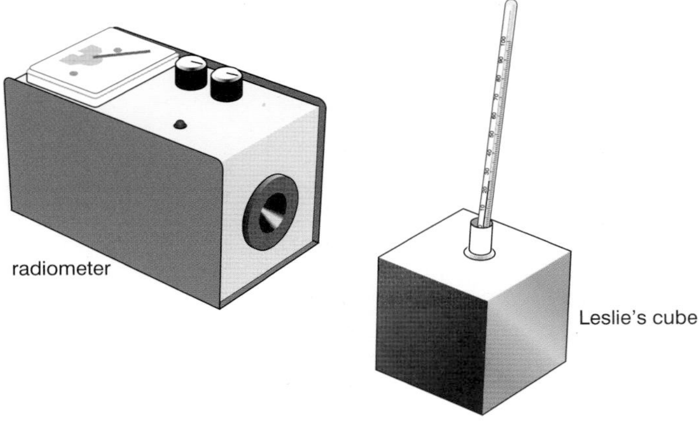

△ Fig. 4.21 A Leslie's cube.

QUESTIONS

1. How does radiation differ from conduction and convection?
2. Which radiates more infrared: a hot object or a cold object?
3. State two factors that affect the amount of thermal radiation emitted by an object.
4. Which side of the Leslie's cube will emit thermal radiation at the greatest rate?

Everyday examples of thermal transfer

Hot-water radiators are used to heat houses in some countries that have cool winters. A radiator transfers some thermal energy by radiation, and if you stand near a hot radiator your hands can feel the infrared radiation being emitted. However, this is only around one quarter of the thermal energy being transferred by the radiator. *Three quarters* of the thermal energy is transferred to the room by the hot air that rises from the radiator. Colder air from the room flows in to replace this hot air, and a convection current is formed as shown.

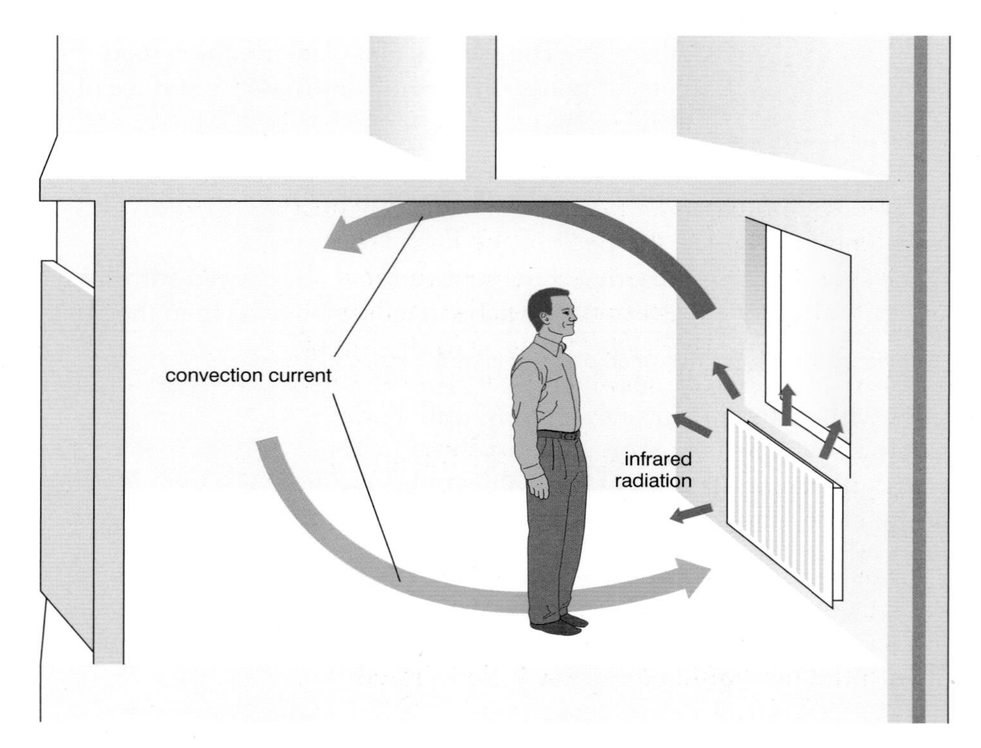

Δ Fig. 4.22 This shows a side view of a room with a hot-water radiator underneath the window.

You will note that the convection current is far more efficient at heating the top of the room than it is at heating the person standing in front of the radiator.

Another example of thermal transfers in everyday life is a vacuum flask. A vacuum flask keeps a drink hot or cold for hours by almost completely eliminating the flow of thermal energy out or in.

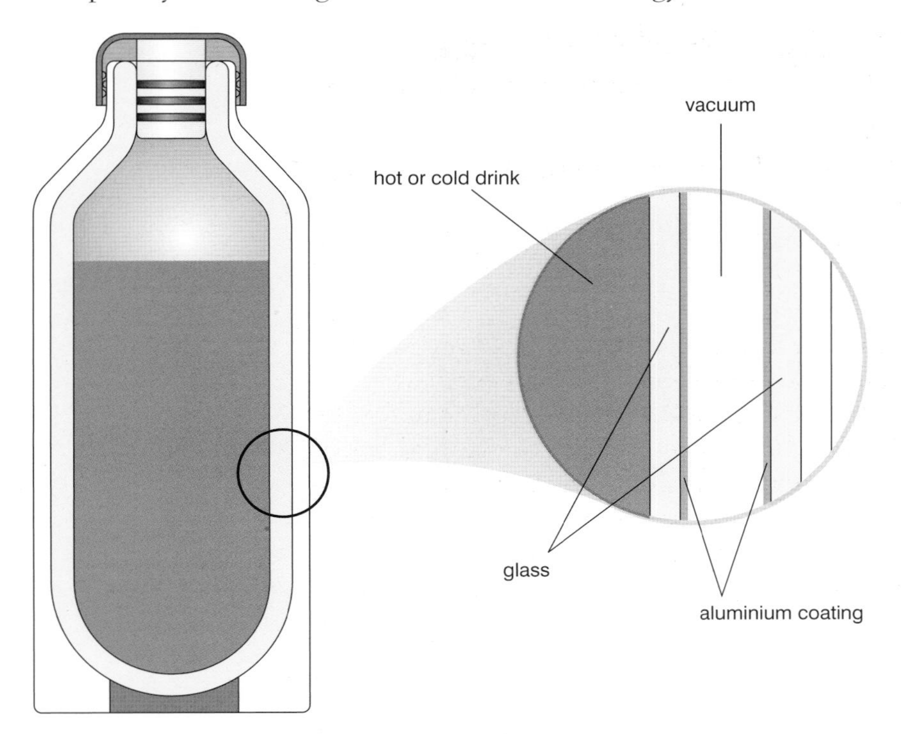

△ Fig. 4.23 A vacuum flask.

Conduction is almost entirely eliminated by making sure that any heat flowing out must travel along the glass of the neck of the flask. The path is a long one, the glass is thin, and glass is a very poor thermal conductor. Energy cannot be lost by conduction across the vacuum space between the two walls of the glass flask. The bung in the top of the flask must also be a very poor thermal conductor: cork or expanded polystyrene is good.

Convection is eliminated because the space between the inner wall and the outer wall of the flask is evacuated so that there is no air to form convection currents.

When the contents are hot, radiation is greatly reduced because the inner walls of the flask are coated with pure aluminium. Because the aluminium is in a vacuum, it stays extremely shiny forever, and so the wall in contact with the hot liquid emits very little infrared radiation.

Developing investigative skills

A student heats water in a beaker until it is just boiling. As the water cools, the student measures its temperature.

Devise and plan investigations

❶ Why should the student record the room temperature as well?

❷ Why is it important to stir the water a few seconds before taking the temperature?

Analyse and interpret

The student obtained the following graph.

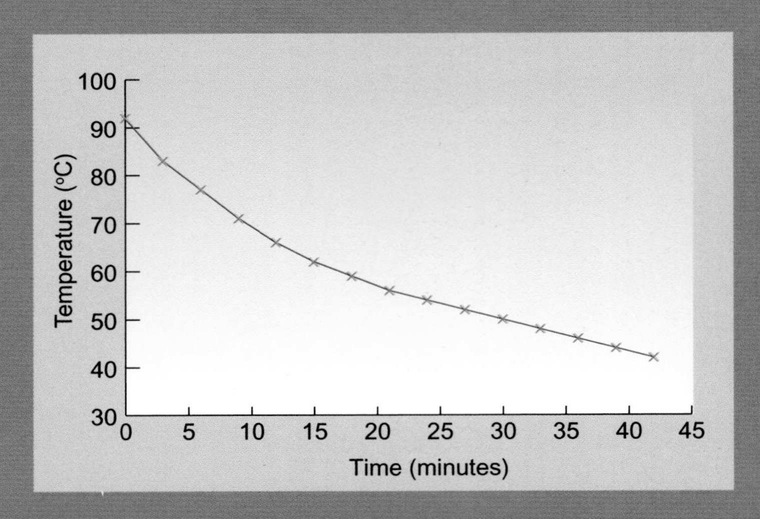

Δ Fig. 4.24 Student's measurements of heat.

❸ How often did the student measure the temperature of the water?

❹ How would the graph have changed if the student had measured the temperature a) more frequently b) less frequently? c) On different surfaces – such as wood and metal?

❺ Compare and explain the rate of cooling from 90°C to 70°C and from 70°C to 50°C.

Evaluate data and methods

❻ How could the student have reduced the rate of cooling of the water?

❼ How could the student investigate the hypothesis 'the bigger the temperature difference between an object and its surroundings, the greater the rate of energy transfer between them'?

QUESTIONS

1. What is the main method of heat transfer in a radiator?

2. Which part of the room is heated most efficiently by a radiator?

3. For a vacuum flask, describe which features reduce the energy transfer by **a)** conduction, **b)** convection and **c)** radiation.

4. Explain why a vacuum flask is good at keeping hot drinks hot AND cold drinks cold.

Using insulation to reduce unwanted energy transfers

There are a number of ways of reducing wasteful energy transfers in a house. The diagram below shows some of them.

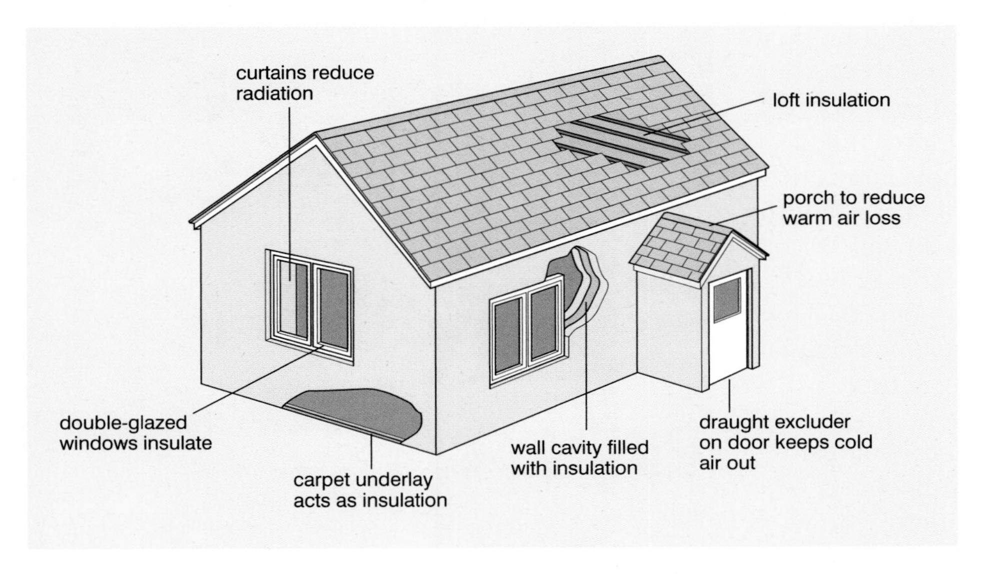

Δ Fig. 4.25 Ways of reducing wasteful energy transfers in a house in a cold climate.

Source of energy wastage	% of energy wasted	Insulation technique
Walls	35	Cavity wall insulation. In the UK, most modern houses have cavity walls, that is, two single walls separated by an air cavity. The air reduces energy transfer by conduction but not by convection as the air is free to move within the cavity. Fibre insulation is inserted into the cavity to prevent the air from moving and so reduces convection.
Roof	25	Loft insulation. Fibre insulation is placed on top of the ceiling and between the wooden joists. Air is trapped between the fibres, reducing energy transfer by conduction and convection.

Floors	15	Carpets. Carpets and underlay greatly reduce unwanted energy transfers by conduction and convection. In some modern houses foam blocks are placed under the floors.
Draughts	15	Draught excluders. Cold air can get into the home through gaps between windows and doors and their frames. Draught excluder tape can be used to block these gaps.
Windows	10	Double glazing. Energy is transferred through glass by conduction and radiation. Double glazing has two panes of glass with a layer of air between the panes. It reduces energy transfer by conduction but not by radiation. Radiation can be reduced by drawing the curtains.

△ Table 4.2 Insulation reduces heat loss.

Mountaineers and other people who need clothing to protect them from extreme cold know that they need to wear several layers of clothing, with each layer full of trapped air. The whole aim is to have a thick layer of air around the body, because air is a poor conductor of heat. The air is trapped in small air pockets. This means that the air cannot circulate easily and form convection currents. Air is generally a good insulator anyway but if you can stop it convecting it becomes an extremely good insulator.

△ Fig. 4.26 Mountaineers need well-insulated clothing.

The fibres of clothes, especially the newer extremely fine spun-polyester Polartec™ fibres, do a very good job at stopping the air from moving, thus preventing convection.

Marathon runners, at the end of a race, wrap themselves in foil blankets to prevent them from cooling down too quickly. The shiny metal surface reflects radiation back to the body. Mountaineers do not use metallised layers to prevent radiation, because such layers would trap perspiration and could interfere with movement. However, metallised plastic layers are used in the emergency survival bags that mountaineers carry.

Developing investigative skills

A student put equal amounts of hot water at 90°C into four identical glass boiling tubes. Three of the tubes were covered in different materials. The other tube was not covered. The temperature of the water in each tube was measured every 5 minutes.

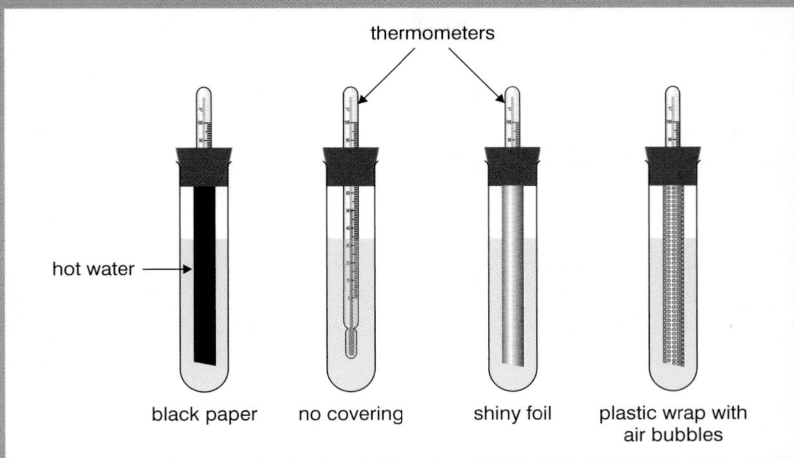

Δ Fig. 4.27 Apparatus for experiment.

His data are shown in the table.

Time (minutes)	Water temperature (°C)			
	Black paper	No covering	Plastic wrap with air bubbles	Shiny foil
0	90	90	90	90
5	57	64	74	81
10	40	49	62	74
15	32	39	54	68
20	26	32	46	63
25	21	26	39	59
30	20	22	34	55
35	20	20	34	52
40	20	20	29	50

Devise and plan investigations

❶ Boiling water can be very dangerous. Suggest two safety precautions that should be followed.

❷ Why has one of the tubes got no black paper, foil or insulation?

Analyse and interpret results

❸ On one set of axes plot a graph of how the temperature of the water changes with time. Draw a line of best fit for each set of data. Label each graph.

❹ Compare the rate of cooling for each of the tubes.

❺ All four test tubes cool fastest in the first five minutes. Explain why.

❻ Shiny foil helps reduce thermal energy transfer. Explain how.

❼ The boiling tube covered with black paper cools quickly. Explain how. In your answer use ideas about absorbing and emitting radiation.

Evaluate data and methods

❽ The student forgot to measure room temperature. Look at the graphs and suggest a value for the room temperature.

❾ Suggest how the temperature measurements could be made more reliable.

❿ If the room temperature had been 30°C, suggest how the cooling curves would have changed.

QUESTIONS

1. If you are insulating a house, explain why the walls and the roof are the most important features to deal with.
2. Air is a bad thermal conductor (a good insulator). Describe how this is used to reduce energy loss from a house.
3. Describe how trapping a layer of air near the body helps a person keep warm.
4. Explain why loose clothing can be an advantage in hotter climates.

End of topic checklist

A moving object stores **kinetic energy**. Raising an object increases its store of **gravitational energy**. Stretching or compressing an object increases its store of **elastic energy**.

Energy stored in chemical bonds is **chemical energy**. Batteries and fuels contain stored chemical energy.

The other **energy stores** are thermal, magnetic, electrostatic and nuclear.

The **law of conservation of energy** says that energy cannot be created or destroyed.

A **Sankey diagram** shows the relative amounts of output energy transferred by different processes.

A **thermal conductor** is a material that transfers thermal energy easily.

A **thermal insulator** is a material that does not transfer thermal energy easily.

Convection is thermal energy transfer in a liquid or gas – when particles in a warmer region gain energy and move into cooler regions carrying this energy with them.

Conduction is the transfer of thermal energy through a material.

Radiation is thermal energy that is carried mainly by infrared radiation.

The facts and ideas that you should understand by studying this topic:

- ❍ Know that energy can be stored and also transferred from one store to another.
- ❍ Know that energy is transferred mechanically (when a force is applied), electrically, by heating and by radiation (light and sound).
- ❍ Know and use the relationship linking efficiency, useful energy output and total energy input:

  $$\text{efficiency} = \frac{\text{useful energy output}}{\text{total energy input}} \times 100\%$$

- ❍ Describe a variety of energy transfers using Sankey diagrams.
- ❍ Describe the processes of thermal energy transfer by conduction, convection and radiation.
- ❍ Know that emission and absorption of radiation from an object depend on its surface and temperature.
- ❍ Understand how insulation is used to reduce unwanted energy transfers.
- ❍ Understand that energy is conserved.

End of topic questions

1. Why are several thin layers of clothing more likely to reduce thermal transfer than one thick layer of clothing? **(3 marks)**

2. The diagram shows a cross-section of a steel radiator positioned in a room next to a wall.

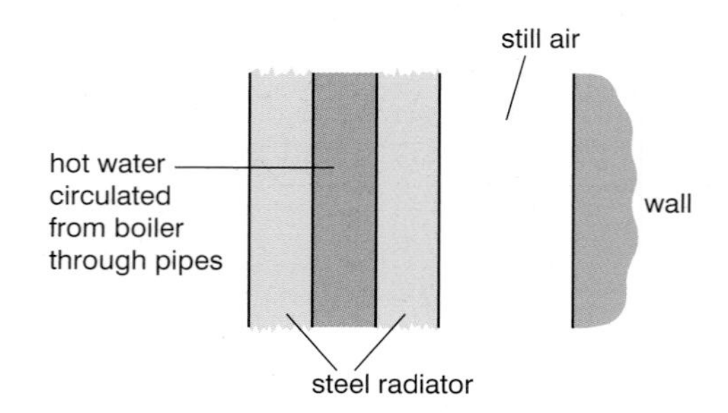

Describe how energy from the hot water reaches the wall behind the radiator. **(3 marks)**

3. The input to an electric motor is 5000 J. 1500 J of the output heats the engine and the surrounding air.

a) How much energy is usefully transferred? **(2 marks)**

b) What is the efficiency of the motor? **(2 marks)**

4. a) The input to an electric fan is 50 J. 30 J is used to spin the fan, 15 J is output as thermal energy and 5 J is output by sound. How much energy does the fan waste? **(2 marks)**

b) What is the efficiency of the fan? **(2 marks)**

5. Some energy transfers are wasteful – not all the energy is transferred usefully.

Copy the table and fill in the gaps. **(12 marks)**

Object	Input energy	Useful energy output	Wasted energy	Efficiency
light bulb	100 J	10 J		
torch	70 J	55 J		
radio	250 J	210 J		

6. Explain what happens to liquids and gases when they are warmed. **(3 marks)**

7. Explain why solids transfer energy mainly by conduction. **(3 marks)**

8. Describe how a convection heater warms a room. **(3 marks)**

9. Using conduction, suggest why serving dishes are usually made from glass or china. **(3 marks)**

10. Which factors affect the rate at which an object transfers energy? **(3 marks)**

11. Draw a Sankey diagram for a device that is 25% efficient. **(3 marks)**

12. Suggest why using oil as a lubricant might improve the efficiency of an engine. **(2 marks)**

13. If a crane does 650 J of useful work for every 1000 J of energy put in, what is its efficiency? **(2 marks)**

14. An engine has 700 J energy input. It has an efficiency of 40%. How much useful energy is output? **(2 marks)**

15. A device is 40% efficient. The total energy input is 800 J. How much energy is wasted? **(2 marks)**

16. An electric fan has an energy input of 80 J. It has an efficiency of 69%. 15 J is wasted by heating. The remainder of the wasted energy by sound. How much energy is wasted by sound? **(2 marks)**

17. Which is the least efficient of these two engines: a petrol engine that wastes 70 J for every 100 J input, or a steam–diesel hybrid engine that gives 20 J useful output energy for every 80 J input? **(3 marks)**

18. Look at the diagram of the single-glazed window. Energy is transferred from the warm air inside the room to the cool air outside. Explain in detail how particles transfer this energy **(6 marks)**

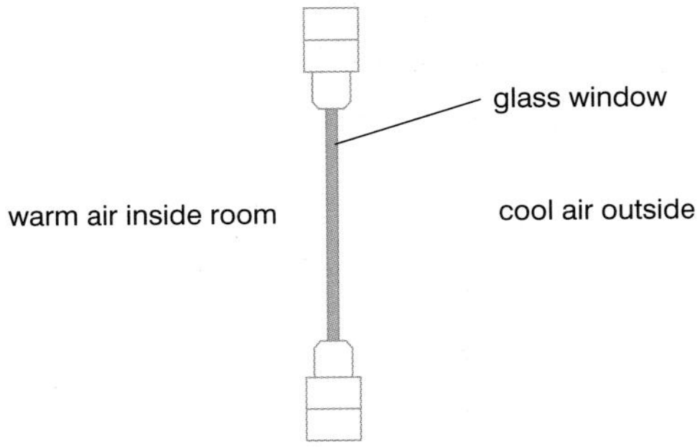

Δ Fig. 4.28 One 'horsepower' transfers energy at a rate equal to 'one horse' – this is taken to be 750 W.

Work and power

INTRODUCTION

How many times do you use the word 'work' in one day? You probably use the word 'power' less often, but, like 'work', it has several different meanings in everyday life, depending on the context. However, in physics these two words have very specific meanings. Work is related to force and distance moved in the direction of the force, while power is the rate of doing work.

KNOWLEDGE CHECK

- ✓ Know that energy can be transferred from one store to another.
- ✓ Know that total energy is always conserved.
- ✓ Know the effects that forces can have on objects.

LEARNING OBJECTIVES

- ✓ Know and use the relationship between work done, force and distance moved in the direction of the force: work done = force × distance moved.
- ✓ Know that work done is equal to energy transferred.
- ✓ Know and use the relationship between gravitational potential energy, mass, gravitational field strength and height: gravitational potential energy = mass × gravitational field strength × height.
- ✓ Know and use the relationship: kinetic energy = ½ × mass × speed^2.
- ✓ Understand how conservation of energy produces a link between gravitational potential energy, kinetic energy and work.
- ✓ Describe power as the rate of transfer of energy or the rate of doing work.
- ✓ Use the relationship between power, work done (energy transferred) and time taken: power = work done ÷ time taken.

WORK

Work is done when the application of a force results in movement. Work can only be done if the object or system has energy. When work is done energy is transferred. The unit of work is the same as the unit of energy, **joules** (J).

Work done is equal to the amount of energy transferred. It can be calculated using the following formula:

work done = force × distance moved in the direction of the force
= energy transferred

$$W = F \times d = E$$

where W = work done in joules (J),

F = force in newtons (N),

d = distance moved in the direction of the force in metres (m) and

E = energy transferred in joules (J)

1 joule of work is done when a force of one newton is moved through a distance of one metre.

In Figure 4.29 the gymnast is not doing any work against his body weight – he is not moving. (He will be doing work pumping blood around his body, however.)

Δ Fig. 4.29 A gymnast hanging still.

The gymnast in Figure 4.30 is doing work. He is moving upwards against his weight. Energy is being transferred as he does the work.

Δ Fig. 4.30 A gymnast lifting himself up is doing work against his own body weight.

WORKED EXAMPLES

1. A cyclist pedals along a flat road. He exerts a force of 60 N on the road surface and travels 150 m. Calculate the work done by the cyclist.

Write down the formula: $W = F \times d$

Substitute the values for F and d: $W = 60\ \text{N} \times 150\ \text{m}$

Work out the answer and write down the unit: $W = 9000\ \text{J}$

2. A person does 3000 J of work in pushing a supermarket trolley 50 m across a level car park. What force was the person exerting on the trolley?

Write down the formula with F as the subject: $F = \frac{W}{d}$

Substitute the values for W and d: $F = \frac{3000\text{ J}}{50\text{ m}}$

Work out the answer and write down the unit: $F = 60\text{ N}$

QUESTIONS

1. Calculate the work done when a 50 N force moves an object 5 m.
2. Calculate the force required to move an object 8 m by transferring 4000 J of energy.
3. Calculate the work done when a force of 40 N moves a block 2 m.
4. How far does an object move if the force on it is 6 N and the work done is 300 J?
5. What force is needed to move a piano a distance of 2 m when the work done is 800 J?

EXTENSION

When something slows down because of friction, work is done. The kinetic energy of the motion is transferred to the thermal energy store of the object and its surroundings as the frictional forces slow the object down. For example, when a cyclist brakes, the work done by friction produces an energy transfer. The kinetic energy store of the cyclist decreases and the thermal energy store of the brake pads increases.

Space shuttles were operated by NASA between 1981 and 2011. A space shuttle was a manned, reusable spacecraft that was launched vertically by rocket but landed horizontally on a runway. The space shuttle used friction to do work on its motion upon re-entry into the Earth's atmosphere.

1. A space shuttle has 8.45×10^{12} J of energy to transfer over an 8000 km flight path. What force is applied by the atmosphere?

Use the internet or books to research the landing of a space shuttle and answer the following questions.

2. What happens to the transferred energy on landing?
3. What temperatures are generated by the work being done? How does this relate to the material used for the underside of the shuttle surface – for example, why was it not made from aluminium or iron?

Δ Fig. 4.31 A space shuttle in orbit.

GRAVITATIONAL POTENTIAL ENERGY AND KINETIC ENERGY

If a load is raised above the ground, it increases its store of gravitational potential energy (GPE). If the load moves back to the ground, the stored potential energy decreases and is transferred to the store of kinetic energy (KE).

Gravitational potential energy can be calculated using the formula:

gravitational potential energy = mass × gravitational field strength × height

$$\text{GPE} = mgh$$

where GPE = gravitational potential energy in joules (J), m = mass in kilograms (kg), g = gravitational field strength of 10 N/kg and h = height in metres (m).

The kinetic energy of an object depends on its mass and its speed. The kinetic energy can be calculated using the following formula:

kinetic energy = $\frac{1}{2}$ × mass × (speed)^2 or

$$\text{KE} = \frac{1}{2}mv^2$$

where KE = kinetic energy in joules (J), m = mass in kilograms (kg) and v = speed in m/s.

REMEMBER

An object gains gravitational potential energy as it gains height. Work has to be done to increase the height of the object above the ground. Therefore:

gain in gravitational potential energy of an object = work done on that object against gravity.

WORKED EXAMPLES

1. A skier has a mass of 70 kg and travels up a ski lift a vertical height of 300 m. Calculate the change in the skier's gravitational potential energy.

Write down the formula: $\text{GPE} = mgh$

Substitute values for m, g and h: $\text{GPE} = 70 \times 10 \times 300$

Work out the answer and write down the unit: GPE = 210 000 J or 210 kJ

2. An ice skater has a mass of 50 kg and travels at a speed of 5 m/s. Calculate the skater's kinetic energy.

Write down the formula: $\text{KE} = \frac{1}{2}mv^2$

Substitute the values for m and v: $\text{KE} = \frac{1}{2} \times 50 \times 5 \times 5$

Work out the answer and write down the unit: KE = 625 J

REMEMBER

As a skier skis down a mountain, the loss in gravitational potential energy should equal the gain in kinetic energy (assuming no other energy transfers take place, as a result of friction, for example). Calculations can then be performed using:

loss in GPE = gain in KE

$mgh = \frac{1}{2}mv^2$

QUESTIONS

1. Calculate the gravitational potential energy gained when a 5 kg mass is lifted 2 m.
2. Calculate the kinetic energy of a 2 kg ball rolling at 2 m/s.
3. What is the kinetic energy of a cyclist travelling at 5 m/s if his mass is 65 kg?
4. What is the mass of an animal with kinetic energy 64 J travelling at 4 m/s?
5. What is the speed of an animal of mass 2 kg with kinetic energy 9 J?

ENERGY AT WORK

The law of conservation of energy, as you saw on page 193, states that energy cannot be created or destroyed, but simply transferred from one store to another. Use this to calculate what happens when energy is transferred between kinetic and gravitational stores.

Most energy transfers are not 100% efficient. However, for this section we are going to make the assumption that they are 100% efficient. Assuming there is no energy lost in the transfer (for example, by work done against friction), $mgh = \frac{1}{2}mv^2$.

WORKED EXAMPLE

If a stone thrown vertically upwards reaches a height of 6.5 m above the hand of the thrower, with what speed was it thrown?

Figure 4.32 shows the path of the stone. When the stone is thrown it is given kinetic energy.

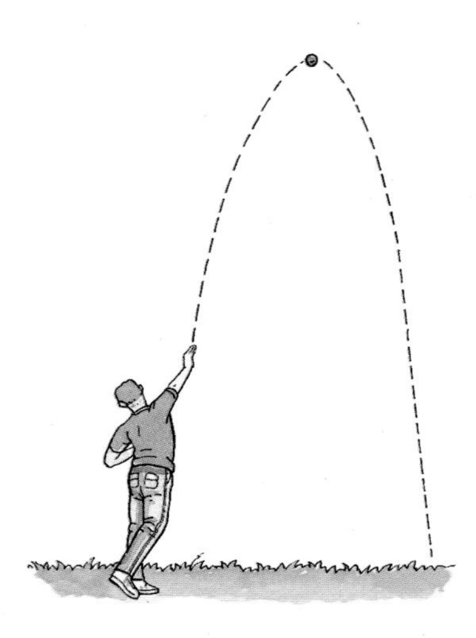

Δ Fig. 4.32 The path of a stone when it is thrown.

At the top of its flight, the stone's store of kinetic energy is much smaller and the store of gravitational potential energy has increased. It still has some kinetic energy because it is moving horizontally. A small amount of energy is lost due to friction between the stone and the air. However, in our calculation we are assuming that the energy transfer is 100% efficient. This means we can say that the decrease in kinetic energy of the stone as it rises equals the increase in the GPE of the stone.

As the final KE of the stone is 0, the initial KE of the stone equals the increase in the GPE of the stone at the top of its flight.

Write down the formula: $\frac{1}{2}mv^2 = mgh$

Note that the mass has cancelled out because the mass does not matter in this case. $\frac{1}{2}v^2 = gh$

Substitute values for g and h: $v^2 = 2gh$

$= 10 \times 6.5 \times 2$

$= 130$

Work out the answer and write down the unit: $v = \sqrt{130}$

$= 11.4$ m/s, or 11 m/s to two significant figures

REMEMBER

Note that in the previous worked example the answer is given to two significant figures. This is the same number of significant figures as the data in the question (6.5). In most physics examples at this level, you should remember to give your answer to two or three significant figures, not all the digits that your calculator might give! You may lose marks in an exam if you give too many significant figures.

QUESTIONS

1. When calculating conversions between gravitational potential energy and kinetic energy, what assumption do you need to make?
2. In practice, what losses of energy could there be?
3. A ball is kicked in the air and reaches a height of 4 m above the ground. What was the speed at which it was kicked?

EXTENSION

Work and kinetic energy (KE) are linked. When objects speed up or slow down, their KE changes. A force always has to act to make this change, so work is also done.

As a revision exercise, you have been asked to come up with a teaching aid that shows how these ideas are linked. Suggested ways to present this are:

- a PowerPoint presentation
- an A3 poster for display
- a worksheet for other students to complete
- a Flash animation
- a podcast.

You should use textbooks or the internet to help you with your research before creating the teaching aid. Any presentations or podcasts should be no more than 5 minutes long.

Key points that should be included are:

- a description and explanation of work
- the work equation
- a description and explanation of KE
- the KE equation
- the work done on a moving object is the same as the change in KE
- a worked example using a suitable situation, such as a thrown cricket ball or a car braking.

POWER

A powerful engine in a car can take you up a road to the top of a mountain more quickly than a less-powerful engine. In the same way, a powerful electric motor on a cooling fan will move the air in the room more quickly; and the 'powerfully built' athlete will, by transferring more kinetic energy to it as it is launched, throw the javelin further.

Power is defined as the rate of doing work or the rate of transferring energy. The more powerful a machine is, the quicker it does a fixed amount of work or transfers a fixed amount of energy.

Since power is the rate of doing work or the rate of transferring energy, power can be calculated using the formula:

$$\text{power} = \frac{\text{work done}}{\text{time taken}} = \frac{\text{energy transferred}}{\text{time taken}}$$

$$P = \frac{W}{t} \text{ or } P = \frac{E}{t}$$

where P = power in joules per second or **watts** (W), E = energy transferred in joules (J), W = work done in joules (J) and t = time taken in seconds (s)

If one joule of energy is transferred in one second this is one watt of power (1 W = 1 J/s).

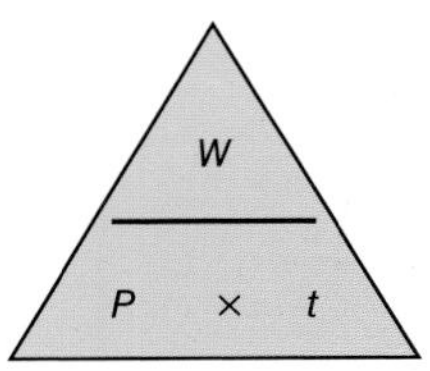

Δ Fig. 4.33 You can use this triangle to help you to rearrange the formula.

WORKED EXAMPLES

1. A crane lifts a 100 kg girder for a skyscraper by 20 m in 40 s. Hence it does 20 000 J of work in 40 seconds. Calculate its power over this time. Note: this calculation tells you the power of electric motor that the crane needs.

Write down the formula: $P = \frac{W}{t}$

Substitute the values for W and t: $P = \frac{20\,000}{40}$

Work out the answer and write down the unit: $P = 500$ W

2. A student with a weight of 600 N runs up the flight of stairs (height 5 m) shown in the diagram (below) in 6 seconds.

Calculate the student's power.

Write down the formula for work done: $W = F \times d$

Substitute the values for F and d: $W = 600 \times 5 = 3000$ J

Write down the formula for power: $P = \frac{W}{t}$

Substitute the values for W and t: $P = \frac{3000}{6}$

Work out the answer and write down the unit. $P = 500$ W

Fig. 4.34 A student running up a flight of stairs.

REMEMBER

The student is lifting his body against the force of gravity, which acts in a vertical direction. The distance measured must be in the direction of the force (that is, the vertical height).

SIMPLE MACHINES

A machine is a device that makes it easier to do work. The law of conservation of energy says that no machine can produce *more* work than you put in. Going further, no machine will even produce *the same* amount of work as you put in, since there will always be non-useful transfers of energy – efficiency will always be less than 100%.

However, machines are still useful since they can:

- decrease the force that has to be applied (the effort) to move the load
- increase the magnitude of the force applied to an object (the load)
- increase the distance that an object (the load) moves
- change the direction of a force.

Notice, of course, that no machine can increase the magnitude of the force AND the distance the load moves.

An example of a simple machine is the inclined plane.

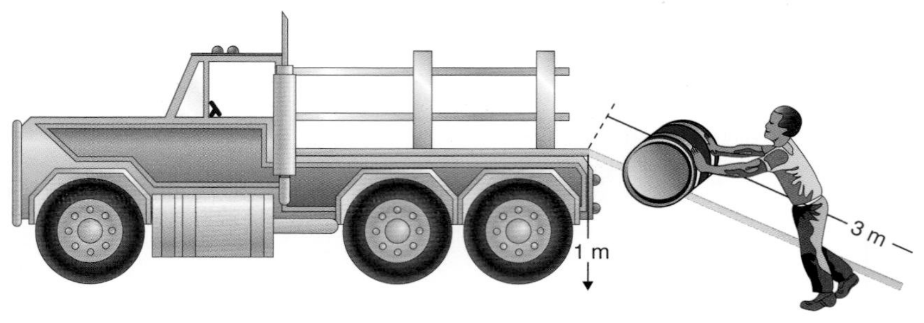

Δ Fig. 4.35 An inclined plane.

Suppose the load has a mass of 120 kg, so its weight is 1200N. To lift it directly onto the back of the truck, the work required would be:

work done = force × distance moved = 1200 N × 1 m = 1200 J

Pushing the load up the inclined plane (the ramp) the same work must be done on it as before since the load ends up in the same place. To find the force required, we can use:

$$\text{force} = \frac{\text{work done}}{\text{distance moved}} = \frac{1200\text{ J}}{3\text{ m}} = 400\text{ N}$$

So using the inclined plane has made it three times easier (three times less force) to move the load onto the truck. In reality the force will be more than 400 N because of friction on the ramp, but this is reduced if the load can roll, like a barrel or by putting it on wheels, rather than sliding it.

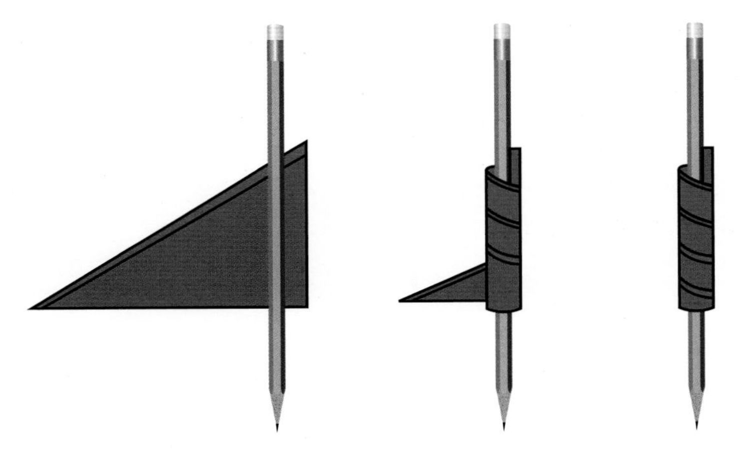

Δ Fig. 4.36 Rolling a triangular piece of paper round a pencil.

Now if you imagine 'rolling' an inclined plane around a cylinder, you will make a screw. This is another machine that reduces the force you have to exert – think of how much force you need to hammer in a nail compared to screwing in a screw. Of course, the *distance* you have to moving is much larger for the screw – you need to turn the screwdriver round a number of times, just like pushing all the way up the ramp – but the *force* you need is much less.

Δ Fig. 4.37 A screw and screwdriver.

DEVELOPING INVESTIGATIVE SKILLS

A student performs several activities to calculate her power when lifting different objects. She lifted different masses through different heights at varying speeds.

Look at her results.

Lifting activity	Mass (kg)	Increase in height (m)	Energy transferred (J)	Time taken (s)	Power (W)
Bag of sugar	1	2		0.6	
Bag of cement	25	0.5		1.6	
Bag of compost	7.5	0.8		0.6	
Watering can	15	1.6		3.2	

Analyse and interpret

❶ Use the equation GPE = $m \times g \times h$ to complete the table (take g to be 10 N kg^{-1}). Give your calculations to an appropriate number of significant figures.

a) How does the power change as the height lifted changes? Give examples to support your conclusions.

b) How confident can you be in these conclusions?

❷ **a)** How does the power change as the mass lifted changes? Give examples to support your conclusions.

b) How confident can you be in this conclusion?

Evaluate data and methods

❸ Suggest changes to this method that would improve the accuracy of the measurement for time.

❹ It is difficult to draw firm conclusions from this data.

How would you improve the experiment to investigate the following:

a) How a change in height affects the power.

b) How a change in mass affects the power.

QUESTIONS

1. Calculate the power of a motor that transfers 1200 J every 5 seconds.
2. A man (70 kg) and a boy (35 kg) run up a set of stairs in the same time. Explain why the man is twice as powerful.
3. When a machine is called 'powerful', what does it mean?
4. What is the unit of power?
5. a) A crane lifts a mass of 60 kg to a height of 5 m. How much work does it do?

   b) The crane takes 1 minute to do this. Calculate the power of the crane.

End of topic checklist

Work is done when a force is applied to an object and the object moves. When work is done, energy is transferred to the object.

Power is defined as the rate of doing work or the rate of transferring energy.

The facts and ideas that you should understand by studying this topic:

- ❍ Know and use the relationship between work, force and distance moved in the direction of the force:
  force × distance moved in the direction of the force.
- ❍ Know and use the relationship between gravitational potential energy, mass, *g* and height:
  gravitational potential energy = mass × *g* × height.
- ❍ Know and use the relationship between kinetic energy, mass and speed:
  kinetic energy = $\frac{1}{2}$ × mass × $(\text{speed})^2$.
- ❍ Know and use the relationship between power, work done and time taken:
  power = work done ÷ time taken.
- ❍ Understand the link between work done and energy transferred, that is, they are equal.
- ❍ Understand how conservation of energy provides a link between work, potential energy and kinetic energy.
- ❍ Understand that power is the rate of energy transfer.

End of topic questions

1. 50 000 J of work is done as a crane lifts a load of 400 kg. How far did the crane lift the load? (Gravitational field strength, g, is 10 N/kg.) **(2 marks)**

2. A student is carrying out a personal fitness test.

   She steps on and off the 'step' 200 times. She transfers 90 J of energy each time she steps up.

   a) Calculate the energy transferred during the test. **(2 marks)**

   b) She takes 3 minutes to do the test. Calculate her average power. **(2 marks)**

3. A child of mass 35 kg climbed a 30 m high snow-covered hill.

   a) Calculate the increase in the child's gravitational potential energy. **(2 marks)**

   b) The child then climbed onto a lightweight sledge and slid down the hill. Calculate the child's maximum speed at the bottom of the hill. (Ignore the mass of the sledge.) **(2 marks)**

   c) Explain why the actual speed at the bottom of the hill is likely to be less than the value calculated in part **b)**. **(3 marks)**

4. Calculate the missing numbers in the table using the appropriate formula. **(6 marks)**

Work done (J)	Force (N)	Distance (m)
	50	4
300		150
4500	60	
	150	600
1500		15
120 000	300	

5. Calculate the increase in potential energy of a piano of mass 300 kg lifted through a vertical height of 9 m. **(2 marks)**

6. Calculate the height climbed up a ladder when the person's mass is 70 kg and the gravitational potential energy gained is 2800 J. **(2 marks)**

7. A 1500 kg helicopter has a potential energy of 1.35 MJ. Calculate its height. **(2 marks)**

End of topic questions continued

8. Calculate the missing numbers in the table. **(6 marks)**

Kinetic energy (J)	Mass (kg)	Speed (m/s)
	85	9
196		1.4
32	1	
	950	13
93 750		250
750	3000	

9. What is the kinetic energy of a bird of mass 200 g flying at 6 m/s? **(2 marks)**

10. What is the speed of a car of mass 1500 kg with kinetic energy of 450 kJ? **(2 marks)**

11. a) A skateboarder of mass 60 kg is travelling at 1 m/s. He is at the top of a slope with a vertical height of 3.15 m. What is his kinetic energy? **(2 marks)**

b) What is the gravitational potential energy of the skateboarder in part **a)**? **(2 marks)**

c) What is the total energy (kinetic + gravitational) of the skateboarder in parts **a)** and **b)**? **(2 marks)**

d) Assuming that no energy is lost in the descent, show that the skateboarder is travelling at about 8 m/s as he reaches ground level after the descent down the 3.15 m slope. **(3 marks)**

12. Look at the data in the table of stopping distances for cars.

Speed of car (m/s)	Thinking distance (m)	Braking distance (m)	Total stopping distance (m)
8	6	6	12
16	12	24	36
32	24	96	120

Describe and explain what happens to the braking distance when the speed is doubled. In your answer use ideas about kinetic energy and its relationship with speed. **(6 marks)**

Human influences on the environment

INTRODUCTION

Over the next century, our management of energy resources will be critically important. Fossil fuels have provided vast amounts of energy since the Industrial Revolution of the 19th century, but the supplies are finite. Competition for these resources continues to grow, and ever-decreasing resources become increasingly expensive. Alongside this has come the growing realisation of the long-term harm done to the Earth's climate by burning these fuels.

△ Fig. 4.38 Wind turbines and solar cells are renewable energy resources.

Alternative sources of energy are available, but they may – or may not – permit current levels of consumption to continue. Everyone should have an understanding of where we get our energy from and the choices that we will have to make.

KNOWLEDGE CHECK

✓ Know how to describe energy transfers using Sankey diagrams.
✓ Know that total energy is always conserved, but that energy transfers are never 100% efficient.

LEARNING OBJECTIVES

✓ Be able to describe the energy transfers involved in generating electricity using wind, water, geothermal resources, solar heating systems, solar cells, fossil fuels and nuclear power.
✓ Be able to describe the advantages and disadvantages of methods of large-scale electricity production from various renewable and non-renewable resources.

FOSSIL FUELS

Most of the energy we use is obtained from **fossil fuels** – coal, oil and natural gas.

Many power stations use fossil fuels to generate electricity that is supplied to homes and factories. There are many other power stations that burn alternative fuels to generate this electricity, but the basic method of producing power is nearly always the same:

- fuel is burned and steam is produced in a boiler
- the steam turns a turbine

- the turbine drives a generator
- the generator produces electricity
- the electricity is supplied.

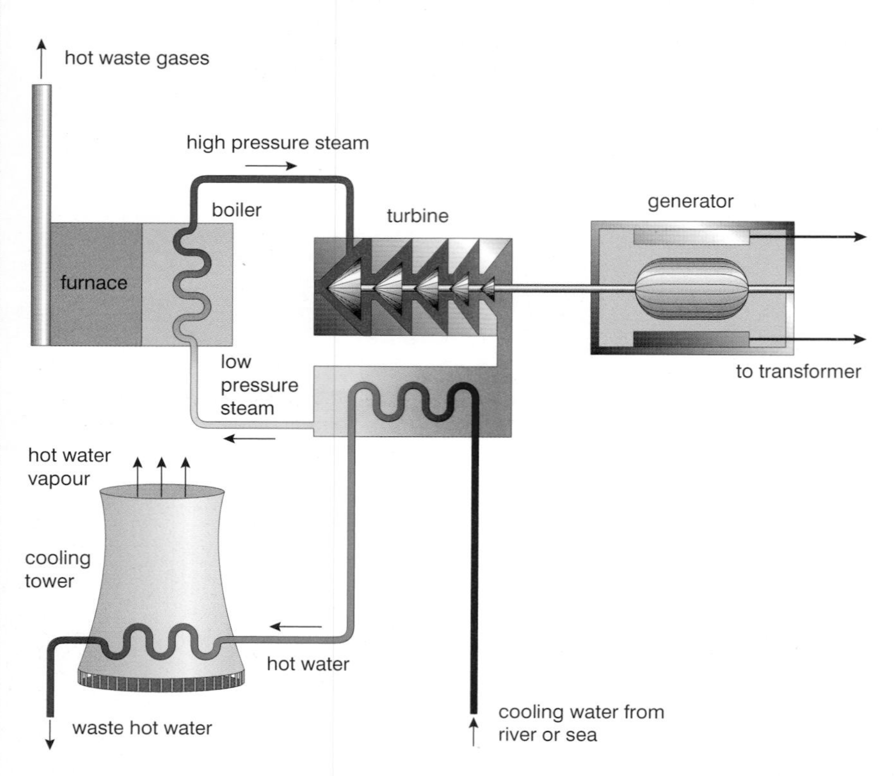

△ Fig. 4.39 The most common fuels used in power stations are still coal, oil and gas.

Nuclear power stations do not 'burn' nuclear fuel as combustion is not involved at all. The nuclear fuel undergoes a nuclear reaction called fission. This releases large amounts of thermal energy which is used to heat water to produce steam. This drives a turbine as in a conventional power station.

A typical nuclear power station produces 3 m^3 of radioactive waste per year. People disagree over whether this radioactive waste is more hazardous than the polluting gases (sulfur dioxide and the greenhouse gas, carbon dioxide) emitted by the 4 million tonnes of coal burned by a single coal power station each year.

Once supplies of these fuels have been used up, they cannot be replaced – they are **non-renewable**. At current levels of use, oil and gas supplies will last for about another 40 years, and coal supplies for no more than a few hundred years from now. The development of **renewable** sources of energy is therefore becoming increasingly important.

RENEWABLE SOURCES OF ENERGY

The wind is used to turn windmill-like turbines which generate electricity directly from the rotating motion of their blades. Modern wind turbines are efficient, but it takes about one thousand of them to generate the same amount of energy as a modern gas, coal or oil-burning power station. If there is not much wind, they are less efficient than this.

Wind is a renewable resource and it does not pollute the atmosphere like fossil fuels do. But if we relied on wind power only we may have enough electricity when it is windy days but no there would be no electricity generated on days when there is little or no wind. The gap on windless days would have to be filled with electricity generated from other sources.

◁ Fig. 4.40 On a windy day a wind turbine generates 2000 kW of electricity. That's enough for 1200 families.

Developing investigative skills

You are going to plan an investigation to evaluate wind power as an energy source. You have the following equipment:

- model wind turbine
- multimeter to measure the voltage generated
- anemometer to measure wind speed
- hair dryer to generate wind power, set on cold
- metre rule to measure distance.

Demonstrate and describe techniques

❶ Plan your experiment, describing clearly the following:

a) the aim of your investigation
b) what you will measure
c) the number and range of readings that you will take
d) the independent variable
e) the dependent variable
f) the control variables
g) how you will make your experiment a fair test.

Make observations and measurements

❷ Draw out a results table that you would use in your investigation.

Evaluate data and methods

❸ Write an evaluation identifying aspects of your experiment where modifications are possible.

The up and down motion of **waves** can be used to move large floats and generate electricity. A very large number of floats is needed to produce a significant amount of electricity.

Dams on tidal estuaries trap the water at high tide. When the water is allowed to flow back at low tide, **tidal power** can be generated. This obviously limits the use of the estuary for shipping and can cause environmental damage along the shoreline.

Δ Fig. 4.41 Artist's impression of wave energy converters.

EXTENSION

The River Severn Barrage is a proposed project to build a huge dam on the estuary of the River Severn in the UK. The cost of the project is estimated to be £15 billion and is projected to produce a clean sustainable source of electricity for the next 120 years.

In the area behind the dam, there are huge areas of mud that are exposed at low tide. These mud flats contain many small animals and are a significant source of food for many species of birds. If the dam is built, these mud flats could be disrupted and it may not be easy for the birds to feed on the small animals in the mud.

Imagine that you are called as an expert witness as part of an environmental group to evaluate the benefits, disadvantages and environmental impact of constructing the barrage. Write a report in preparation for a press release. It should be approximately 200 words long.

Dams can be used to store water, which is allowed to fall in a controlled way that generates electricity. This is particularly useful in hilly regions for generating **hydroelectric power**. When demand for electricity is low, surplus electricity can be used to pump water back up into the high dam for use in times of high demand.

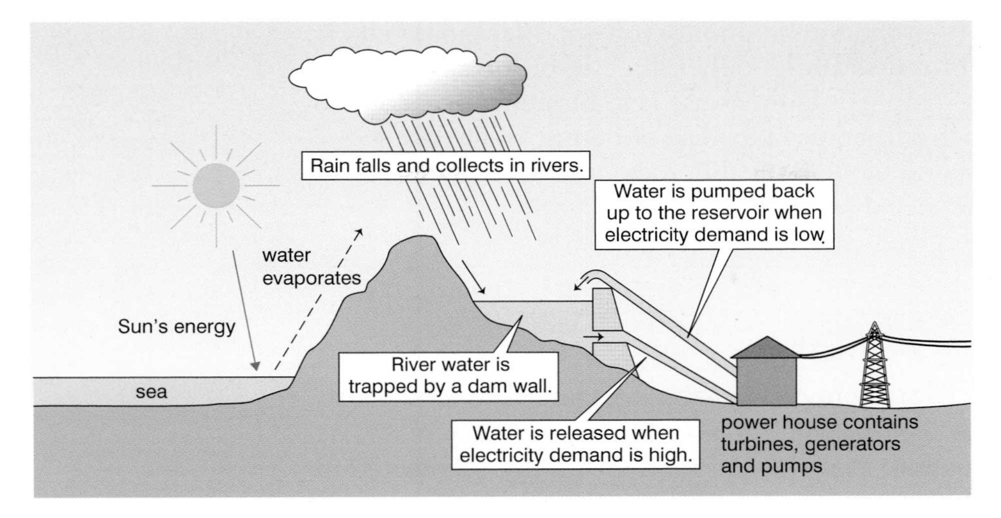

Δ Fig. 4.42 A pumped storage hydroelectric power station.

Solar cells can be used to convert light from the Sun directly into electricity. This electricity can be stored in batteries, to be used when convenient. In some countries where there is a national electricity distribution system, houses can output the electricity generated from solar cells to the 'grid'. In this way householders can sometimes sell more electricity to the electricity company than they buy from it. Electricity generated in this way is renewable, but it is only really effective in a sunny location, so it is not a solution for everything. More efficient solar cells that generate electricity even when the angle of sunlight is low are becoming more affordable.

Δ Fig. 4.43 Solar cells on the roof of a house.

The energy from the Sun can also be used simply to heat water that is pumped through a panel, often on the roof of a house (Figure 4.44). Heating the water this way reduces the demand on other energy resources. Again, the energy can be stored in the water for later use, although this system needs lots of sunlight to be fully effective.

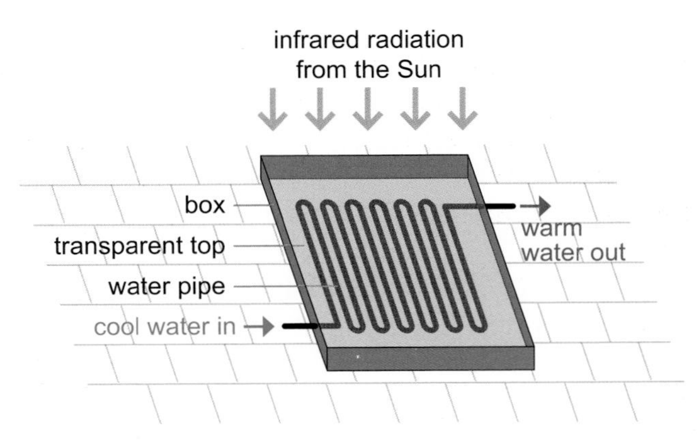

Δ Fig. 4.44 How a solar heating system works.

REMEMBER

Make sure you don't confuse the two systems for using the Sun's energy. Solar panels heat water, solar cells (also called **photovoltaic cells**) generate electricity.

Plants use energy from the Sun in photosynthesis. Plant material can then be used as a **biomass fuel** – either directly, by burning it, or indirectly. A good example of indirect use is to ferment sugar cane to make ethanol, which is then used instead of petrol. Waste plant material can be used in biodigesters to produce methane gas. The methane is then used as a fuel.

ENERGY SUPPLIES FOR THE FUTURE

The world demand for energy is now about 50 times higher than it was 200 years ago, as shown in Figure 4.45. Most of this increase has taken place since about 1950. The demand for energy is expected to increase still further in the future.

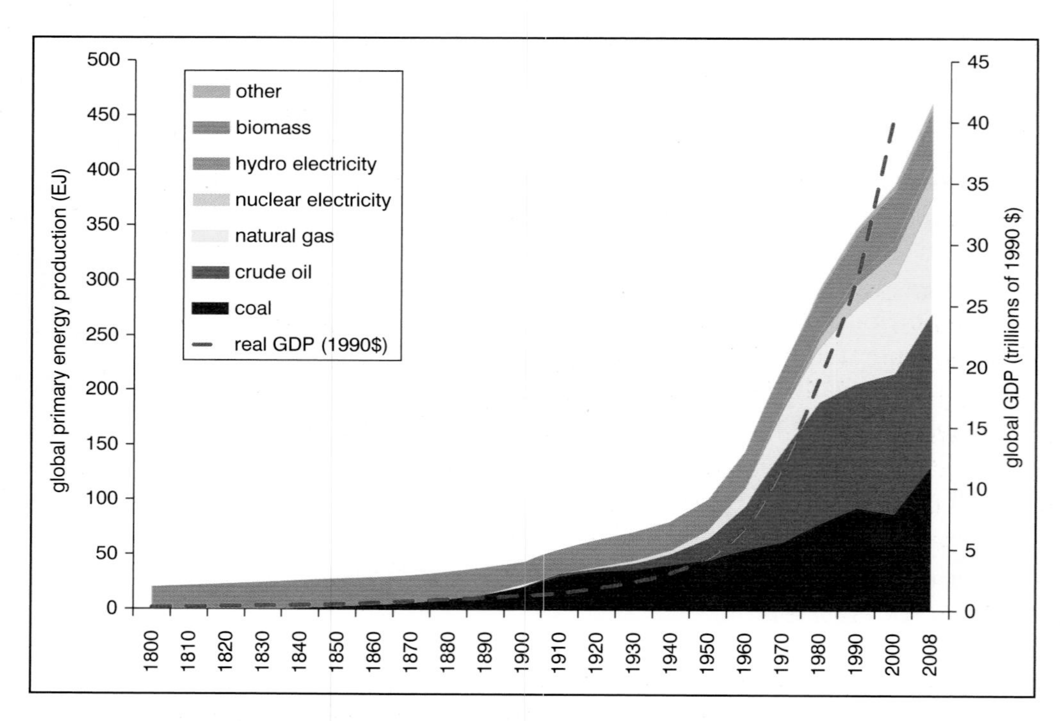

Δ Fig. 4.45 World demand for energy from 1800 to 2008.

Most of the huge increase in energy demand has been for the main fossil fuels – coal, oil and natural gas. These are non-renewable resources. We cannot predict easily or accurately how large the reserves are. The coal reserves will last for several hundred years, and the natural gas and oil for perhaps some tens of years. In 2010, the International Energy Agency (IEA) produced the prediction of what the sources of oil and gas will be in the future, as shown in Figure 4.46.

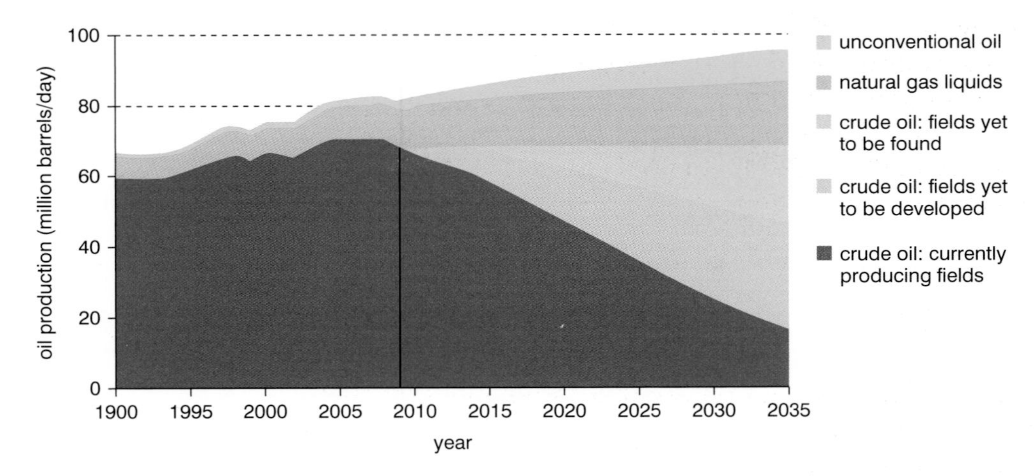

△ Fig. 4.46 IAE prediction of future energy sources.

According to this prediction, the production of oil is expected to increase further. But there is a gap to be filled with 'fields yet to be found'. Even if this oil is found (in a form that can be exploited), it does not alter that fact that oil will – one day quite soon – run out. Oil will only become more expensive in the long term. The 'unconventional oil' refers to oil that is currently uneconomic to extract.

There are additional pressures because of governments' commitments to reduce the amount of carbon dioxide their countries release into the atmosphere. Fossil fuels are major contributors to these emissions. However, it will not be easy to balance reduced carbon dioxide emissions against further increases in burning fossil fuels. Fuel for vehicles can now be made from vegetable oils, and there has been a large increase in this process over recent years. This brings its own problems as land is shifted away from food production to grow crops for fuel. This could have an impact on the amount of food available for us to eat.

Governments are investing in renewable energy sources. In Europe, countries such as Germany and Denmark lead the way in the use of solar cells, even though their northern climate is not ideal. However, major shifts of emphasis are needed if renewable energy sources are to make a significant difference to the balance of world energy supply.

There are difficult decisions to be made with regard to energy sources. The supply of energy in 50 or 100 years time may look very different to today.

Geothermal power is obtained using thermal energy transferred from the interior of the Earth. In certain parts of the world, underground water is heated to form hot springs which can be used directly for heating buildings. Water can also be pumped deep into the ground to be heated.

SELECTING ENERGY SOURCES TO MEET DEMAND

All energy sources have advantages and disadvantages that need to be considered when planning to meet the demand for energy. The choice of energy source is also highly dependent on location. Far from a coastline, tidal power is clearly not a viable option.

Wind turbines may be an ideal solution in a windy area. However, it is unlikely that it will be windy 100% of the time so, if these were the only source of power, no electricity would be produced on windless days. There is a case for storing some of the generated energy in batteries, but this would be very expensive given the size and quantity of batteries that would be needed.

In general, a mix of energy sources is needed that can supplement each other to maintain a supply of electricity that meets demand.

In the small-scale example shown in Table 4.3, it would be too expensive to run a petrol generator to provide electricity at all times. Petrol has the further disadvantages that it is a non-renewable, polluting fossil fuel. The use of wind turbines, solar cells and wave generators would be better for the environment. But on a dark windless night no electricity could be generated. In such a situation, a solution may be to use the three renewables most of the time but use the petrol generator as a back-up when necessary.

Energy collecting device	Source of energy	Energy output	Advantage	Disadvantage
wind turbine	wind	electricity	renewable	only works when windy
solar cells	sunlight	electricity	renewable	only works in daylight
wave power	wind	electricity	renewable	only works when windy
petrol generator	fossil fuel	electricity	can be used at any time	fuel is expensive

Δ Table 4.3 Example of energy solutions for an island community.

QUESTIONS

1. What energy transfers take place in a solar cell?
2. What energy transfers take place in a wind turbine?
3. Describe the process used to generate electricity in power stations.
4. How is electricity produced from geothermal sources?

End of topic checklist

Energy resources are classified as **renewable** or **non-renewable**. Both can be used to generate electricity.

Fossil fuels are fuels such as coal, oil and natural gas.

The facts and ideas that you should understand by studying this topic:

❍ Describe energy transfers involved in generating electricity using:

- wind
- water
- geothermal resources
- solar heating systems
- solar cells
- fossil fuels (coal, oil, gas)
- nuclear power.

❍ Describe advantages and disadvantages for each of these resources.

End of topic questions

1. **a)** What is meant by a non-renewable energy source? **(2 marks)**

   b) Name three non-renewable energy sources. **(3 marks)**

   c) Which non-renewable energy source is likely to last the longest? **(2 marks)**

2. Look at the graph, which shows the amount of energy from different sources used in the OECD (Organisation for Economic Cooperation and Development) nations between 1980 and 2001.

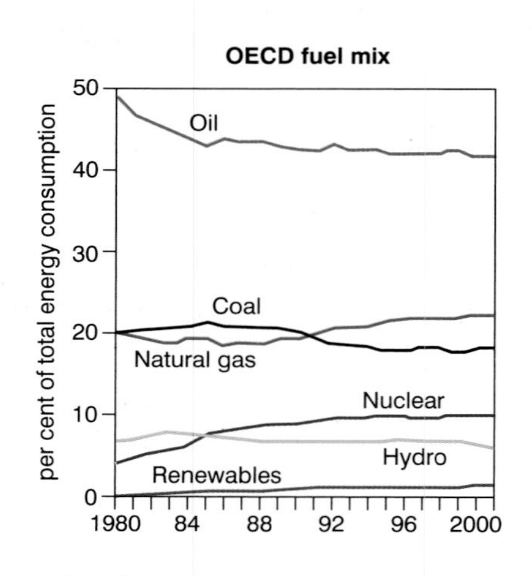

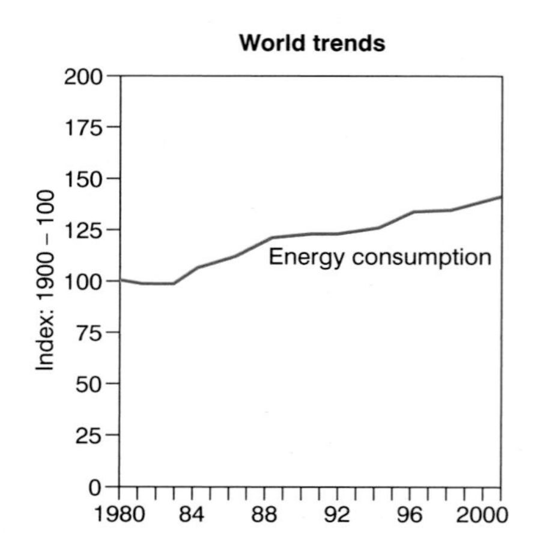

   a) Describe the trend in the total energy used in this period. **(3 marks)**

   b) Describe the main changes in the sources of this energy. **(3 marks)**

   c) What do you expect to have happened to these graphs since 2000? Give reasons where possible. **(3 marks)**

3. A site has been chosen for a wind farm (a series of windmill-like turbines).

   a) Give two important factors in choosing the site. **(2 marks)**

   b) Give one advantage and one disadvantage of using wind farms to generate electricity. **(2 marks)**

4. Draw up two tables, one for renewable energy resources and one for non-renewables.

   Add columns to your table to describe at least one advantage and one disadvantage of each energy resource when it is used to provide large-scale electricity production. **(6 marks)**

5. **a)** Explain why wind turbines do not always produce electricity. **(3 marks)**

   b) Explain why wind turbines are useful in remote areas. **(3 marks)**

6. Fossil fuels produce sulfur dioxide when they are burned.

a) Which types of power station can pollute the air with sulfur dioxide? **(3 marks)**

b) What are the advantages and disadvantages of this type of power station? **(2 marks)**

7. Compare the effects on the environment of coal-fired power stations and nuclear power stations.

a) Carbon dioxide is a greenhouse gas. Which of these power stations release greenhouse gases? **(2 marks)**

b) Which of the fuels used in these power stations, will run out one day? **(2 marks)**

8. Power stations need to be located on suitable sites. Write down three factors that a company may consider before choosing a site for a coal-fired power station. **(3 marks)**

9. It has been decided to build a new power station. The choice is between a large hydroelectric plant and an oil-fired power station.

a) What affects the total cost of electricity produced? **(2 marks)**

b) How do you expect the fuel costs for each to change over the next 50 years? **(2 marks)**

c) What happens to some of the energy stored in the oil and water? **(2 marks)**

d) Which type of power station is more efficient? **(2 marks)**

e) What are the advantages and disadvantages of these types of power station? **(2 marks)**

10. Imagine that the energy resources in your area are being changed.

Write a short report making recommendations on what types of energy resources should be used.

Remember to explain why you are recommending each energy source is used. **(5 marks)**

Exam-style questions
Sample student answers

Question 1

The diagram shows the main features of a power station that uses fossil fuels.

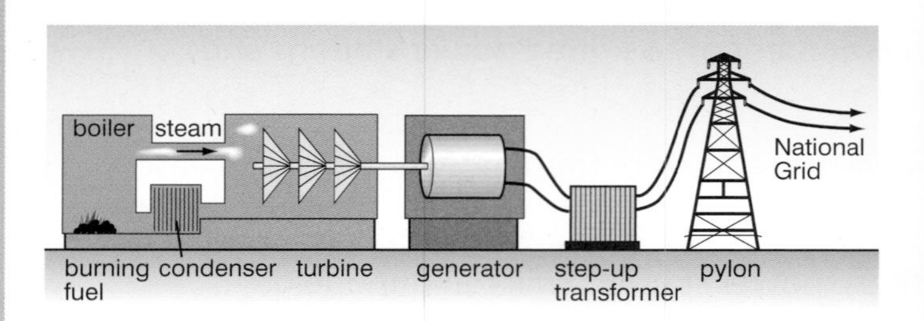

a) Give an example of a fossil fuel.

(1)

b) Describe the energy transfers that take place in

i) the boiler

The fuel is burned to make heat energy in the steam. ✗ (1)

ii) the generator

The generator spins and the kinetic energy is turned into electrical energy. ✓ 1 (1)

c) A typical power station of this type might be 20% efficient.

What does this statement mean?

It means that some of the energy is wasted, for example as heat in the cooling towers and when the steam is turned back into water so that it can go back to the boiler to be heated again and turned into steam again. ✓ ✗ 1 (2)

EXAMINER'S COMMENTS

a) Correct. Alternatives would be oil or natural gas – but be careful if you shorten this to 'gas', because oxygen is a gas, for example, but it is not a fossil fuel!

b) The student clearly knows what is going on, but in the first part has not been careful enough in answering the actual question. To gain the mark, the initial energy store (chemical energy in the fuel) should have been mentioned.

c) Again, the student has clearly learned a lot about the systems in a power station, but they have not kept their response to answering the question. In their enthusiasm to describe everything they know about the situation they have forgotten to explain what 20% efficient means – which was the point of the question. Always restrict your answer to what you need to put, not everything you know.

d) i) Fossil fuels running out may well be true, but is not relevant to the answer.

ii) Be careful whenever using economic reasons ('cheap' or 'expensive'). Often, this can be used to support either side of an argument depending on what factors are included in working out the cost. Only use these answers if you are absolutely clear about the reasons behind them.

e) Here, the student has benefitted from having a good knowledge of the topic. Although the answer could have been broken up and set out a little better, there are plenty of points to gain all five marks. The student has given examples of renewable energy sources and then described advantages AND disadvantages – as the question required.

Exam-style questions continued

d) Using fossil fuels in power stations can cause environmental problems.

i) Describe one example.

Burning fossil fuels releases carbon dioxide which is a greenhouse gas. Fossil fuels are also running out and soon they will be gone. ✓ ✓ ✗ (2) **(3)**

ii) Fossil fuels are widely used in power stations despite environmental concerns.

Give two advantages of using fossil fuels in power stations.

1. *They are cheap.* ✗

2. *You get a lot of energy from them compared to other sources.* ✓ (1) **(2)**

e) Many people would like to see greater use of renewable energy resources in generating electricity.

Describe advantages and disadvantages of using renewable sources to generate electricity.

Renewable energy sources, such as wind energy and solar energy, will not run out, ✓ *so once the power stations have been built, they provide a cheap source of energy.* ✓ *They do not produce harmful gases, such as carbon dioxide and sulfur dioxide, so they have less harmful effects on the environment.* ✓ *However, renewable sources can sometimes be unreliable (it is not always windy or sunny)* ✓ *and the energy output is small compared to the energy from a coal power station.* ✓ (5) **(5)**

(Total 15 marks)

Exam-style questions continued

Question 2

A student is investigating the efficiency of water heating.

She makes a heater coil from a piece of resistance wire and then uses it to heat up some water.

She heats the water for 10 minutes.

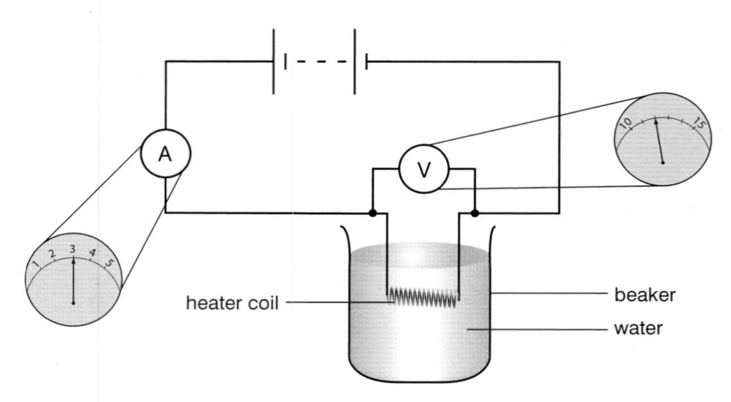

a) i) Copy the table. Use the information in the diagram to complete the table. **(2)**

Current in A	Voltage in V

ii) Use the equation $E = V \times I \times t$ to calculate the energy transferred to the heater during the 10 minutes. **(2)**

b) By measuring the temperature rise of the water, the student is able to calculate that 18500 J of energy was needed to raise the temperature of the water.

i) State the equation linking efficiency, useful energy output and total energy input. **(1)**

ii) Calculate the efficiency of the heating system. **(2)**

iii) Suggest why the system is less than 100% efficient. **(2)**

c) i) Copy and complete the sentence below. **(4)**

In this experiment, energy from the electrical supply was transferred to useful energy in the water and non-useful energy in the

ii) Draw a Sankey diagram for this situation. **(3)**

(Total 16 marks)

Exam-style questions continued

Question 3

A child climbs to the top of a slide at a playground.

The child has a mass of 30 kg and the top of the slide is 3.0 m above the ground.

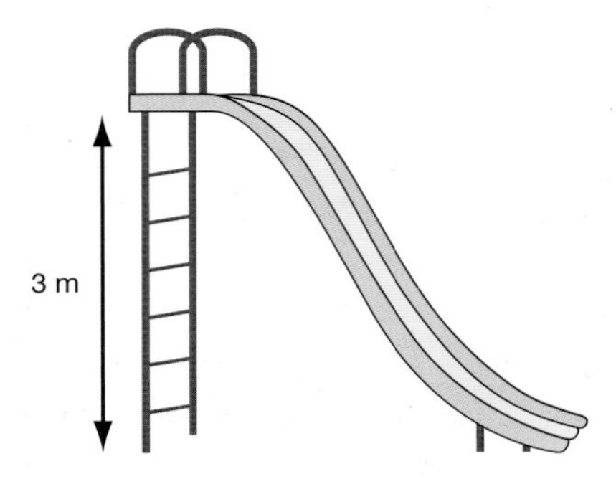

a) i) State the equation linking gravitational potential energy, mass, *g* and height. **(1)**

ii) Calculate the gravitational potential energy gained by the child when he climbs to the top of the slide. **(2)**

iii) State the link between the gravitational potential energy gained and the work done by the child. **(1)**

iv) To calculate the power of the child as they climb to the top of the slide, what other measurement would be needed? **(1)**

b) The child rides down the slide.

Assuming there are no energy losses, calculate the speed of the child at the bottom of the slide. **(4)**

(Total 9 marks)

Exam-style questions continued

Question 4

A student investigates the effect of insulation on cooling.

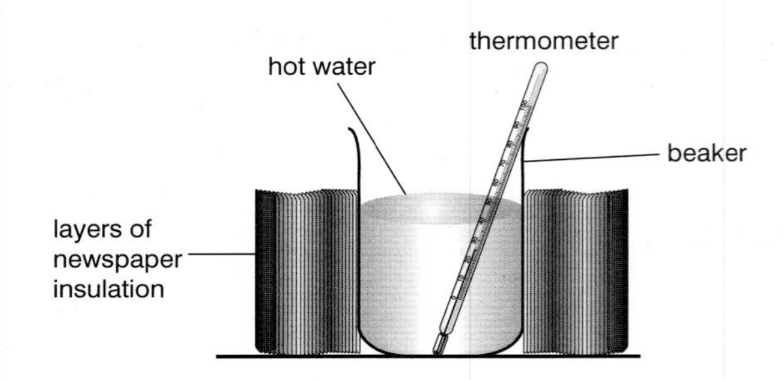

The student puts some hot water into a beaker and measures the temperature drop in 20 minutes.

He repeats the experiment using layers of paper as insulation.

His results are shown in the table.

Number of layers of insulation	Temperature drop in 20 minutes (°C)
0	21
5	20
10	18
15	17
20	18

a) i) Draw a graph of these results. **(5)**

ii) The student concludes that the graph shows that thicker insulation reduces the heat loss.

Is this a correct conclusion from this data?

Explain your answer. **(2)**

Exam-style questions continued

b) The student did not keep the starting temperature constant.

He gave his reason as 'I was only interested in how much it went down, so it didn't really matter where I started'.

Comment on the student's reason, explaining whether or not the student is correct. **(3)**

c) Look again at the diagram of the experiment.

The method used by the student is unlikely to give valid results.

i) What are 'valid' results? **(1)**

ii) Use ideas about energy transfer by particles to explain why the method used is unlikely to give valid results. **(4)**

(Total 15 marks)

Materials can exist in the form of a solid, a liquid or a gas. These are the three main states of matter. Substances can change from a solid to a liquid in a process called melting and from liquid to gas in a process called evaporation. Gases can change to liquids by condensation and liquids change to solids in solidification.

You are probably most familiar with these changes for water. It is possible for the states to exist at the same time: for example, there is a temperature at which ice, water and steam are all present. This is called the triple point of water and is used to define the kelvin scale of temperature, which you will meet later in this section.

STARTING POINTS

1. Describe how particles are arranged in **a)** a solid **b)** a liquid and **c)** a gas.
2. What happens to the particles when a solid melts?
3. Explain what happens when a solid dissolves in a liquid.
4. How are density, mass and volume related?
5. Are evaporation and boiling the same thing? Give a reason for your answer.
6. What happens to the speed of gas molecules as temperature increases?

CONTENTS

5 Solids, liquids and gases

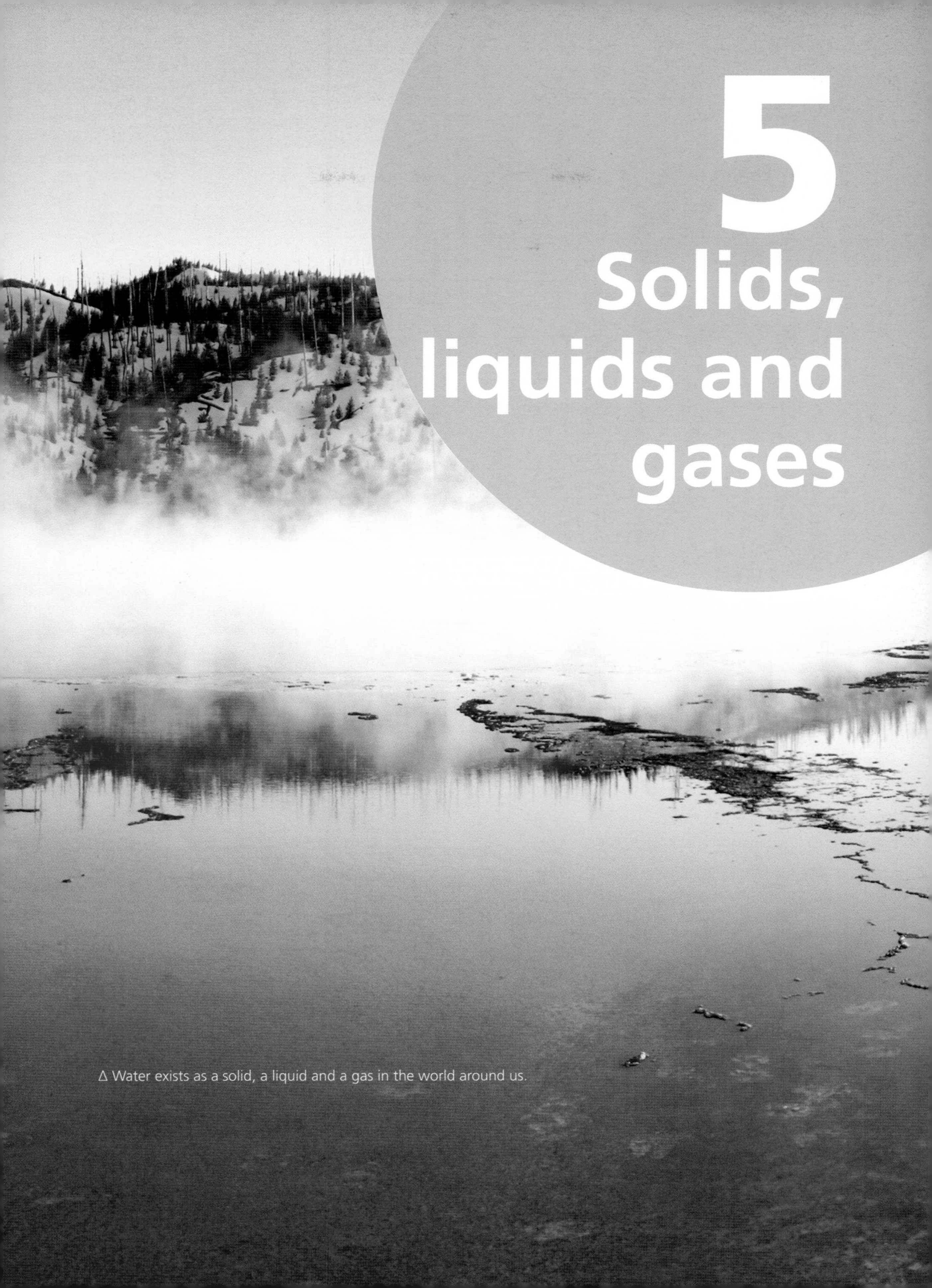

Δ Water exists as a solid, a liquid and a gas in the world around us.

Units

INTRODUCTION

For the topics included in solids, liquids and gases, you will need to be familiar with:

Quantity	Unit	Symbol
mass	kilogram	kg
length	metre	m
area	metre squared	m^2
volume	metre cubed, centimetre cubed, millilitre	m^3, cm^3, ml
time	second	s
speed	metre per second	m/s
acceleration	metre per second squared	m/s^2
force	newton	N
pressure	pascal, newton per metre squared	Pa, N/m^2
temperature	degrees Celsius, kelvin	°C, K
specific heat capacity	joules/kilogram degree Celsius	J/kg °C

Some of these you have met previously but the new units will be defined when you meet them for the first time.

LEARNING OBJECTIVES

✓ Use the following units: degree Celsius (°C), kelvin (K), joule (J), kilogram (kg), kilogram/metre^3 (kg/m^3), metre (m), metre^2 (m^2), metre^3 (m^3), metre/second (m/s), metre/second^2 (m/s^2), newton (N) and pascal (Pa), joules/kilogram degree Celsius (J/kg °C).

✓ Use the following unit: joules/kilogram degree Celsius (J/kg °C).

Density and pressure

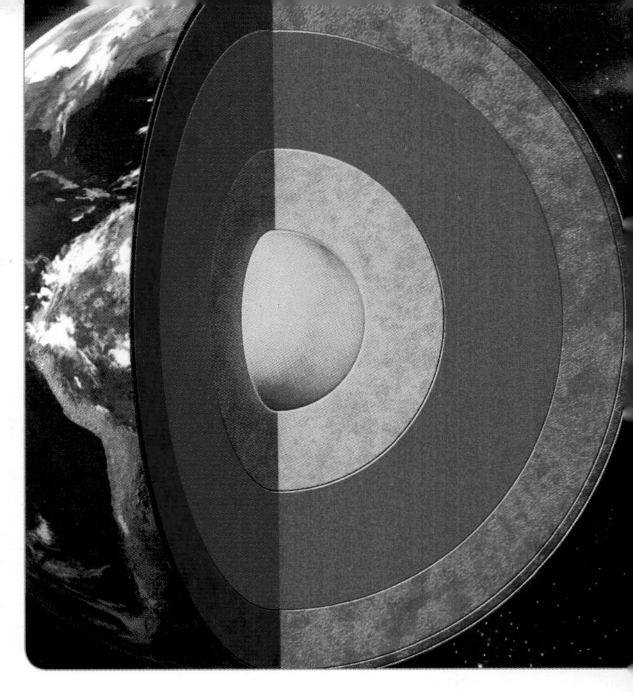

△ Fig. 5.1 The Earth has a layered structure. The crust and the lower mantle are solid but the upper mantle contains a layer of semi-molten rock. Convection currents in this layer cause the crust above to move very slowly.

INTRODUCTION

Which is heavier, a tonne of feathers or a tonne of iron? This is a trick question, of course: they have the same weight. But there would be a big difference if you loaded each one onto a truck; the feathers would take up much more space. Ideas about density help to describe the difference between the feathers and the iron, and ideas about pressure help to describe the effect on the truck.

Density and pressure are useful measures that give an insight into other areas of physics. For example, they help to explain why objects float, which leads on to describing convection currents. This leads us to think about the movement of the continents on the surface of the Earth and the very structure of the Earth itself.

KNOWLEDGE CHECK

✓ Know how to calculate areas of regular shapes, such as squares and rectangles.
✓ Know how to calculate the volume of regular objects, such as cubes and cylinders.

LEARNING OBJECTIVES

✓ Be able to calculate densities using the relationship between density, mass and volume: density = mass ÷ volume.
✓ Be able to investigate density using direct measurements of mass and volume.
✓ Be able to calculate pressures using the relationship between pressure, force and area: pressure = force ÷ area.
✓ Understand how the pressure at a point in a gas or liquid at rest acts equally in all directions.
✓ Be able to use the relationship for pressure difference: pressure difference = height × density × gravitational field strength.

DENSITY

You must have noticed that the weight of objects can vary greatly. A plastic teaspoon weighs less than a metal one, and a gold ring weighs twice as much as a silver one, even if the objects are exactly the same size.

The **density** of a material is a measure of how 'squashed up' it is, and a heavy object contains more mass than a light object of the same size.

Density is calculated using this formula:

$$\text{density} = \frac{\text{mass}}{\text{volume}}$$

or $\rho = \frac{m}{v}$

where m = mass in g or kg, V = volume in cm^3 or m^3 and ρ = density in g/cm^3 or kg/m^3

Note that in this equation you must use g and cm throughout, or you must use kg and m. And note that if you measure the weight in N, you must convert it into g or kg.

Δ Fig. 5.2 Gold is one of the densest metals. A block the size of one-litre carton of milk would have a mass of almost 20 kg and would be very hard to pick up.

The density of a regularly shaped object

You can use this formula to work out the density of a block.

WORKED EXAMPLE

A brick has dimensions 20.0 cm × 9.0 cm × 6.5 cm.

Weight of brick = 22.2 N

What is the density of the brick?

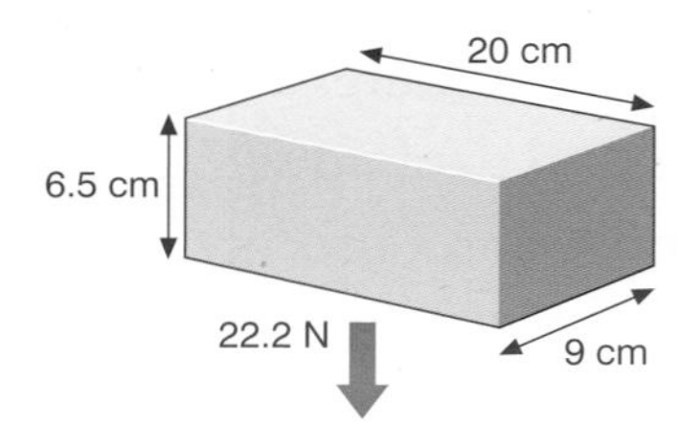

Δ Fig. 5.3 Finding the density of a brick.

Mass of brick: $m = \frac{w}{g}$

$= \frac{22.2}{10}$

$= 2.2$ kg or 2220 g

(Remember that 1 kg = 1000 g)

Volume of brick (V) = $20 \times 9 \times 6.5 = 1170$ cm^3

Density of brick: $\rho = \frac{\text{mass}}{\text{volume}}$

$= \frac{2220}{1170}$

$= 1.9$ g/cm^3 to two significant figures

The density of water is 1.0 g/cm^3, and the rule is that an object of greater density will sink in a liquid of lower density. So it is not very surprising that a brick sinks in water. But will it sink in mercury? To decide whether it will sink in mercury, you need to compare the density of the brick (1.90 g/cm^3) to the density of mercury using the same units. From the table, you can see that the density of mercury is 13.6 g/cm^3, which is very much greater than that of the brick, so the brick will float on liquid mercury.

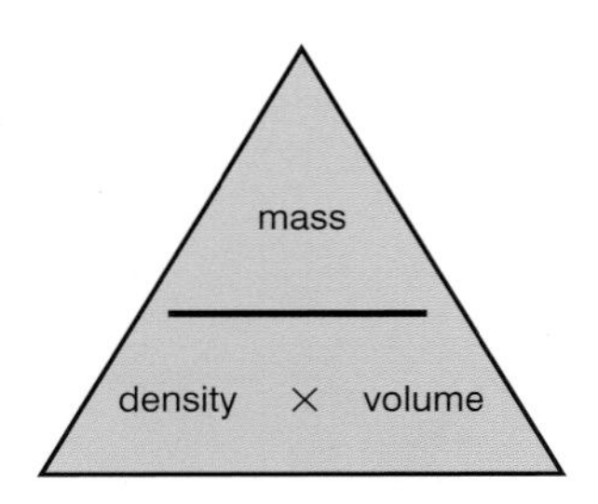

Δ Fig. 5.4 The equation triangle for mass, density and volume.

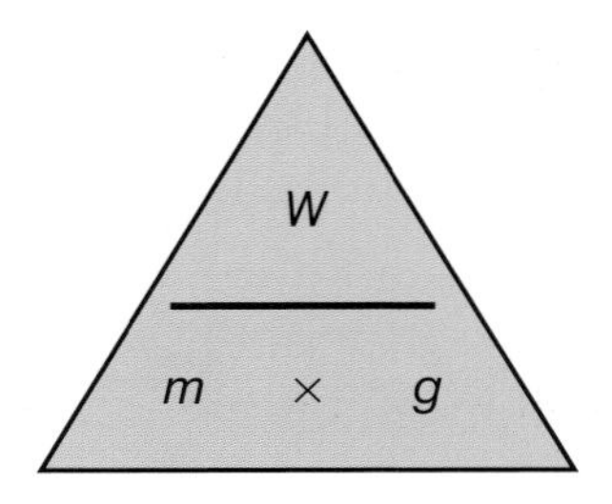

Δ Fig. 5.5 The equation triangle for weight, mass and acceleration due to gravity.

	Density (g/cm³)	Density (kg/m³)
Vacuum	0	0
Helium gas	0.00017	0.17
Air	0.00124	1.24
Oil (Petroleum)	0.88	880
Water	1.0	1000
Sea water	1.03	1030
Plastic	0.9–1.6	900–1600
Wood	0.5–1.3	500–1300
Magnesium	1.74	1740
Aluminium	2.7	2700

Δ Table 5.1 Densities of a variety of substances.

QUESTIONS

1. A small rectangular block of steel measures 2 cm × 4 cm × 5 cm and has a mass of 312 g.

   a) Calculate its volume.

   b) Calculate its density.

2. Why is a loaf of bread usually less dense than a root vegetable?

3. A block of wood floats on sea water. What can you say about the density of the block of wood?

Why do materials have different densities?

If you use a modern electron microscope to look inside a block of gold and inside a block of aluminium, you will see that the atoms are almost exactly the same size (the gold atoms are just a little bit bigger). Most of an atom is empty space, and its mass is concentrated in the nucleus, which is much smaller than the atom. The gold has extra density because the nuclei of its atoms are far more massive than the nuclei of the aluminium atoms.

Many materials have a lower density because they contain large bubbles or other voids inside them. Bread has a lower density than most cakes; and expanded polystyrene cups have a lower density than other cups.

∆ Fig. 5.6 A bag of popcorn (left) has a far lower density than the same bag filled with corn that has not been popped (right).

An aircraft is another example of a lower density. Although aircraft are made of aluminium and other light metals, there is no way that they could fly if they were made of solid aluminum. In fact, the average densities of all aircraft are so low that if they make an emergency landing on water, they can easily float for long enough for everyone to escape.

AIRCRAFT CAN FLOAT

In January 2009, an aircraft took off from LaGuardia Airport, New York, bound for Charlotte, North Carolina. About three minutes into the flight, the aircraft struck a flock of Canada geese, which resulted in a complete loss of thrust from both engines. The pilot realised that he could not safely reach any airfield to land, so he decided to ditch the aircraft on the Hudson River in New York City. He ditched safely about three minutes after losing power. The pilot said later that he had chosen the location to be as close as possible to boats to maximise the chance of rescue.

Immediately after the aircraft ditched, the crew began to evacuate the passengers through the emergency exits. A panicking passenger opened a rear door, which could not be resealed. This made the aircraft fill with water more quickly than it would otherwise have done. However, all 155 passengers and crew safely evacuated. The aircraft was almost completely intact but partially submerged and slowly sinking. Passengers and crew were quickly rescued by nearby ferries and other boats.

The survival of all on board was due to the pilot's quick thinking, everybody's co-operation, and the fact that the density of the aircraft allowed it to stay afloat l ong enough to evacuate.

Δ Fig. 5.7 The ditched aircraft floating in the Hudson River.

WORKED EXAMPLE

What is the mass of a block of expanded polystyrene that is 1 m long, 0.5 m wide and 0.3 m high? The density of this block of polystyrene is 40 kg/m^3.

Volume of block:	$V = 1.0 \times 0.5 \times 0.3$
	$= 0.15\ \text{m}^3$
Write down the formula:	$m = \rho \times V$
Substitute values for ρ and V:	$m = 40 \times 0.15$
Work out the answer and write down the units:	$m = 6\ \text{kg}$

Measuring the density of an irregular object

This method involves submerging an object in a liquid and measuring the volume of the liquid that is displaced. It only works if the object is denser than the liquid used so that it sinks. It does not work if the object absorbs the liquid, nor if it is damaged by the liquid.

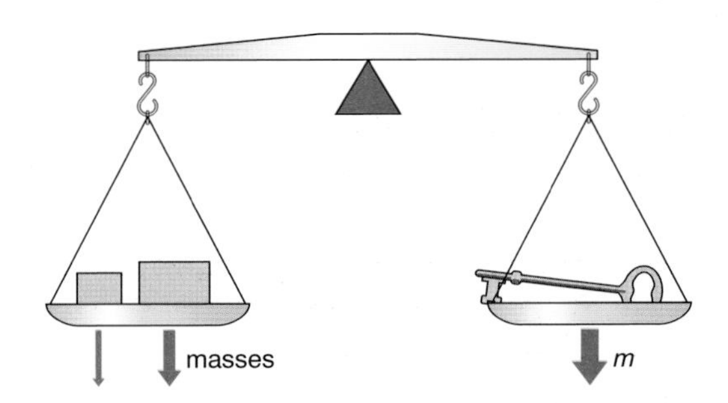

Δ Fig. 5.8 Using a balance to find the mass of an object.

1. Use a balance to weigh the object in question, as shown in Figure 5.8, and find its mass, m.
2. Choose a measuring cylinder that is wide and deep enough to hold the object. A narrower cylinder will give a more accurate answer than a wider one. Add liquid to the cylinder to fill it to a deep enough level so that the object will be completely submerged, and then measure the volume of liquid V_1 (see Figure 5.9). The exact amount of liquid that you use is not at all critical. Water is often the liquid used.

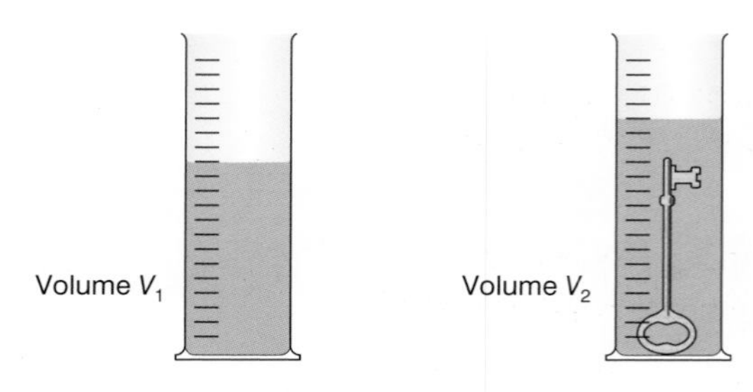

△ Fig. 5.9 Measuring the volume of an object.

3. Lower the object into the liquid (without splashing) and measure the new reading V_2 (as shown in Figure 5.10). This is the volume of the object and the liquid. The volume of the object is $V_2 - V_1$.

From the mass and the volume you can calculate the density of the object.

△ Fig. 5.10 Measuring the volume of a copper object by displacing water.

WORKED EXAMPLE

A small metal statue is measured to have a mass of 90 g.

A measuring cylinder is filled with water to the 82 cm³ mark. The statue is lowered into the measuring cylinder and the water rises to the 91 cm³ mark. What material could the statue be made of?

Volume of the statue: V $= 91 - 82$

$= 9\ \text{cm}^3$

Write down the formula: $\rho = \dfrac{m}{V}$

Substitute the values for m and V: $\rho = \dfrac{90}{9}$

Work out the answer and write down the units: $= 10\ \text{g/cm}^3$

Table 5.1 suggests that the metal could be silver.

An experiment of this type is never perfectly accurate, so the density that you measure will never be exactly the same as the values given in tables.

QUESTIONS

1. In which situations does the method of finding volume by displacing a liquid not work?
2. What features of the measuring cylinder would allow you to make your measurements as accurate as possible?

3. Calculate the density of a block which has a mass of 25 g and where the change in volume when it is placed in a measuring cylinder of water is 20 cm³.

4. Calculate the mass of a steel key that causes a change in volume in a measuring cylinder of water of 10 cm³. Density of steel = 7800 kg/m³

5. A plastic key ring has a mass of 10 g. It has a density of 1.4 g/cm³. It is lowered into a measuring cylinder which contains water. What will the change in the volume reading be?

Developing investigative skills

A student is finding the density of some different materials. The samples she has are all regular shapes.

The student has a ruler, marked in mm, and an electronic balance that measures to the nearest 0.1 g. The student finds the mass and the volume of each sample. Her data is shown below.

Δ Fig. 5.11 Regular shapes for the experiment.

Sample	Mass (g)	Volume (cm³)	Density (g/cm³)
Aluminium	97.2	36	
Brass	302.4	36	
Copper	321.5	36	
Iron	282.6	36	

Devise and plan investigation

❶ Describe how the student should use the ruler to find the volume of each sample.

❷ The student checks that the balance reads zero before she puts the sample on. Does this improve the accuracy or the precision of the experiment? Explain your answer.

Analyse and interpret

❸ Use the equation density = mass / volume to complete the table. Include the units at the top of the 'density' column.

❹ How many significant figures should you give your values of density? Explain your answer.

Evaluate data and methods

❺ How could the method be changed to find the density of objects with an irregular shape?

PRESSURE

△ Fig. 5.12 The 'skis' underneath stop the snowmobile from sinking into deep soft snow.

The snowmobile in the picture can travel over soft snow because its skis spread its weight over a large area of snow. If the rider got off and stood on the snow, he would probably sink into it up to his knees, even though he is much lighter than the snowmobile.

If a pair of shoes has small pointed heels, the wearer can easily damage a wooden floor by putting indentations into it. And you can push a drawing pin into a notice board by the pressure of your thumb.

In every case the question is not just *what force* is used, but also *what area it is spread over*. Where there is a large force over a small area, there is a high **pressure**, and a small force over a large area gives a low pressure.

Pressure is measured in newtons per square metre (N/m^2), usually called **pascals** (Pa). So 1 Pa = 1 N/m^2.

In order to measure how 'spread out' a force is, use this formula:

$$\text{pressure} = \frac{\text{force}}{\text{area}}$$

$$p = \frac{F}{A}$$

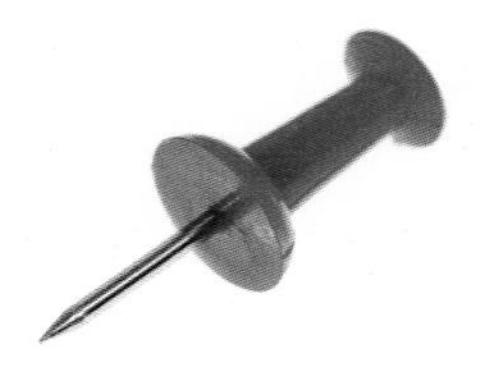

△ Fig. 5.13 When a drawing pin is placed pin side down, the pressure on the surface is greater than when it is placed head side down.

Where p = pressure in pascals, Pa

(or newtons per square metre, N/m^2)

F = force in newtons, N

A = area in m^2

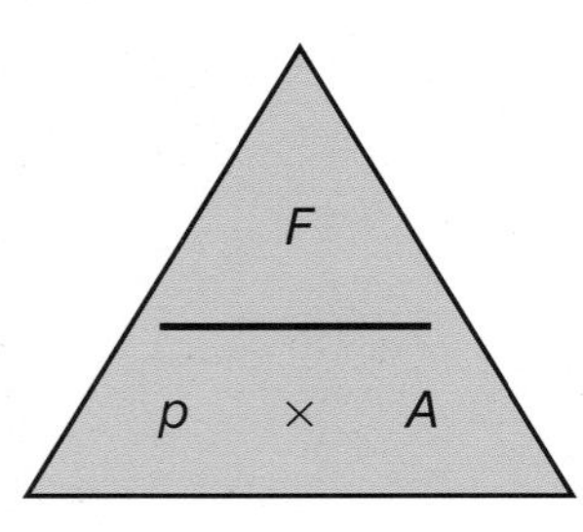

◁ Fig. 5.14 The equation triangle for force, pressure and area.

WORKED EXAMPLE

What pressure on the snow does a snowmobile make if it has a weight of 800 N and the runners have an area of 0.2 m²?

Write down the formula: $p = \frac{F}{A}$

Confirm that F is in N and A is in m^2.

Substitute the values for F and A: $p = \frac{800}{0.2}$

Work out the answer and write down the units: $p = 4000$ Pa or 4 kPa

Note that 4 kPa is a very low pressure. If you stand on the ground in basketball shoes, the pressure on the ground will be around 20 kPa. The wheel of a car creates a pressure on the ground of around 200 kPa. Pressures can be quite high, and so the unit kPa is often used.

QUESTIONS

1. Why can you push a drawing pin into a surface using your thumb when you can't push your thumb into the same surface?
2. Calculate the pressure exerted by a 100 N force acting on an area of 0.2 m^2.
3. A pressure of 40 Pa is exerted over an area of 2 m^2. Calculate the force involved.
4. A force of 500 N produces a pressure of 640 Pa. Over what area is the force acting?

Developing investigative skills

A student has been reading about how scientists can gain information about the mass of dinosaurs from the depth of their fossilized footprints.
He decides to investigate how far a wooden block sinks into sand when the pressure on it changes. The student loads 100g masses onto the block one at a time and measures how deeply the block is pushed into some sand.

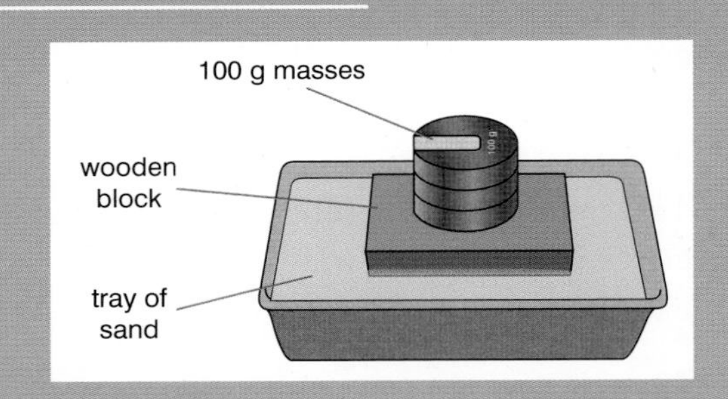

△ Fig. 5.15 Apparatus for experiment.

The student finds very little evidence of a pattern in his measurements. He feels this is because the block tends to tip over, rather than standing straight, which means that the sand is not equally pushed down. He also feels that the sand does not push down very much anyway, it just gets pushed to the side.

To extend his experiment, the student has an idea about investigating if the 'wetness' of the sand makes a difference to the way the block behaves, but he has not yet devised a plan to test this.

Devise and plan investigation

❶ Explain how using the block in different ways and using different numbers of 100 g masses allows the student to test a variety of different pressures.

❷ Devise a method to measure the depth to which the block sinks in the sand. You should name any equipment needed. Remember that the block may not sink equally in all directions.

❸ What was the independent variable in this investigation? What was the dependent variable?

Analyse and interpret

❹ Use ideas about particles to explain the student's observation that the sand 'just gets pushed to the side'.

Evaluate data and methods

❺ Suggest a way that the student could measure the 'wetness' of the sand in a reliable way.

❻ Give an example of a situation where the idea of pressure can be used to explain why an object does not sink into a material such as sand.

Pressure in fluids

Because particles in a liquid or gas (for example, in fluids) are constantly in random motion, they are constantly colliding with each other and the walls of the container.

This causes a force on the other particles and the container walls. Usually this force is described in terms of the pressure it causes on a particular area.

The pressure at a point in a gas or liquid which is at rest acts equally in all directions.

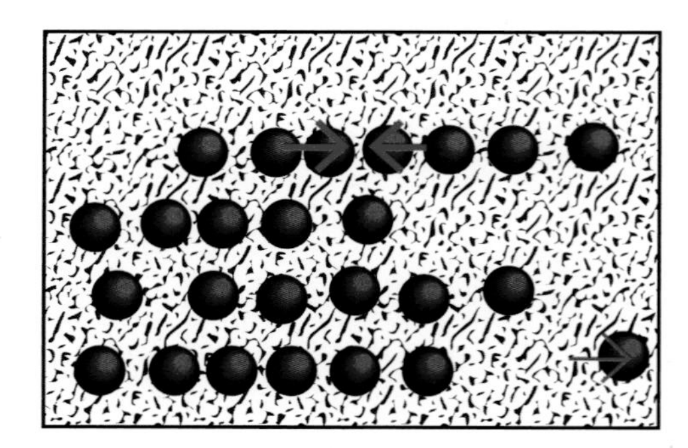

Δ Fig. 5.16 The particles in a fluid are constantly colliding with each other and the walls of the container.

Atmospheric pressure

Because we have spent all of our lives living in the atmosphere of the Earth, we seldom think that we have a height of 20 km or so of air pressing on us. We do not feel the pressure because it does not just push *down*, it pushes us *inwards* from all sides. Our lungs do not collapse, because the same air pressure flows into our lungs and presses outwards. It would be a very different story if our lungs did not contain any air and there was a vacuum inside them.

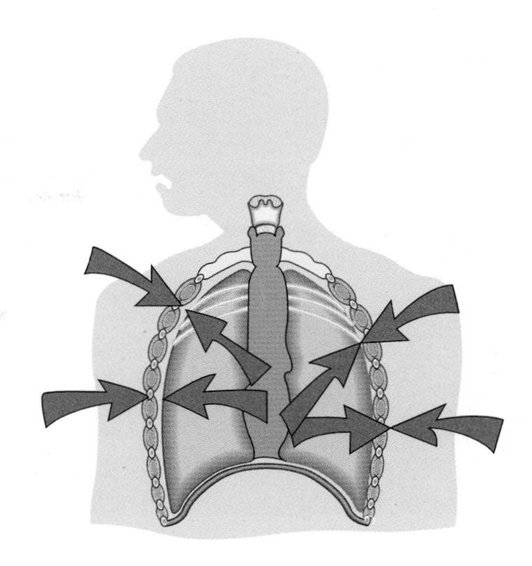

△ Fig. 5.17 Because the pressure inside is the same as the pressure outside, our lungs do not collapse.

If an aluminium soft drink can is filled with 1 tablespoon of water and heated over a Bunsen burner until the water boils, then, as seen in Figure 5.18, the can is grasped with tongs, turned upside down, and dipped into a beaker of cold water, the can will collapse almost instantly due to the change in air pressure. Boiling the water drives the air out of the can and replaces it with water vapor. When the water vapor condenses, the pressure inside the can is much less than the air pressure outside, causing the can to collapse.

△ Fig. 5.18 Collapsing can experiment.

Similarly, if a plastic bottle has same pressure inside and out as shown in Figure 5.19, the bottle will be as normal. If the pressure inside the bottle is reduced, so that the pressure outside is greater, as shown in the middle diagram, then the bottle will start to collapse. If the pressure inside is completely removed, as shown on the right, then the bottle will collapse.

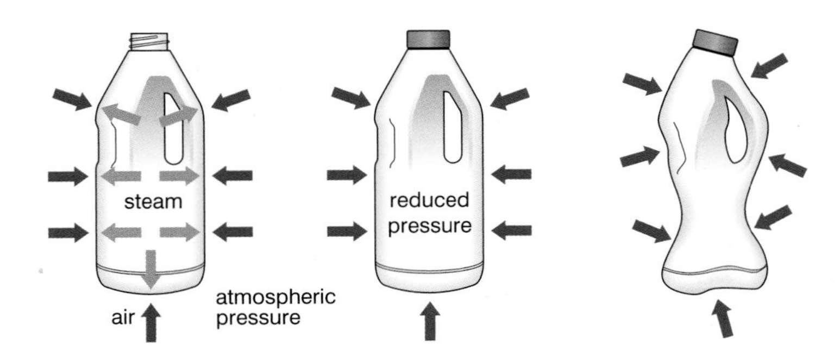

△ Fig. 5.19 How a bottle filled with steam collapses when it cools down.

Atmospheric pressure is approximately 100 kPa. This value is a pure coincidence. In fact it is around 101.3 kPa, though it increases and decreases by 5 per cent or so depending on the weather. But in the same way that we often take g to be 10 m/s^2 on the Earth when it is more accurately 9.8 m/s^2, we often choose to take atmospheric pressure to be 100 kPa.

Pressure may also be measured in bar and millibar. Normal atmospheric pressure is approximately 1 bar. The pressure on a scuba diver's cylinder of air can easily be 200 bar. You will see millibar used in some weather forecasts. Atmospheric pressure is approximately 1000 mbar.

QUESTIONS

1. What causes the pressure in a fluid in a container?
2. Why do our lungs not collapse under atmospheric pressure?
3. Why does a plastic bottle collapse if air is removed from inside it?

Pressure difference, height and density

If you dive below the water, the height of the water above you also puts pressure on you. At a depth of 10 m of water, the pressure has increased by 100 kPa, and for each further 10 m of depth the pressure increases by another 100 kPa. The rapid increase in pressure explains why scuba divers cannot go down more than 20 m without taking extra safety precautions.

PRESSURE AND SUBMARINES

Early military submarines had propulsion systems that could not operate well when submerged, so these submarines spent most of their time on the surface, with hull designs that balanced the need for a relatively streamlined structure with the ability to move on the surface. Late in World War II, technological advances meant that longer and faster submerged operations were possible.

Δ Fig. 5.20 Submarines have two hulls to withstand the water pressure at depths down to about 300 m.

Submarines actually have two hulls. The external hull, which forms the shape of the submarine, is sometimes called the light hull. (This term is particularly appropriate for Russian submarines, whose external hull is usually made of steel that is only 2 to 4 millimetres thick.) The pressure hull, which is inside the external hull, is designed to withstand the pressure outside it from the water around the submarine. It has normal atmospheric pressure inside, which allows the submarine crew to breathe normally. The dive depth (the maximum depth at which the submarine can operate) depends on the strength of the hull. Submarines used in World War I had hulls made of carbon steel and could not dive below 100 m. In World War II, high-strength alloyed steel was used, and the dive depth increased to 200 m. This is still the main material used today, with a current limit of 250–300 m dive depth. A few submarines have been built with titanium hulls and the deepest diving submarine was the Soviet *Komsomolets*, which dived to about 1000 m.

The increase in pressure below the surface of a liquid depends on the depth below the surface and the density of the liquid. The pressure is much higher at a certain depth below the surface of mercury than it is below the same depth of water. It does not depend on anything else, and note in particular that the pressure does not depend on the width of the water.

If a diver goes to inspect a well, the pressure 10 m below the surface is the same as the pressure 10 m below the surface of a large lake. This explains why an engineer who is designing a dam needs to make it the same thickness whether the lake is going to be 100 m long or 100 km long.

Scuba divers breathe compressed air at high pressure to prevent their lungs collapsing due to the high pressure from the water above them. This is a safe sport, but only because new divers are trained to a very high standard before they are allowed to dive.

△ Fig. 5.21 Scuba divers.

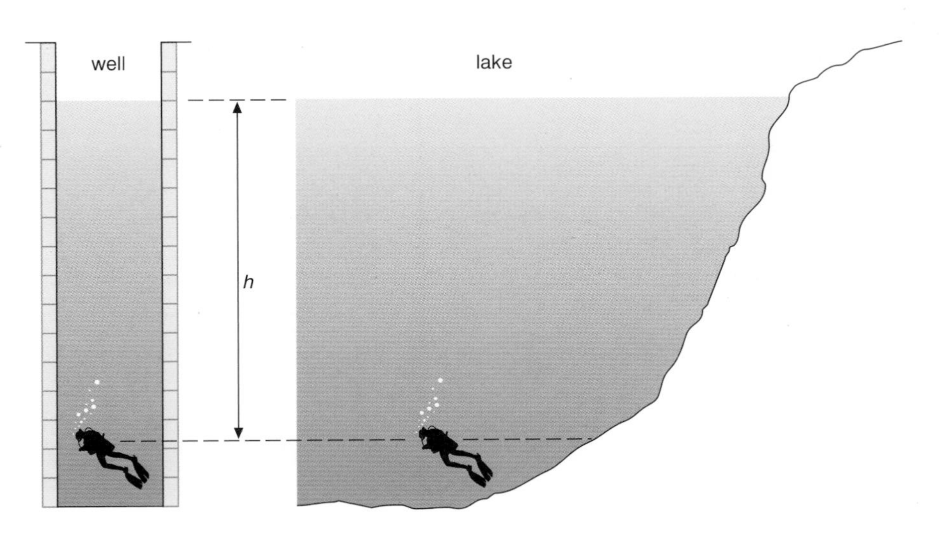

△ Fig. 5.22 The pressure on the diver is the same in the well and in the lake. In both cases it depends only on the density of the liquid and his depth, h.

The pressure below the surface of a fluid – and between any two points in the fluid – can be calculated by the following equation:

pressure difference = height × density × gravitational field strength

$$p = h \times \rho \times g$$

where p = pressure difference in pascals (Pa), ρ = density in kilograms per cubic metre (kg/m^3) and g = gravitational field strength (N/kg).

REMEMBER

Note that the density ρ must be in kg/m^3. If it is quoted in g/cm^3, you must convert it. Remember that 1 g/cm^3 = 1000 kg/m^3.

Note that there is one major cause of confusion. Consider the pressure on scuba divers. Before they jump in, the pressure on them is already 100 kPa (or 1 bar). When they have dived down 10 m, the pressure on them increases by 100 kPa, so the total pressure on them is now 200 kPa (2 bar). The pressure is coming 100 kPa from the air above them, and 100 kPa from the water above them. At 20 m, the total pressure on them is 300 kPa, and so on.

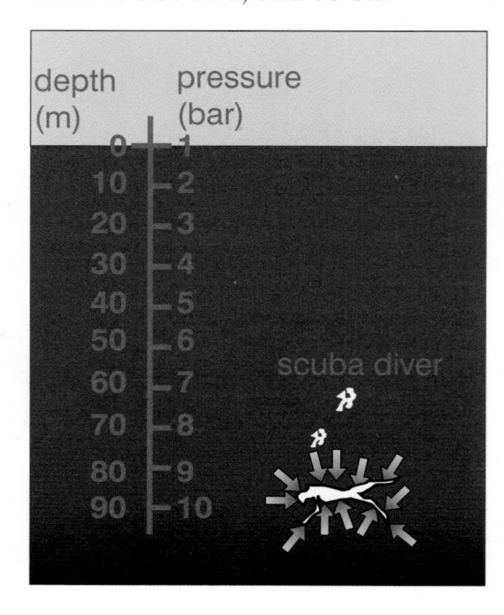

◁ Fig. 5.23 How the pressure on a diver varies with depth.

WORKED EXAMPLE

An aquarium has a tunnel through a tank of water at a depth of 5 m below the surface. The manufacturer guarantees the tunnel to a pressure difference of 200 kPa. Is the tunnel safe?

Write down the equation: $p = h \times \rho \times g$

Substitute the values into the equation: $p = 5 \times 1000 \times 10$

Work out the answer and write down the unit: p = 50 000 Pa or 50 kPa

The tunnel is safe.

Note that the total pressure on the outside of the tunnel is 50 kPa from the water, plus 100 kPa from the air pushing on top of the water, giving 150 kPa. However, the tunnel is also full of air, which is pushing outwards with a pressure of 100 kPa. So the tunnel only has to stand a pressure difference of 50 kPa.

QUESTIONS

1. What factors does the pressure in a fluid depend on?
2. Calculate the increase in pressure as you dive from the surface of a lake to a depth of 8 m. (Density of water = 1000 kg/m^3 and g = 10 N/kg).
3. What is the pressure difference at 100 m below sea level? (Density of sea water = 1030 kg/m^3)
4. What is the pressure on a scuba diver at a depth of 30 m in fresh water?
5. At the bottom of a column of mercury of height 15 cm, what is the pressure? (Density of mercury = 13 600 kg/m^3)

End of topic checklist

Density is the mass of a body divided by its volume. Its units can be g/cm^3 or kg/m^3.

Pressure is the effect of a force spread out over an area. Its units are N/m^2 or pascals (Pa).

The facts and ideas that you should understand by studying this topic:

- ❍ Know and use the relationship between density, mass and volume: density = mass/volume
- ❍ Know and use the relationship between pressure, force and area: pressure = force/area
- ❍ Know and use the relationship between pressure difference, density, g and height difference: pressure difference (in a fluid) = height difference × density × g
- ❍ Be able to investigate density using direct measurements of mass and volume.
- ❍ Understand that at a point in a gas or liquid at rest the pressure acts equally in all directions.

End of topic questions

1. State whether each of the following objects will sink or not.

   a) wood in oil

   b) wood in mercury

   c) plastic in oil

   d) steel in mercury

   e) silver in air

   f) gold in mercury (if tested by experiment it must be done rapidly as the gold will dissolve very quickly)

   g) helium balloon in air

   The answer may be that 'It depends on what sample of the material you choose'.

   (7 marks)

2. Write out the worked example on page 260 but with the lengths in metres and the mass in kilograms. Give the answer in kg/m³. **(1 mark)**

3. A king who has studied physics believes that his jeweller has given him a crown that is a mixture of gold and silver, not the 1.93 kg of pure gold that he paid for. He weighs the crown in a balance and finds that it has the correct mass of 1.93 kg. He then immerses it in a measuring jug where the water volume was 800 cm³. If the crown is pure gold, what will the new water volume be? What will happen to the water level if the jeweller has cheated? **(4 marks)**

4. a) Calculate the pressure on the floor caused by:

   i) an ordinary shoe heel (person of mass 40 kg, heel 5 cm × 5 cm) when all the person's weight is on one heel

   ii) an elephant (of mass 500 kg, one foot area of 300 cm²) when all four feet are on the ground

   iii) and a high-heeled shoe (worn by a person of mass 40 kg, heel of area 0.5 cm²) when all the person's weight is on one heel. **(6 marks)**

   b) Which ones will damage a wooden floor that starts to yield at a pressure of 4000 kPa? (Note: to convert from cm² to m² you need to divide by 10 000.) **(1 mark)**

5. The density of fresh water is 1000 kg/m³. The pressure gauge on a submarine in a river was reading 100 kPa when it was at the surface.

   a) A sailor notices that the gauge is now reading 250 kPa. How deep is the submarine? **(2 marks)**

End of topic questions continued

b) How would this answer change if the submarine was diving in sea water that is slightly denser than fresh water? **(1 mark)**

6. A diver on Saturn's moon Titan is 50 m below the surface of a lake of liquid methane. The density of liquid methane is 0.42 g/cm^3. The gravitational field strength on Titan is 1.4 N/kg.

a) What is the increase in pressure on him due to his depth in the methane?

b) The pressure of the atmosphere on Titan is 1.6 bar. What is the total pressure on the diver (in kPa)? **(2 marks)**

7. A skater glides on one skate. If the mass of the skater is 65 kg, and the area of the skate is 9×10^{-4} m^2, what pressure is exerted on the ice by the skate? **(2 marks)**

8. An oil well is 1500 m deep and is filled with a fluid of density 960 kg/m^3. What is the pressure due to the fluid at the bottom of the well? **(2 marks)**

9. A diver is exploring a sunken ship and notes that the pressure is 2.96×10^5 Pa at the ship compared to 1.00×10^5 Pa at the surface. Taking the density of water to be 1000 kg/m^3, calculate the depth that the diver is at. **(3 marks)**

10. A helium balloon has a volume of 0.04 m^3. The density of helium is 0.18 kg/m^3. What mass of helium does each balloon contain? **(2 marks)**

11. a) How does the pressure in a liquid depend on the depth of the liquid and the density of the liquid? **(2 marks)**

b) A container has an area of 4 m^2 and is filled with water of density 1000 kg/m^3. What is the pressure of the water at a point 0.5 m below the surface? **(2 marks)**

12. The air pressure at the base of a mountain is 1.01×10^5 Pa. At the top of the mountain, the air pressure is measured at 0.8×10^5 Pa. Given that the density of air is 1.2 kg/m^3, calculate the height of the mountain. **(2 marks)**

13. The density of water in a lake is 1.02×10^3 kg/m^3. Atmospheric pressure is 1.01×10^5 Pa. What is the total pressure at a depth of 12 m below the surface of the lake? **(2 marks)**

14. A fish is swimming at a depth of 10.4 m in water of density 1.03×10^3 kg/m^3. Calculate the pressure at this depth caused by the water. **(2 marks)**

15. A small child of weight 120 N can stand with one bare foot on a Lego brick of surface area 2 cm^2 without feeling pain. When an adult weighing 600 N stands with one foot on the same brick it could be painful. Explain why. **(5 marks)**

Change of state

INTRODUCTION

One of the most powerful ideas in science is the idea that all matter is made of tiny particles arranged in various ways depending on the energy available. To link together such a wide variety of materials – from the hardest, most dense metals to a breath of wind, and from great icebergs to the steam from a kettle – in such a simple way is a great achievement and a story that has been pieced together over many years.

Δ Fig. 5.24 Molten iron being poured from a furnace.

Materials science is still an area of expanding research, but it remains based upon the study of solids, liquids and gases and the changes that happen when energy is added or removed.

KNOWLEDGE CHECK

✓ Be able to classify substances as solid, liquid or gas.
✓ Be familiar with some of the simple properties of solids, liquids and gases.
✓ Know that all substances are made up of particles.

LEARNING OBJECTIVES

✓ Explain why heating a system will change the energy stored within the system and raise its temperature or produce changes of state.
✓ Describe the changes that occur when a solid melts to form a liquid, and when a liquid evaporates or boils to form a gas.
✓ Describe the arrangement and motion of particles in solids, liquids and gases.
✓ Be able to obtain a temperature–time graph to show the constant temperature during a change of state.
✓ Know that specific heat capacity is the energy required to change the temperature of an object by one degree Celsius per kilogram of mass (J/kg °C).
✓ Use the equation: change in thermal energy = mass × specific heat capacity × change in temperature, $\Delta Q = m \times c \times \Delta T$.
✓ Be able to investigate the specific heat capacity of materials including water and some solids.

STATES OF MATTER

Almost all matter can be classified as a solid, a liquid or a gas. These are the three **states of matter**.

◁ Fig. 5.25 The main body of this rocket is filled with liquid oxygen and liquid hydrogen, which have to be kept at extremely low temperatures to prevent them from heating up and turning back into gas. If the fuel were made colder it would turn into a solid.

Atoms and molecules exert forces on each other when they are close together. The atoms in a solid are locked together by the forces between them. But even in a solid, the particles are not completely still. They vibrate constantly about their fixed positions. If the material is heated, it is given more energy, and the particles vibrate faster and further.

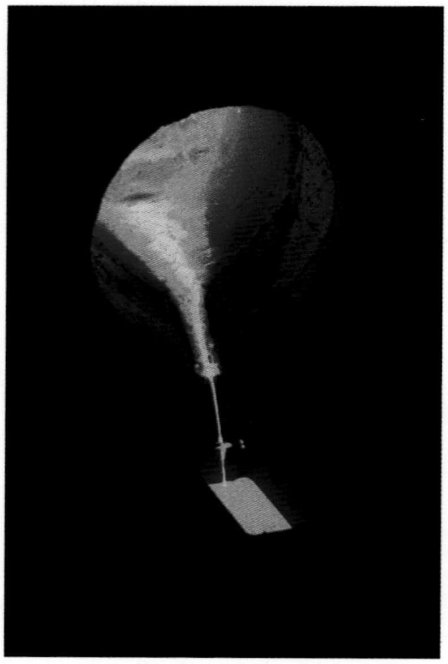

If the temperature is increased more, the vibrations of the particles increase to the point at which the forces are no longer strong enough to hold the structure together in the rigid order of a solid. The forces can no longer prevent the atoms moving around, but they do prevent them from flying apart from each other. This is what makes a liquid. The volume of the liquid is the volume occupied by the particles of which it is made.

If the temperature is increased even more, then the particles do fly apart. They now form a gas. The particles fly around at high speed – several hundred kilometres per hour! If they are in a container, they travel all over it, bouncing off the walls. The volume of a gas is not fixed; it just depends on the size of the container that the gas is put in. We use the **kinetic molecular model** to explain the behaviour of solids, liquids and gases. The following table summarises this model.

△ Fig. 5.26 The molten iron can be poured into a mould before it cools down and turns back into a solid.

Solids, liquids and gases

	Solid	Liquid	Gas
Arrangement of particles	Regular pattern, closely packed together, particles held in place by the forces of attraction between them	Irregular, closely packed together, particles able to move past each other	Irregular, widely spaced, particles able to move freely
Diagram	Δ Fig. 5.27 Arrangement of particles in a solid.	Δ Fig. 5.28 Arrangement of particles in a liquid.	Δ Fig. 5.29 Arrangement of particles in a gas.
Motion of particles	Vibrate in place within the structure	'Slide' over each other in a random motion	Random motion, faster movement than the other states

Δ Table 5.2 Arrangement and motion of particles in the three states of matter.

QUESTIONS

1. What happens to the motion of atoms as the temperature increases?

2. Explain why it is easier to compress a gas than a liquid.

3. Describe the arrangement of particles in:

a) a solid

b) a liquid

c) a gas.

4. What does the volume of a gas depend on?

Melting and boiling

If you take a thermally insulated beaker that contains pieces of extremely cold ice and warm it up with an electrical heater, with a current through it that is measured by an ammeter and a voltage across it that is measured by a voltmeter, you can measure the temperature of the ice (and then water) every few seconds. You can then plot a graph of the temperature of the ice and water as the contents of the beaker warm up. The graph will look like Fig. 5.31.

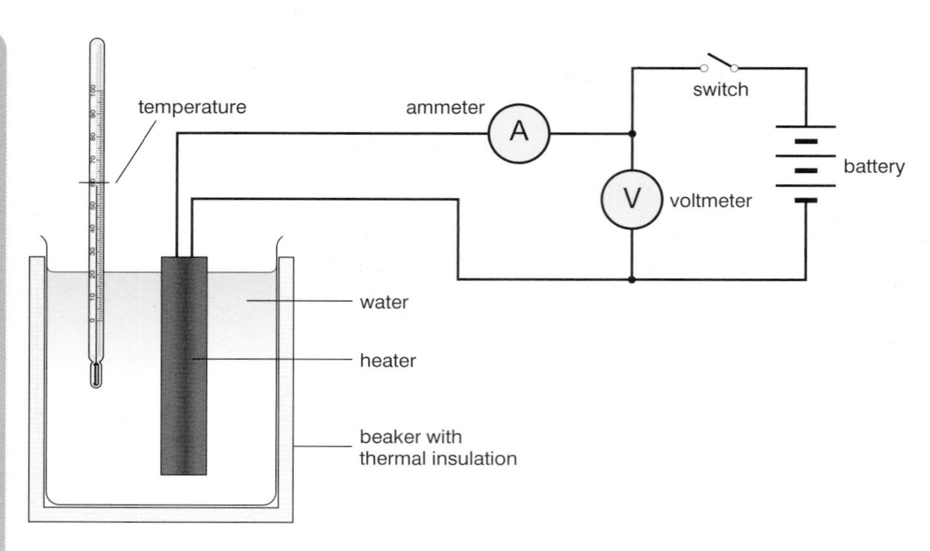

Δ Fig. 5.30 Warming water using an electrical heater.

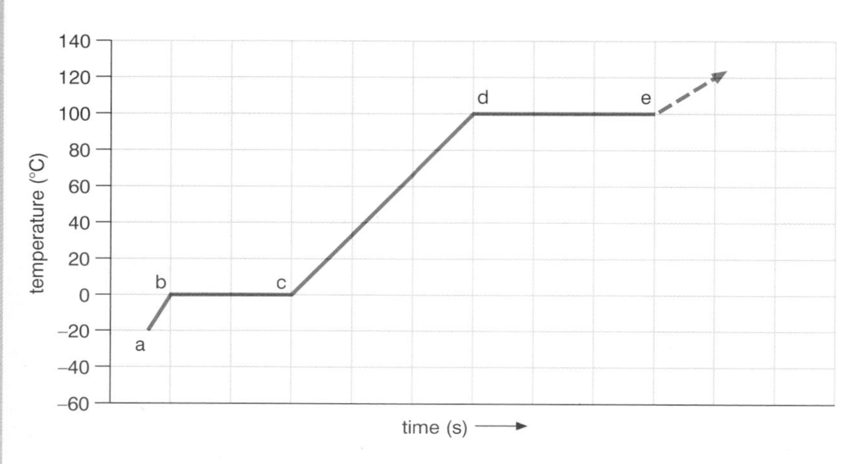

Δ Fig. 5.31 Graph of temperature against time when heating ice and water.

From point a to point b the ice is warming up, but it is not starting to melt. (This is similar to the behaviour of ice-cream after you take it out of the freezer, as it slowly takes heat out of the air and warms up without melting to begin with.)

Along line a–b, the heater is increasing the energy stored in the ice, and this is shown by the increase in temperature.

From point b to point c, the beaker contains a mixture of ice and water. The temperature stays constant, at 0 °C, and the heater melts the ice.

This makes a very important point. The energy transferred from the heater has gone into the beaker, and so the energy stored by the contents of the beaker has gone up, but the temperature has not gone up. The energy has been used to separate the particles in the ice to melt it, changing it from solid to liquid, and this energy is stored in the water. The arrangement of the particles has changed from a solid (Fig. 5.27) to a liquid (Fig. 5.28).

In fact you have to put in almost as much energy to melt the ice as you will do in the next step, to raise the water from freezing point to boiling point. The same amount of energy must be removed again to turn the water back into ice. This is why it takes a freezer so long to freeze water.

From point c to point d, the input of heat energy into the water raises its temperature from 0 °C to 100 °C, and at point d the water boils.

In boiling, every particle in the liquid has enough energy to break away. This happens at a particular temperature – the **boiling point**. At the boiling point, the energy added to the material will be breaking the particles apart – the temperature does not change.

So from d to e the temperature of the boiling water stays constant, at 100 °C.

Condensation is the reverse of boiling, where the gas turns into a liquid, and **solidification** is the reverse of melting.

QUESTIONS

1. Explain why the temperature–time graph in Figure 5.31 shows a steady temperature as ice melts.
2. Why does boiling happen at one particular temperature?

Developing investigative skills

A student put some solid wax into a boiling tube. She placed the boiling tube in a hot water bath which melts the wax. She removed the boiling tube and allowed the wax to cool. The student measured the temperature of the wax every minute using a thermometer. Her data is shown below.

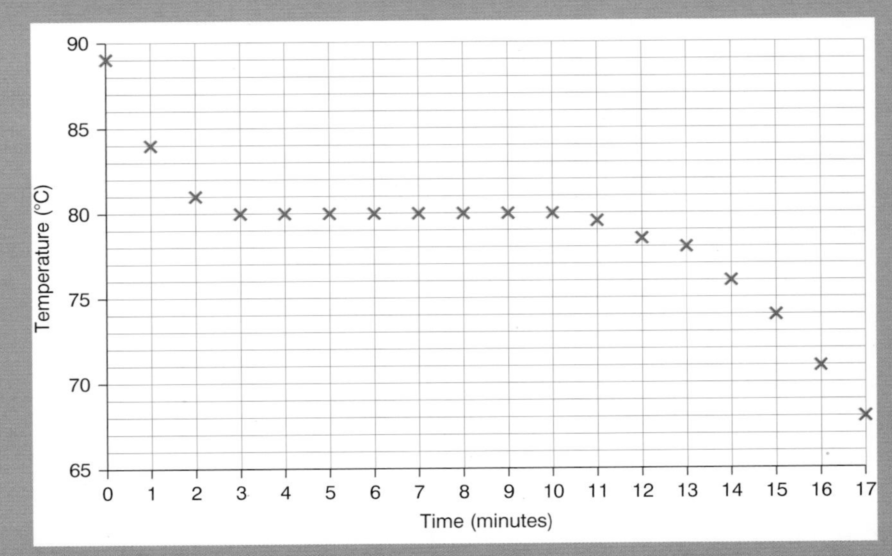

Δ Fig. 5.32 Temperature-time graph from the student's experiment.

Devise and plan investigation

❶ Why was the boiling tube heated in a water bath rather than with a Bunsen burner?

Analyse and interpret

❷ What is the melting point of the wax used in this experiment?

❸ The wax cools quicker between 0 and 1 minute than between 16 and 17 minutes. Explain why.

❹ Describe and explain what is happening to the wax between 4 and 11 minutes. Use ideas about particles in your answer.

Evaluate data and methods

❺ **a)** Suggest a way the student could measure the temperature more continuously.

b) What effect would this have on the answer obtained in question 2?

❻ The student decides to repeat the experiment by using a much larger mass of candle wax. Suggest how this would affect the graph. Explain your answer.

Evaporation

When particles break away from the surface of a liquid and form a **vapour**, the process is known as **evaporation**.

The more energetic molecules of the liquid escape from the surface.as shown in Figure 5.33. This reduces the average energy of the molecules remaining in the liquid and so the liquid cools down.

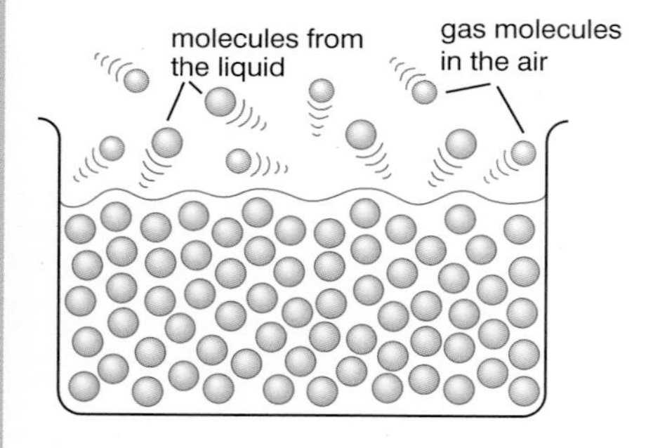

Δ Fig. 5.33 Particles break away from the surface of a liquid during evaporation.

Evaporation is increased at higher temperatures. It is also increased by a strong flow of air across the surface of the liquid, as in this way the evaporating molecules are carried away quickly. A certain amount of water will also evaporate more quickly if you increase its surface area. Tea or coffee in a short, wide cup cools down much more quickly than in a tall narrow mug, because the large surface area of the bowl allows more evaporation.

◁ Fig. 5.34 Clouds are formed from invisible water vapour that evaporates from the sea and is carried away by the wind. When the water vapour cools at high altitude, it turns back into the small droplets of water that you can see as these clouds.

QUESTIONS

1. What factors increase the rate of evaporation?
2. Why does tea in a narrow mug cool down more slowly than tea in a wide mug?

EXTENSION

Imagine you are a particle that has experienced evaporation. Write a letter to your friend describing the experience. Your letter should answer the following questions.

1. What change of state did you go through?
2. How close to your neighbours were you in your original state?

   What was given to you to make you change state?
3. How did you change state?

SPECIFIC HEAT CAPACITY

Different materials need different amounts of energy to change their temperature by the same amount, for the same mass of material. It depends on their **specific heat capacity**.

For example, many houses use liquid-filled radiators to heat a cold room. If one radiator is filled with oil and another is filled with the same amount of water, the water-filled radiator needs more energy to heat it from, say 15°C to 55°C, than the oil-filled radiator.

Each kilogram of water needs 4200 J of energy to change its temperature by 1°C. Each kilogram of oil needs only 1500 J of energy to heat it up by 1°C. This is because water has a higher specific heat capacity. The high specific heat capacity of water is very useful in storing thermal energy in heating systems. The water stores more energy than oil for the same mass so it has more energy to transfer to the cold room.

Material	Specific heat capacity (J/kg °C)
Oil	1500
Water	4200
Aluminium	913
Copper	330
Concrete	880
Iron	460

△ The specific heat capacities of some everyday materials.

Energy calculations using specific heat capacity

The specific heat capacity is the energy required to change the temperature of an object by one degree Celsius per kilogram of mass (J/kg °C).

The change in thermal energy when a system is heated can be calculated using:

change in thermal energy = mass × specific heat capacity × change in temperature

$\Delta Q = m \times c \times \Delta T$

WORKED EXAMPLE

An iron pan has a mass of 3 kg. It contains 3 kg of water. Calculate the energy needed to raise the temperature of the water **and** the pan by 80°C.

For water:

Write down the equation: $\Delta Q = m \times c \times \Delta T$

Substitute the values: $\Delta Q = 3 \times 4\,200 \times 80$

Calculate the answer: $\Delta Q = 1\,008\,000$ J

For iron:

Write down the equation: $\Delta Q = m \times c \times \Delta T$

Substitute the values: $\Delta Q = 3 \times 460 \times 80$

Calculate the answer: $\Delta Q = 110\ 400$ J

Energy needed to heat water and iron pan = 1 008 000 + 110 400
= 1 118 400 J.

Developing investigative skills

A student investigates the specific heat capacity of different metal blocks. He compares blocks of the same mass, 1 kg.

The student heats up an aluminium block with a 50 W electrical heater. The heater slots into the larger hole in the aluminium block and the thermometer slots into the smaller hole. He uses a stopwatch to measure the time taken for the temperature to rise by 10°C.

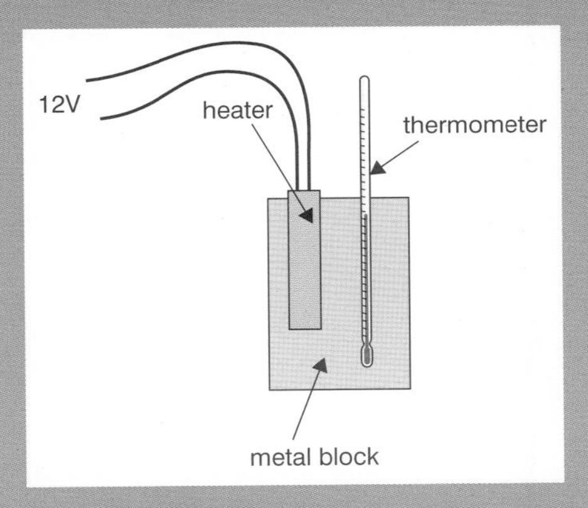

Δ Fig. 5.35 Apparatus used in the experiment.

He repeats the experiment for different metals. His data is shown in the table.

Type of metal	Temperature rise (°C)	Power of heater (W)	Time taken to increase the temperature by 10°C (s)	Energy transferred by the heater (J)
Aluminium	10	50	221.0	
Copper	10	50	90.0	
Lead	10	50	30.0	
Steel	10	50	120.0	

Devise and plan investigation

❶ The heater has a power rating of 50 W or 50 J/s. Use the data to calculate the energy transferred by the heater for each metal. Complete the table with your answers.

❷ A small amount of oil is put in the hole to surround the thermometer. Suggest reasons why oil is used.

❸ Name two control variables in this experiment.

Analyse and interpret

❹ Determine the specific heat capacity of each material.

Evaluate data and methods

❺ The student did some research to find out the specific heat capacity of some materials. Look at his findings:

Metal	Specific heat capacity (J/kg°C)
Aluminium	921
Copper	377
Lead	126
Steel	502

How do his experimental results compare to the accepted values? Suggest reasons for any differences.

❻ Suggest how the student could improve his method to get more accurate results.

❼ The student decides to use the heater to find the specific heat capacity of water. Design a method that the student could use. Include in your description the measurements and calculations he should make. Why should he stir the water at the end?

End of topic checklist

The three **states of matter** are solid, liquid and gas.

The **kinetic molecular model** is the theory describing the movement of particles in solids, liquid and gases.

Melting and **boiling** require an energy input but take place without a change in temperature.

Evaporation only takes place at the surface and can occur at any temperature. Boiling takes place throughout the liquid and occurs only at the boiling point.

Specific heat capacity is the energy needed to raise the temperature of 1 kg of a substance by 1°C. The units of specific heat capacity are J/kg °C.

The facts and ideas that you should understand by studying this topic:

- ❍ Be able to explain why heating a system will either raise its temperature or produce changes of state.
- ❍ Be able to describe the changes that occur when a solid melts to form a liquid, and when a liquid evaporates or boils to form a gas.
- ❍ Be able to describe the arrangement and motion of particles in solids, liquids and gases.
- ❍ Describe an experiment to obtain a temperature–time graph during a change of state.
- ❍ Be able to use the equation: change in thermal energy = mass × specific heat capacity × change in temperature, $\Delta Q = m \times c \times \Delta T$.
- ❍ Know that different materials have different specific heat capacities.
- ❍ Describe how to investigate the specific heat capacity of materials including water and some solids.

End of topic questions

1. Use ideas about particles to explain why:

a) solids keep their shape, but liquids and gases do not **(2 marks)**

b) solids and liquids have a fixed volume, but gases fill their container. **(2 marks)**

2. Water at 60 °C is changing from a liquid to a gas. What is this change of state called? **(1 mark)**

3. Name the three factors that determine how much energy is needed to increase the temperature of an object. **(3 marks)**

4. Using ideas about particles, explain the difference between evaporating and boiling. **(3 marks)**

5. Explain, using ideas about particles, what happens when a solid melts. **(3 marks)**

6. A metal pan has a mass of 2 kg and required 80 320 J of energy to increase its temperature from 20°C to 100°C. Calculate its specific heat capacity. **(5 marks)**

7. Explain why, in cold weather, you sometimes see droplets of water on the inside of windows. **(3 marks)**

8. The bonds between particles in liquid A are stronger than those in liquid B. Suggest which liquid will evaporate most easily and say why. **(3 marks)**

9. Explain what happens to a pan of boiling water if it continues to be heated. **(3 marks)**

10. A bath contains 250 kg of water at 50°C. The water has a specific heat capacity of 4200 J/kg°C. Calculate the energy used to heat the water from 20°C. **(2 marks)**

11. 2.5 kg of cold water is heated to boiling point using 892 500 J of energy. Calculate the temperature of the cold water. **(4 marks)**

12. Using your knowledge about particles and energy, explain why lower air pressures make liquids evaporate more quickly. **(3 marks)**

Ideal gas molecules

△ Fig. 5.36 The balloon rises because the gas inside it is less dense than the surrounding air.

INTRODUCTION

You can start to apply your knowledge of forces and motion to the molecules of a gas. This helps to build up a theory – the kinetic theory of gases – to see if you can predict the behaviour of gases and match this against experiments with gases.

This topic begins with quite simple theory, looking only at gases whose particles are generally separated from each other. However, these ideas have proved to be remarkably successful in describing the behaviour of other materials. You will learn that the behaviour of gases at different temperatures and pressures can be summarised by one very simple equation.

KNOWLEDGE CHECK

- ✓ Be able to describe how particles are arranged and move in solids, liquids and gases.
- ✓ Know how to define and calculate pressure.
- ✓ Know some everyday properties of gases – for example that they expand when heated and they exert a pressure on container walls.

LEARNING OBJECTIVES

- ✓ Be able to explain how molecules in a gas have random motion and that they exert a force and hence a pressure on the walls of a container.
- ✓ Be able to explain, for a fixed amount of gas, the qualitative relationship between:
  - pressure and volume at constant temperature
  - pressure and Kelvin temperature at constant volume.
- ✓ Use the relationship between the pressure and Kelvin temperature of a fixed mass of gas at constant volume:

  $$\frac{p_1}{T_1} = \frac{p_2}{T_2}$$

- ✓ Use the relationship between the pressure and volume of a fixed mass of gas at constant temperature:

  $$p_1V_1 = p_2V_2$$

- ✓ Understand why there is an absolute zero of temperature which is –273 °C.
- ✓ Describe the Kelvin scale of temperature and be able to convert between the Kelvin and Celsius scales
- ✓ Understand why an increase in temperature results in an increase in the average speed of gas molecules.
- ✓ Know that the Kelvin temperature of a gas is proportional to the average kinetic energy of its molecules.

KINETIC THEORY OF GASES

The **kinetic molecular model** helps to build up a set of ideas from the basic idea that a gas is made of many tiny particles, called molecules, which are always moving at high speed and in random directions. These ideas give a picture of what happens inside a gas.

Observed feature of a gas	Related ideas from the kinetic theory
Gases have a mass that can be measured.	The total mass of a gas is the sum of the masses of the individual molecules.
Gases have a temperature that can be measured.	The individual molecules are always moving. The faster they move (the more kinetic energy they have), the higher the temperature of the gas.
Gases have a pressure that can be measured.	When the molecules hit the walls of the container they exert a force on it. It is the total force, divided by the surface area of the container, that is observed when measuring pressure.
Gases do not have a fixed volume, but take the volume of their container.	Although the volume of each molecule is only tiny, they are always moving about and spread out throughout the container. This volume can be measured.
Temperature has an **absolute zero**.	As temperature falls, the speed of the molecules (and their kinetic energy) becomes less. At absolute zero (–273 °C) the molecules would have stopped moving.

Δ Table 5.3. Kinetic theory and gases.

Temperature and pressure of a gas

The kinetic model says that the pressure on the walls of a container is caused by the collisions with the speeding molecules. You can feel this pressure if you try to hold a bicycle pump in the pushed-in position while blocking the air outlet with your finger as shown in Figure 5.37. (If the pump is broken and allows the air to escape, this does not work.)

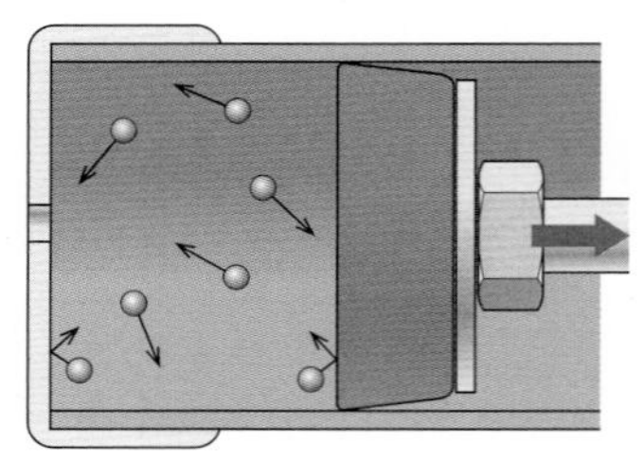

Δ Fig. 5.37 The molecules of the gas are colliding with the piston and trying to push it out.

In the diagram, the piston is not moving. However, there is a force trying to push it out. It is clear that if the molecules travel faster then they will hit the piston in the pump more often and harder. The pressure on the piston and on the walls will go up. This is exactly what will happen if the air gets hotter.

Note that, of course, the molecules will hit each other as well as the walls of the container. At normal pressures they travel a lot less than 1 mm between collisions. This does not affect the way that the model works.

◁ Fig. 5.38 The inner tube from a tyre has been pumped up with air before use as a toboggan. It is the pressure caused by the movement of the air molecules that keeps it inflated. Because the temperature is low, the inner tube will have needed more air. On a hot day this tube could burst.

QUESTIONS

1. How does the kinetic model of matter explain pressure on the walls of a container?
2. Why does the pressure in a pump increase if the molecules move faster?

These ideas help to explain some experimental results:

- If the volume of the gas stays constant, then the pressure of the gas increases as temperature increases. This is because at a higher temperature the molecules move more quickly and so the force on the walls will be higher.
- If the temperature of the gas stays constant, then the volume of the gas decreases as the pressure increases. This is because if the volume of the gas is reduced the molecules hit the walls more frequently, increasing the pressure.

QUESTIONS

1. Explain how the kinetic theory explains the measurable volume of a gas.
2. Describe the link between the pressure of a gas and its volume at constant temperature.
3. How is this linked to kinetic theory?

ABSOLUTE ZERO AND THE KELVIN SCALE OF TEMPERATURE

As a material is cooled down, its molecules vibrate less. If the substance is a liquid or a gas, the molecules move more slowly. If you keep cooling the material down, it eventually reaches a temperature so low that even gases such as nitrogen and oxygen have turned into solids –all movement of the atoms in the solid has stopped. This temperature is the lowest that can be reached. It is known as absolute zero,
and it has been shown to be –273 °C. There can be no lower temperature, as the molecules cannot do less than not move!

Because there is an absolute zero, there is an alternative system of measuring temperature in which the temperature at absolute zero is given the number 0. All other temperatures are then higher than this. The 'steps' are the same size as in the Celsius scale, which means you have to go up 273 degrees to reach the melting point of ice, and another 100 degrees to reach the boiling point of water. This scale is known as the Kelvin scale of temperature.

The two temperature scales work as shown in Table 5.5.

Scale	Absolute zero	Melting point of ice	Boiling point of water	Melting point of gold
Celsius	–273 °C	0 °C	100 °C	1064 °C
Kelvin	0 K	273 K	373 K	1337 K

Δ Table 5.5 The Kelvin and Celsius scales.

So to convert a temperature in degrees Celsius to kelvin you add 273.

To convert a temperature in kelvin to degrees Celsius, you subtract 273.

Note that the unit for the Celsius scale is always called 'degree Celsius' and written as °C, though you will sometimes see degC. The unit for the Kelvin scale is kelvin and is always written as K.

QUESTIONS

1. Convert to kelvin:

a) 50 °C

b) 2000 °C.

2. Convert to degrees Celsius:

a) 100 K

b) 1500 K.

3. At absolute zero, what happens to the particles in the solid?

Energy and temperature

You have already seen that the individual molecules of a gas are always moving. The higher the temperature of the gas, the faster they move (the more kinetic energy they have).

Although all of the molecules in a constant volume of gas are travelling at different speeds, it is possible to calculate the average speed of the molecules and hence their average kinetic energy. We know that the average kinetic energy of the molecules will increase as the temperature increases, but it is perhaps surprising to discover that the average kinetic energy is exactly proportional to the temperature of the gas in kelvin. So, for example, if the temperature of a gas is doubled from 273 K to 546 K, then the average kinetic energy of the molecules will exactly double as well.

We can now express the relationship between the pressure, volume and temperature of a gas in a quantitative way, and explain this using the kinetic model.

Experimental gas law	Name of law	Link to the kinetic theory
If the volume of the gas stays constant, then the pressure of the gas is proportional to its absolute temperature.	**Pressure law**	A higher temperature means the molecules move more quickly, so the force on the walls will be higher. If the volume of the gas is constant (which means the surface area will stay constant), then the pressure (= force/area) must increase.
If the temperature of the gas stays constant, then the volume of the gas is inversely proportional to the pressure.	**Boyle's law**	The temperature stays constant, so the average speed of the molecules stays constant. If the volume of the gas is reduced by half, then the molecules make the same number of collisions with half the surface area of wall, so the pressure (= force/area) must be doubled. This is **inverse proportionality.**

△ Table 5.4. The experimental laws of gases.

Gases only follow these rules if *three* conditions are met:

1. The mass of the gas must remain constant (that is, no particles move in or out of the system).
2. The temperature must be measured using the **Kelvin scale** (see page 000).
3. Gases are **ideal**, that is, do not liquefy or solidify. (In such an ideal gas, the particles are assumed to be so far apart that there are no attractive forces between them.)

The rest of this topic looks at these ideas in more detail.

Pressure and temperature, at constant volume

We know that the pressure in a sealed container increases as the temperature goes up. The container has a fixed volume and a fixed mass of gas. Experiments show that equal temperature increases cause equal pressure increases, but using the Celsius scale the graph does not pass through the origin (Figure 5.39). However, if you are using the Kelvin scale, the pressure is proportional to the temperature. So if the temperature doubles, the pressure doubles. To put this into an equation:

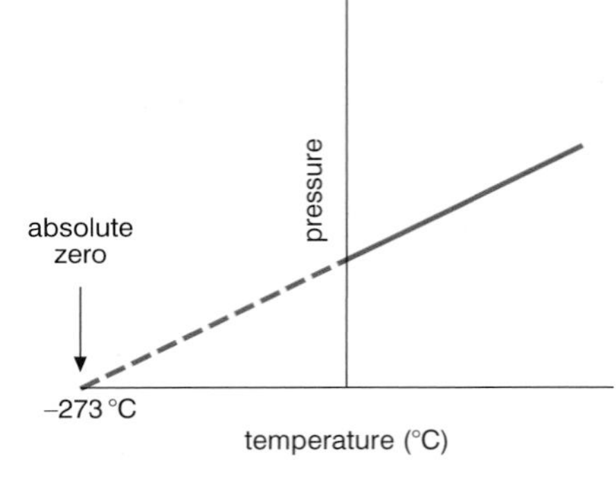

Δ Fig. 5.39 Absolute zero is the lowest possible temperature.

$$p \propto T$$

$$p = kT$$

where p is the pressure in the gas, T is the temperature in kelvin and k is a constant that depends on the size of the container and the amount of gas in it.

To solve a question, you need to know the equation:

$$\frac{p_1}{T_1} = \frac{p_2}{T_2}$$

where p_1 is the initial pressure and p_2 is the final pressure. These pressures must be in the same units, but they can both be in Pa or kPa or any other pressure unit. T_1 and T_2 (the initial and final temperatures) must be in kelvin.

WORKED EXAMPLE

A car tyre is filled to a pressure of 3 bar at 20 °C. After a long journey, the tyre reaches a temperature of 55 °C. What is the pressure now?

First convert the temperatures to kelvin. You could convert the pressures to pascal, but this time, choose to leave them in bar, where 1 bar is atmospheric pressure.

The initial temperature T_1 is (20 + 273) K = 293 K.

The final temperature T_2 is (55 + 273) K = 328 K.

The initial pressure p_1 is 3 bar.

Write down the equation: $\frac{p_1}{T_1} = \frac{p_2}{T_2}$

Substitute values into the equation: $\frac{3}{293} = \frac{p_2}{328}$

Rearrange the equation, and multiply both sides by 328 to find p_2:

$$\frac{p_2 \times 328}{328} = \frac{3 \times 328}{293}$$

$$p_2 = \frac{3 \times 328}{293}$$

Work out the answer and write down the units:

$p_2 = 3.36$ bar

Note that the pressure does not go up very much, since (as measured in the Kelvin scale) the temperature has not gone up very much. Tyre pressures should be measured with the tyre cold. They are designed to have a higher pressure when they are hot.

Δ Fig. 5.40 When you have finished pumping up a bicycle tyre, the part of the tube that is attached the valve on the wheel can get hot.

QUESTIONS

1. Describe exactly the link between the temperature and the average kinetic energy of the molecules in a gas.

2. If you use the equation linking pressures and temperatures, the temperatures must be measured in kelvin. What is the rule for the units of the pressure?

3. A car tyre contains a fixed mass of air. When the air in the tyre is at a temperature of 16 °C, the pressure is 305 kPa.

a) What is the temperature of the tyre on the kelvin scale?

b) After use, the temperature of the air in the tyre increases to 28 °C. What is the pressure in the tyre?

Pressure and volume, at constant temperature

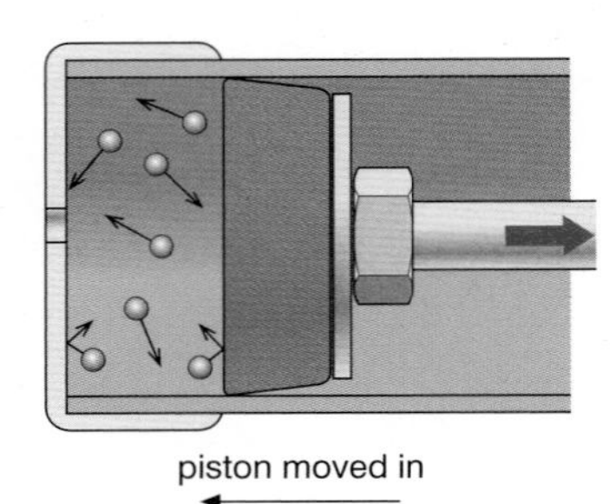

Δ Fig. 5.41. The pressure increases when the volume is reduced.

If the piston of a bicycle pump shown in Figure 5.40 is pushed in with the air outlet blocked, then the more that you push it in, the harder and harder it gets to push it further. This is because the pressure in the container goes up.

The molecular model says that there are the same number of molecules in the container travelling at the same speed. However, because the molecules are now packed in more densely, there will be more collisions with the walls and with the piston per second. If the volume is halved, then the number of collisions with the walls and with the piston will double, and the pressure on the piston will double. This law is often called Boyle's law. It only applies if the temperature of the gas (in this case, air) does not change.

A fixed amount of gas in a sealed container at constant temperature obeys the following equation:

pressure × volume = constant

$pV = \text{constant}$

where p = pressure in Pa (or N/m^2) and V = volume in m^3

Pascals and newtons per square metre are the same thing. Apart from them, you can use whichever units you like so long as you stick with them.

The constant will be a constant for a particular sample of gas in a particular container. So, in an experiment (or an exam question) you can write that the initial values of pressure and volume multiplied together, $p_1 \times V_1$ are constant. The final values of pressure and volume multiplied together, $p_2 \times V_2$, are constant.

This is the same constant in both cases. So:

$$p_1V_1 = \text{constant} = p_2V_2$$

or

$$p_1V_1 = p_2V_2$$

This equation (Boyle's law) only applies if the temperature stays constant.

Air may heat up if it is compressed quickly. A bicycle pump can get very hot due to this effect. So the law really only applies if you let the gas cool down after compressing it, or you compress it very slowly. If you allow the gas to expand, it cools down, so you have to take precautions here as well.

WORKED EXAMPLE

A bicycle pump contains 400 cm^3 of air at atmospheric pressure. If the air is compressed slowly, what is the pressure when the volume of the air is compressed to 125 cm^3? What happens to the pressure if the air is compressed quickly? (Remember that atmospheric pressure = 100 kPa.)

Write down equation: $p_1V_1 = p_2V_2$

Substitute values into the equation: $100 \times 400 = p_2 \times 125$

$p_2 \times 125 = 40\,000$

Rearrange the equation to find p_2: $p_2 = \frac{40\,000}{125}$

Work out the answer and write down the unit: $p_2 = 320$ kPa

If the air is compressed quickly, it will also heat up to a higher temperature. This will mean that the final pressure will be greater than 320 kPa.

QUESTIONS

1. Describe the quantitative relationship between the pressure of a gas and its volume at constant temperature (Boyle's law).
2. What conditions must be met for gases to follow the gas laws?
3. An aerosol has a volume of 150 cm^3. It contains gas at a pressure of 3.5×10^5 Pa. If the temperature stays constant, what will be the volume of gas if it is allowed to expand at a pressure of 101 kPa?

End of topic checklist

Absolute zero is the lowest temperature that can be reached. It is –273 °C.

The **Kelvin scale** is a temperature scale that starts at absolute zero.

Temperature on Kelvin scale = temperature in degrees Celsius + 273.

The **pressure law** states that if the volume of the gas stays constant, then the pressure of the gas is proportional to its absolute temperature.

Boyle's law states that if the temperature of the gas stays constant, then the volume of the gas is inversely proportional to the pressure.

Two quantities are **inversely proportional** when one quantity increases as the other quantity decreases.

The facts and ideas that you should understand by studying this topic:

- Describe the Kelvin scale of temperature and understand why there is an absolute zero of temperature.
- Understand how the kinetic theory pictures particles in a gas to be in constant random motion.
- Understand that an increase in temperature leads to an increase in the speed and hence kinetic energy of the particles in a gas.
- Understand how the pressure in a gas is related to the volume, if the temperature and mass of the gas remain constant.
- Use the relationship between initial temperature and pressure and final temperature and pressure for a fixed mass of gas at constant volume, $p_1/T_1 = p_2/T_2$. (Temperature must be measured in kelvin.)
- Use the relationship between initial pressure and volume and final pressure and volume for a fixed mass of gas at constant temperature, $p_1V_1 = p_2V_2$

End of topic questions

1. How is the speed of a gas molecule linked to the temperature of the gas? **(2 marks)**

2. How does the kinetic theory explain the fact that gases exert a pressure on the container? **(2 marks)**

3. How does kinetic theory explain the idea of absolute zero? **(2 marks)**

4. **a)** Convert these temperatures from °C to kelvin.

   i) 20 °C **ii)** 150 °C **iii)** 1000 °C **(3 marks)**

   b) Convert these temperatures from kelvin to °C.

   i) 300 K **ii)** 650 K **iii)** 1000 K **(3 marks)**

5. A student blows up a balloon. At room temperature, 20 °C, she measures the volume of the balloon as 1500 cm^3. Then she puts the balloon in a freezer where the temperature is −13 °C. Assuming the pressure stays constant, work out the new volume of the balloon. **(3 marks)**

6. A sample of gas is sealed in a 20 cm^3 metal container at a pressure of 1×10^5 Pa and a temperature of 37 °C. Calculate the new pressure of the gas if the metal container is slowly crushed to a volume of 5 cm^3 with the same temperature. **(3 marks)**

7. A car tyre has a pressure of 5×10^5 Pa at a temperature of 17 °C. The car is then driven for some distance and the temperature of the air in the tyre rises to 52 °C. Calculate the new pressure in the tyre, assuming the volume of the tyre stays the same. **(3 marks)**

8. A gas cylinder contains 5×10^{-3} m^3 of gas at a pressure of 1.5×10^6 Pa. What volume of gas would be released at a pressure of 1×10^5 Pa? **(3 marks)**

9. A gas cylinder contains 9×10^{-3} m^3 at a pressure 2.5×10^6 Pa and a temperature of 17 °C. The cylinder is fitted with a pressure safety valve which opens at 5×10^6 Pa. Calculate the highest temperature to which gas in the cylinder can be raised before the safety valve opens. **(3 marks)**

End of topic questions continued

10. A cylinder with a volume of $0.17\ m^3$ containing 20 kg of compressed air is stored at a temperature of 7 °C. The pressure of the gas in the cylinder is 9.5×10^5 Pa.

a) The compressed air is used in a process in which the air at a temperature of 7 °C has to be fed in at a pressure of 150 kPa and at a rate of $1000\ cm^3$ per minute.

i) What volume would the gas occupy at a pressure of 150 kPa at 7 °C? **(2 marks)**

ii) What volume of air can one cylinder supply to the process? **(2 marks)**

iii) How long can one cylinder supply the process for? **(2 marks)**

b) The safety valve on the cylinder activates when the pressure inside the cylinder is greater than 1×10^6 Pa. What is the greatest temperature at which the cylinders can be stored without the safety valve being activated? **(3 marks)**

Exam-style questions

Sample student answers

Question 1

A bicycle tyre contains air at a constant volume.

a) The air exerts a pressure on the walls of the tyre. Use the kinetic theory to explain how the pressure is exerted. **(2)**

The particles move around quickly and hit each other causing pressure. ✓ (1)

b) What would happen to the pressure in the tyre on a **hot** day? Explain why. **(4)**

The pressure would increase because it is hotter. ✓ (1)

c) What would happen to the pressure in the tyre on a **cold** day? **(1)**

The pressure would decrease because it is colder. ✓ (1)

d) Why do recommended tyre pressures depend on the weight of the rider? **(2)**

Adult cyclists need more pressure. ✗

e) Why is it harder to go fast with under-inflated tyres? **(2)**

Flat tyres slow you down. ✗

EXAMINER'S COMMENTS

a) The student does not say how pressure is exerted. Also, the answer does not mention the walls of the tyre. A better answer would have been that the particles move around quickly and hit the inside walls of the tyre (1 mark). These collisions exert forces on the inside of the tyre which push out the walls. (1 mark)

b) Yes, the pressure would increase but no explanation is given. The answer would have gained more marks if it has also said that on a hot day the particles move faster (1 mark) and collide with the walls of the tyre more often (1 mark) and with more force (1 mark)

c) Correct answer (1 mark)

d) The answer is unspecific. Some adults are lighter than some children. The volume of the tyre remains the same so the area in contact with the road will largely remain the same (1 mark). More pressure is therefore needed to support the extra weight (1 mark).

e) The answer contains no science at all and is just a rewording of the question. A better answer would have referred to an underinflated tyre will continuously change shape when revolving (1 mark) which requires extra energy (1 mark).

Exam-style questions continued

Question 2

Two students want to find the density of clay.

They each have a sample of clay.

a) State the equation linking density, mass and volume. .... **(1)**

The students suggest different methods to find the volume of their sample of clay.

b) The first student shapes his sample of clay into a regular cube shape.

Then he measures the length of the sides.

He finds the volume by doing a calculation with his measurements.

i) How should the student choose his equipment to make his measurement of volume as precise as possible? **(2)**

ii) Describe a feature of this method that may lead to inaccurate results. **(1)**

c) The second student decides to find the volume of her sample of clay using a measuring cylinder.

Describe how she should do this. **(4)**

d) i) Describe how the students can use an electronic balance to find the mass of their sample of clay. **(1)**

ii) If the electronic balance is incorrectly calibrated, how will this affect their measurements? **(1)**

iii) How could the students check the calibration of the electronic balance? **(2)**

(Total 12 marks)

Question 3

a) A student investigates the specific heat capacity of four materials: A, B, C and D.

She heats the same mass of different materials and measures their temperatures.

Look at the graph of her results.

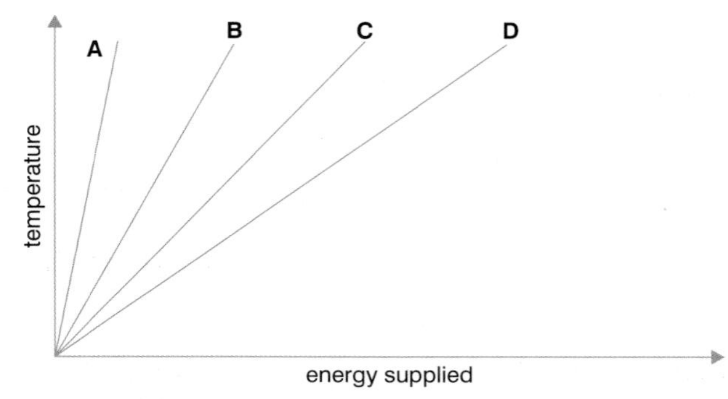

Exam-style questions continued

i) Which material needs **most** energy to raise its temperature by 1°C?

A **B** **C** **D** **(1)**

ii) Which material needs **least** energy to raise its temperature by 1°C?

A **B** **C** **D** **(1)**

iii) Material D has the highest specific heat capacity. Explain how you can tell this from the graph. **(2)**

b) A home owner has two radiators in her home. They are filled with 5 kg of different liquids and heated electrically.

Look at the diagrams.

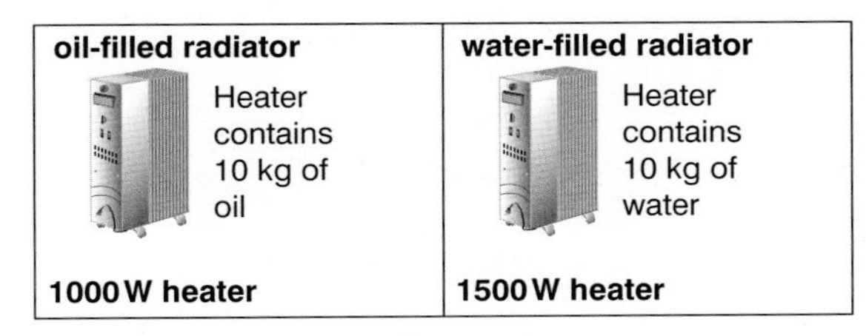

Look at the information about oil and water.

Material	SHC (J/kg °C)	Freezing point (°C)	Boiling point (°C)
Oil	1700	–24	250
Water	4200	0	100

The house has a glass conservatory which can get can get very cold in the winter. Sometimes it can get to –6°C.

i) The homeowner thinks that the oil radiator may be better for the conservatory. Use the data to suggest why. **(1)**

ii) Both radiators have a 'cut-out' which prevents them getting hotter than 60°C.

Suggest a reason why. **(1)**

c) The homeowner does a calculation. She knows that the oil heater produces 1000 J of energy each second.

i) Calculate the energy transferred by the oil heater in 10 minutes. **(1)**

ii) She calculates that 10 kg of oil should change temperature by 40°C when the heater supplies 680 000 J of energy. Complete a calculation which shows this. **(1)**

iii) She plugs in the heater. The mains electricity supplies 680 000 J of energy to the radiator. It only heats up from 20°C to 52°C.

Calculate the temperature rise and explain why the radiator does not reach 60°C. **(4)**

You will probably have experienced magnetism. There is a magnetic field around the Earth which, when it is plotted, looks very similar to that around a bar magnet. However, it is the existence of electromagnetism that really makes a difference to your life. Without electromagnetism, the generation of electricity would not happen in the way that it does. Without electromagnetism, the high voltages transmitted down power lines would not be able to be transformed into the lower voltages that you need in your home.

STARTING POINTS

1. How can electromagnets be made stronger?
2. Describe some similarities and differences between magnets and electromagnets.
3. Describe the form of the electromagnetic field around the Earth.
4. What is the difference between a magnetically 'hard' and a magnetically 'soft' material?
5. How could you investigate the magnetic field pattern for a) a permanent bar magnet and b) the field between two bar magnets?
6. What is a solenoid?

CONTENTS

6 Magnetism and electromagnetism

Δ Electromagnetic effects are used in many devices.

Units

For the topics included in magnetism and electromagnetism, you will need to be familiar with:

Quantity	Unit	Symbol
electric current	ampere (amp)	A
potential difference (p.d.)	volt	V
power	watt	W

LEARNING OBJECTIVES

✓ Be able to use the following units: ampere (A), volt (V) and watt (W).

Magnetism

INTRODUCTION

The property of magnetism has been known for many centuries. Ancient travellers used naturally magnetic rocks such as lodestone, which point towards the north when suspended, to guide their journeys. The development of the compass made it possible to make long sea voyages where previously ships had stayed within sight of land. There is evidence that some animals, such as birds, can sense the magnetic field of the Earth and that they have their own 'in-built compass' which may help navigation during long migrations.

Δ Fig. 6.1 A compass on a boat.

The magnetic field of the Earth has a wider importance. It acts as a shield, protecting the surface of the Earth from many charged particles emitted from the Sun.

A study of magnetism helps us to understand our Earth better and will be crucial when we start to explore electromagnetism later in this section.

KNOWLEDGE CHECK

- ✓ Know that the ends of a bar magnet are called poles and that is where the magnetism is strongest.
- ✓ Know that only some materials are magnetic.
- ✓ Know that the Earth has a magnetic field, which is how compasses are able to point to the North.

LEARNING OBJECTIVES

- ✓ Know that magnets repel and attract other magnets and attract magnetic substances.
- ✓ Be able to describe the properties of magnetically hard and soft materials.
- ✓ Be able to investigate the magnetic field pattern for a permanent bar magnet and between two bar magnets.
- ✓ Know that magnetism is induced in some materials when they are placed in a magnetic field.
- ✓ Describe how to use two permanent magnets to produce a uniform magnetic field pattern.

MAGNETS REPEL AND ATTRACT

If a permanent magnet is suspended and allowed to swing, it will line up approximately north–south. Because of this, the two ends of a magnet (which are the most strongly magnetic parts) are called the north pole and the south pole, often labelled N and S. (Strictly speaking, they are called the north-seeking pole and the south-seeking pole.)

If two north poles from different magnets are brought together, there will be a repulsion between them. This also happens if two south poles are used. However, if a north pole and a south pole are brought together, there will be an attraction. Magnets will also attract magnetic substances such as iron, nickel and cobalt.

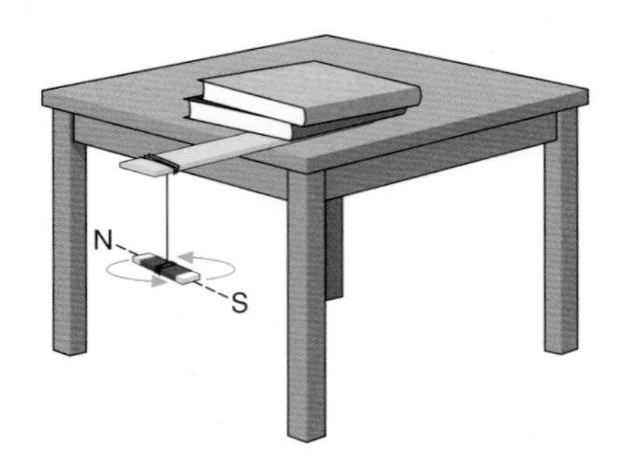

△ Fig. 6.2 A bar magnet aligns itself with the Earth's magnetic field when it is suspended.

QUESTIONS

1. What happens when a north-seeking pole of a magnet is brought close to a south-seeking pole of another magnet?
2. What happens when two like poles are brought close to each other?
3. What happens when a bar magnet is suspended?

MAGNETICALLY HARD AND SOFT MATERIALS

There are several elements that are magnetic, the most important of which are iron, cobalt and nickel. Scientists have developed alloys and ceramics made from combinations of elements to get the exact properties that they want. Some of these materials are **magnetically hard** (such as steel, which is an alloy of iron and other elements such as carbon or tungsten). This means that they stay magnetic once they have been magnetised.

When we refer to a 'magnet', we mean a **permanent magnet** that is made of magnetically hard materials.

Other materials are **magnetically soft** (such as pure iron), which means that they do not stay magnetic – this is particularly useful in devices such as the **electromagnet** and the relay. In these cases, magnetism is only needed under particular circumstances, such as when a switch is closed (see next topic, Electromagnetism).

Alloys are made by melting different metallic elements (iron, aluminium, copper, tungsten, and so on) together. The resulting metal is known as a ferrous metal if it contains significant iron, and as a non-ferrous metal if it does not. Nickel and brass (which is an alloy of copper and tin) are examples of non-ferrous metals.

In the past all magnetic materials were ferrous, but this is no longer true, and the strongest magnets may not contain any iron at all – for example, samarium-cobalt (SmCo), which is used in headphones.

◁ Fig. 6.3 The headphones worn by this radio announcer contain samarium-cobalt (SmCo) magnets.

'Magnetically hard' and 'magnetically soft' materials does not refer to their physical hardness but to their magnetic behaviour. You may have seen rubberised magnetic strips used on notice boards. These strips are permanent magnets, but are physically soft. Microphones like the one shown in Figure 6.3, also contain magnets.

QUESTIONS

1. Describe the difference between magnetically hard and magnetically soft materials.
2. What do we mean when we use the term *magnet*?
3. Where are magnetically soft materials particularly useful?

MAGNETIC FIELD LINES

Magnets have a **magnetic field** around them – a region of space where their magnetism affects other objects. Magnetic fields may be described using **magnetic field lines**. These lines show the path that a free north pole would take: heading away from a north pole and ending up at a south pole. The more concentrated the field lines are, the stronger the magnetic effect.

To show the field lines, place a bar magnet under a thin sheet of plastic, and sprinkle iron filings on to the top of the plastic. The iron filings will arrange themselves into strings of filings along the field lines.

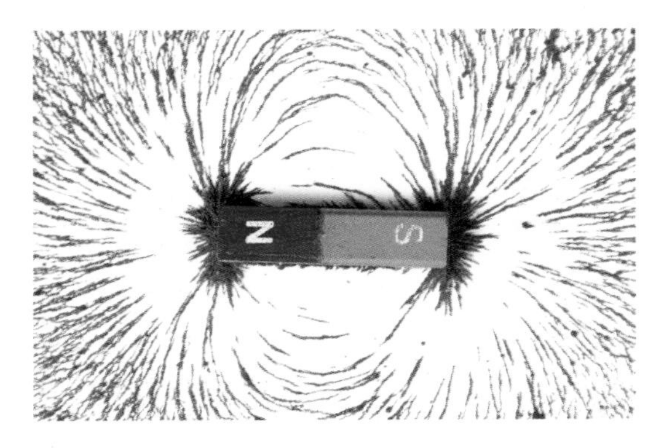

◁ Fig. 6.4 The iron filings show the field lines around a bar magnet.

It is also possible to follow the path of the field lines by placing a small compass (known as a plotting compass) on the plastic in place of the iron filings. If you move the compass in the direction that its north pole is pointing, then it will follow a field line.

THE EARTH'S MAGNETIC FIELD

The Earth has a magnetic field. This fact immediately gives us some evidence about the composition of the Earth itself. There must be magnetic materials – large amounts of them – within the Earth to generate this field. Based on this and other evidence such as the estimated mass of the Earth, geologists believe that the core of the Earth is mostly iron. Also, at least some of this core must be mobile. In this way, the dynamo theory (which is a theory which attempts to explain how the Earth generates its magnetic field) suggests that a magnetic field that extends thousands of miles out into space can be maintained. So, the core must contain lots of iron and at least some of it must be mobile – it is a liquid.

Look at the shape of the magnetic field. By plotting it out as you would use a plotting compass to investigate a bar magnet in the classroom, a familiar pattern emerges.

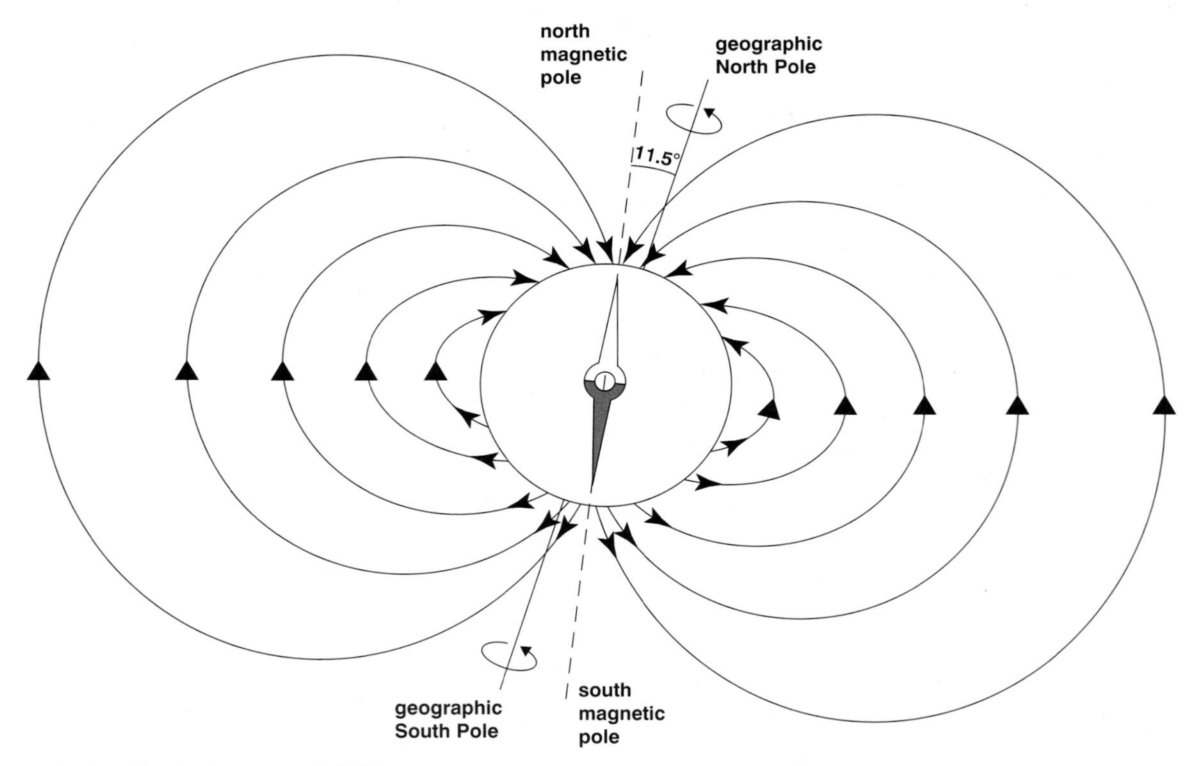

Δ Fig. 6.5 The Earth's magnetic field.

This looks very much like the pattern from a bar magnet, so it can be helpful to think of the Earth as having a giant bar magnet inside it – although of course it doesn't.

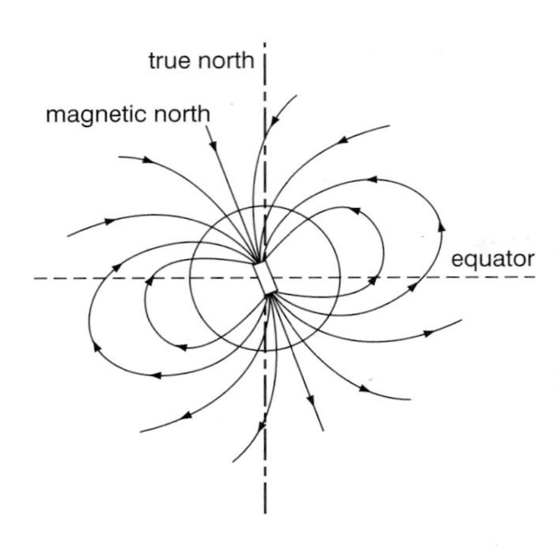

◁ Fig. 6.6 The Earth as a bar magnet.

Igneous rocks contain a relatively high proportion of iron. When liquid magma cools and hardens, the iron particles in igneous rocks 'line up' in the magnetic field of the Earth, just like iron filings, and the rocks become magnetised in the direction of the Earth's magnetic field. When physicists measure the magnetic field over rocks of the ocean floor – particularly in the mid-Atlantic – they find that alternating bands of rock are magnetised in opposite directions. This is evidence that the Earth's magnetic field has changed over time. In fact, the Earth's magnetic field has completely reversed its direction many times. It also appears that the field is overdue for another change. How that will happen is unknown at this time. It may be a cause for concern, however, since the Earth's magnetic field extends out into space. Many charged particles are emitted from the Sun and are deflected by the magnetic field (an example of the motor effect), leading to effects such as the aurora. An aurora is a natural light display in the sky, which is found particularly in the Arctic and Antarctic regions, and is caused by the collision of energetic charged particles with atoms high in the atmosphere. The charged particles, which come originally from the magnetosphere and solar wind are directed by the Earth's magnetic field into the atmosphere.

Δ Fig. 6.7 An aurora.

If the magnetic field weakens, then many more charged particles will strike the Earth, with possibly damaging effects on living tissue.

QUESTIONS

1. What is a magnetic field?
2. What do magnetic field lines show?
3. Describe two methods to how to show magnetic field lines for a bar magnet.

MAGNETIC INDUCTION

If a soft magnetic material is brought near to a magnet it will be attracted. It has had magnetism **induced** in it; it has become **magnetised**. When the magnet is taken away, the material loses its magnetism again. Note that the magnet will continue to attract the soft magnetic material even if the material is turned round. This is the opposite behaviour to two magnets, as two magnets will repel each other in certain orientations. This simple method enables you to work out whether you are holding two magnets or one magnet and one piece of soft magnetic material.

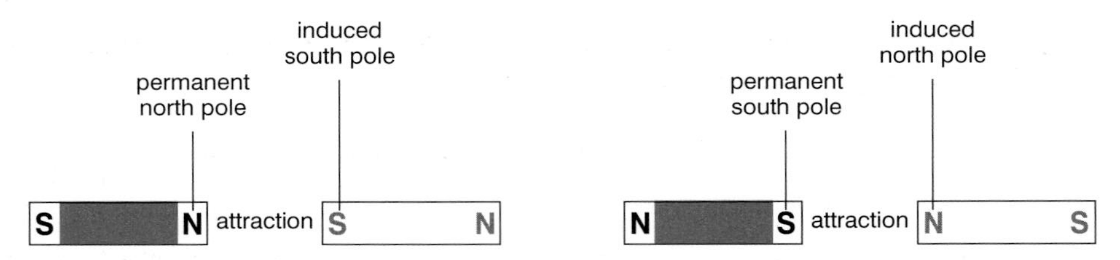

Δ Fig. 6.8 The pole of a permanent magnet always induces the opposite pole in an unmagnetised piece of magnetic material.

QUESTIONS

1. Describe what happens when a soft magnet is brought near to a permanent magnet.
2. If a soft magnet is brought near to a permanent north pole, what pole is induced in the end of the soft magnet nearest the north pole?
3. To see if a piece of metal is a magnet, you should check to see if it repels a known magnet. Explain why attracting a known magnet is not sufficient.
4. In which orientations will two magnets repel each other?

MAGNETIC FIELD PATTERNS

The idea of field lines was first developed in the 19th century by Michael Faraday, the inventor of the electric motor. Note that where the magnets are repelling each other, the field lines do not go from one magnet to the other. Where the magnets are attracting each other, lines do cross from one to the other. This fact will help you to draw the lines more easily. You can investigate the field patterns using the techniques described on page 297.

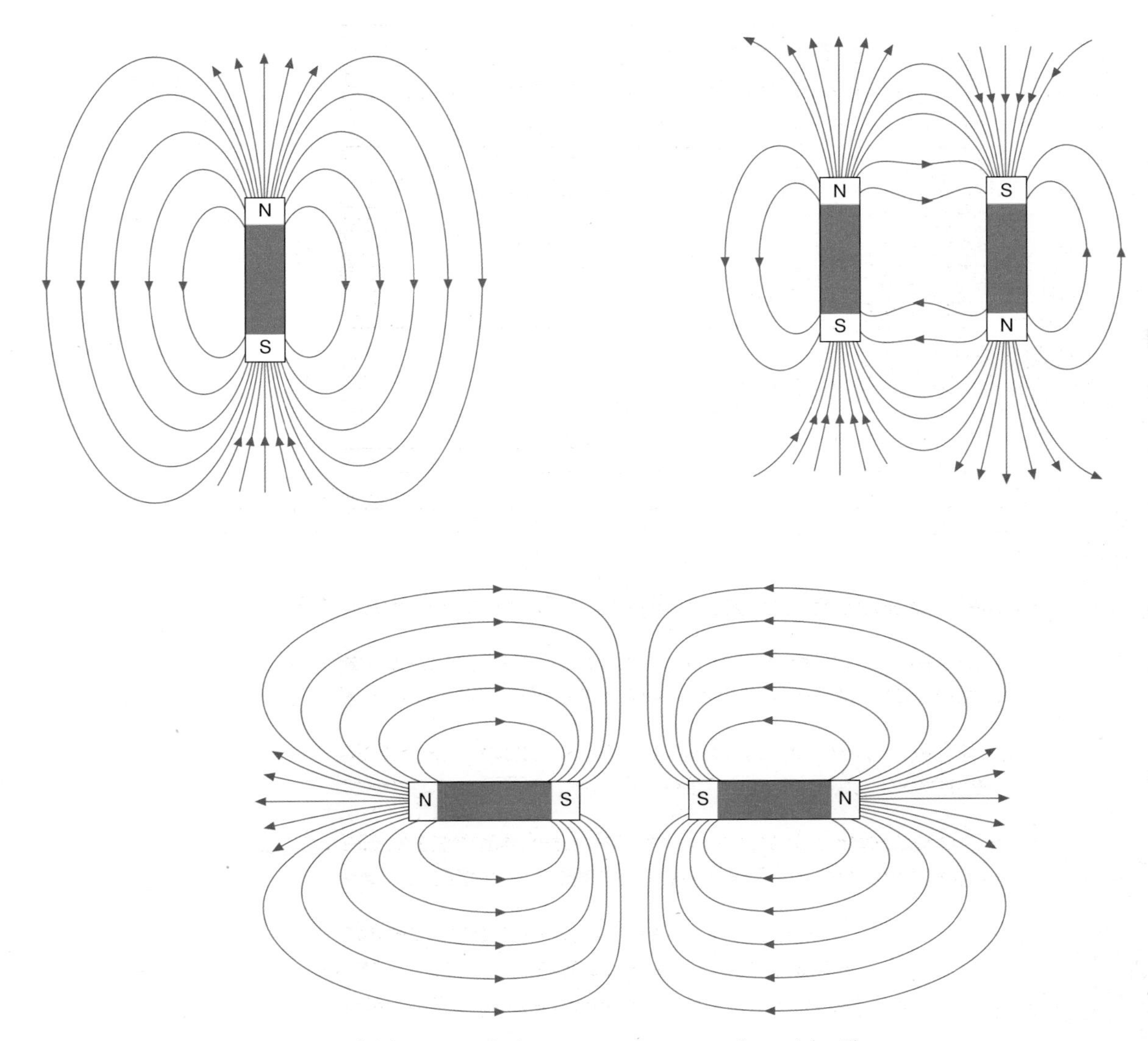

Δ Fig. 6.9 Top left: The magnetic field pattern of a bar magnet. Bottom and top right: The magnetic field pattern between two bar magnets.

MICHAEL FARADAY

Michael Faraday, who named magnetic field lines 'lines of force', was the son of a blacksmith. He was a bookbinder's apprentice who often read the books that were brought in for rebinding. One of these was a copy of the section of the *Encyclopedia Britannica* which included an article about electricity. This interested him so much that he attended lectures given by Humphry Davy, who was the leading chemist of his time in the UK. Faraday kept notes of the lectures and, on the strength of these, when Davy needed an assistant, he gave the job to Faraday. This was the start of Faraday's highly successful career in physics and chemistry. In 1831, while he was working on electromagnetism, he wrote, 'I am busy just now again on Electro-Magnetism and I think I have got hold of a good thing but can't say; it may be a weed instead of a fish that after all my labour I may at last pull up.'

Faraday's discoveries led him to producing a metal cage which blocks out electric fields. It is now known as the Faraday cage, and is used to shield equipment from electric fields such as lightning.

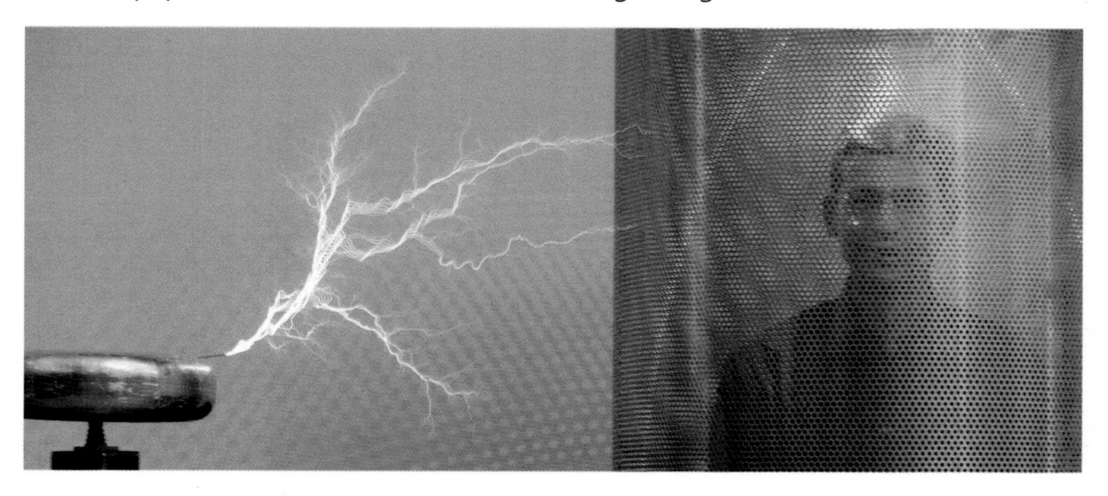

△ Fig. 6.10 A demonstration of how a Faraday cage can shield a person from electric discharges. The person is standing inside a metal cage.

QUESTIONS

1. Explain the magnetic field pattern for a single bar magnet.
2. Why are there field lines between the north and south poles of two magnets that are side by side?
3. Why are there no lines between two like poles?

Uniform magnetic fields

If two bar magnets are arranged with the N pole of one very close to the S pole of the other (Figure 6.11), then some of the field lines will travel straight from the one pole to the other. This will give an approximately uniform magnetic field in the gap between the two poles.

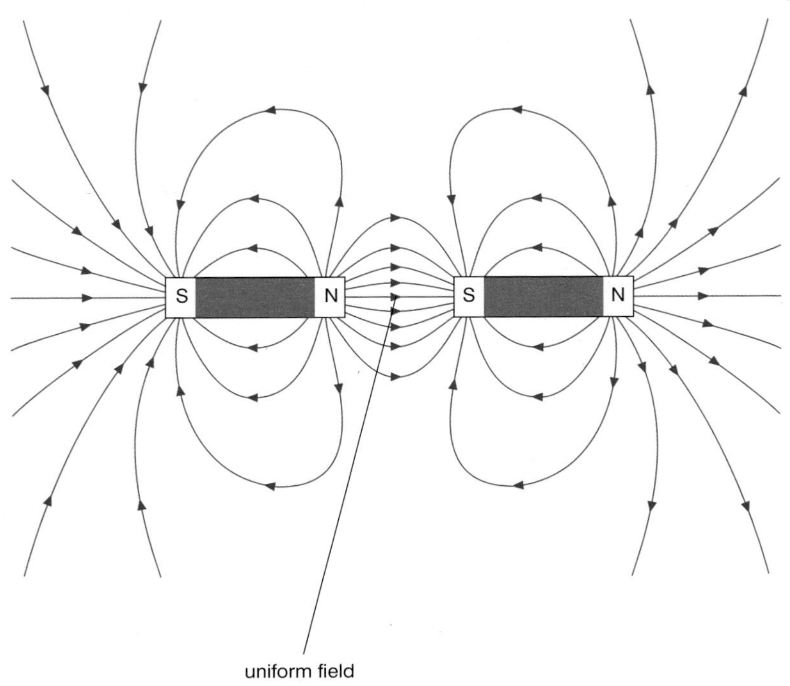

△ Fig. 6.11 Midway between the poles the field is uniform.

You can get a more uniform field by using a single bar magnet that has been bent round until its poles are close together as shown in Figure 6.12. A larger region of uniform field can be produced by using slab magnets, as can be seen in Figure 6.33.

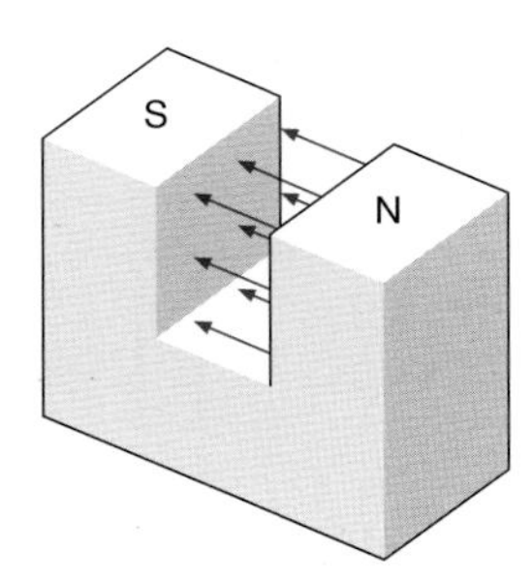

◁ Fig. 6.12 A single bar magnet bent with its poles close together.

QUESTIONS

1. Magnetic field lines never cross. Explain why not.
2. What does 'uniform' mean, as in 'uniform field'?
3. Where will you find a uniform magnetic field?

End of topic checklist

Magnets can attract or repel – like poles **repel**, unlike poles **attract**.

Magnetic materials are materials that are attracted to magnets and can be made into magnets. Iron, cobalt, and nickel are magnetic materials.

Some materials retain their magnetism (**magnetically hard**) while others lose it (**magnetically soft**).

Magnetic field lines show the path a free north pole would follow.

A **permanent magnet** is made of a magnetically hard material.

To **induce** is to affect something without touching it. An electric force can induce charge in a conductor. A changing magnetic force can induce electric current in a wire.

The facts and ideas that you should understand by studying this topic:

- ❍ Describe experiments to investigate the magnetic field around bar magnets and combinations of bar magnets, using iron filings and/or plotting compasses.
- ❍ Describe how two permanent magnets can produce a uniform magnetic field.
- ❍ Understand that like poles repel and unlike poles attract.
- ❍ Understand that testing for repulsion is the only way to test to see if an object is a magnet.
- ❍ Understand the term 'magnetic field line' and use it to describe magnetic field patterns.
- ❍ Understand the terms 'magnetically hard' and 'magnetically soft'.

End of topic questions

1. What is the difference between a magnetically hard material and a magnetically soft material? Give an example of each. **(4 marks)**
2. Ranjit has a piece of metal that he thinks is a magnet. He holds it near another magnet and it is attracted. Ranjit says this proves his metal is a magnet. Explain why Ranjit is wrong. **(3 marks)**
3. Sketch the magnetic field pattern for a single bar magnet. How would the diagram change if the magnet were made stronger? **(2 marks)**
4. You are given two bar magnets. Describe how you would investigate the field pattern between them. **(4 marks)**
5. Describe how to produce a uniform magnetic field pattern using two permanent magnets. **(4 marks)**

Δ Fig.6.13 Part of an electric motor.

Electromagnetism

INTRODUCTION

Almost anything you see where electricity causes something to move – such as electric drills, headphones, CD players, hairdryers, loudspeakers – must have a magnet in it somewhere to work. When magnetism is combined with electricity, a force is created. It is this force, which we call the motor effect, which enables all these devices to work. In this section you will learn how electromagnets are constructed, and how to increase the size of the motor effect.

KNOWLEDGE CHECK

✓ Know how an electric current can be produced in a wire.
✓ Be able to describe the factors that control the size of a current in a wire and the heating effect that can be produced.
✓ Know that electric currents in wires are due to the motion of electrons inside the wire.

LEARNING OBJECTIVES

✓ Know that an electric current in a conductor produces a magnetic field around it.
✓ Be able to describe how to make an electromagnet.
✓ Be able to draw the magnetic field patterns for a straight wire, a flat circular coil and a solenoid when each is carrying a current.
✓ Know that there is a force on a charged particle when it moves in a magnetic field as long as its motion is not parallel to the field.
✓ Understand why a force is exerted on a current-carrying wire in a magnetic field, and how this effect is applied in simple d.c. electric motors and loudspeakers.
✓ Use the left-hand rule to predict the direction of the resulting force when a wire carries a current perpendicular to a magnetic field.
✓ Describe how the force on a current-carrying conductor in a magnetic field changes with the magnitude and direction of the field and current.

ELECTROMAGNETS

An electric current in a conductor produces a magnetic field around it (Figures 6.14 and 6.15). **Electromagnets** are made out of a coil of wire connected to a source of electrical energy. When the current is switched off, the coil does not produce a magnetic field. However, when an electric current is passed through the coil, as shown in Figure 6.16, a magnet is formed with the north pole at one end of the coil and the south pole at the other end. The coil is known as a **solenoid**.

Δ Fig. 6.14 There is no current through the wire, and the compass points in the direction of the Earth's magnetic field.

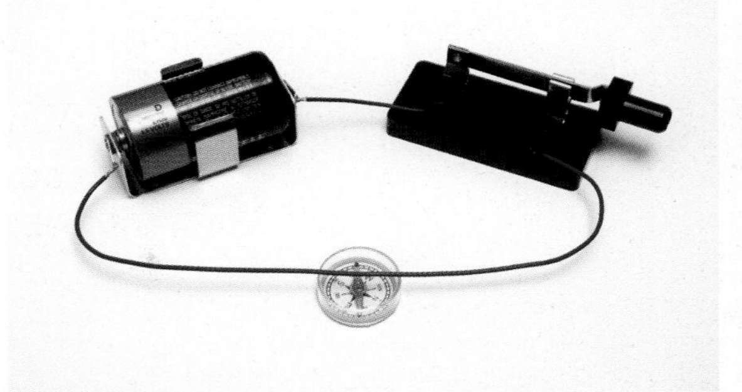

Δ Fig. 6.15 The electric current through the wire has produced a magnetic field around the wire. The compass has lined itself up to the magnetic field of the current-carrying wire.

If the coil is wrapped around a magnetically soft core as shown in Figure 6.16, when the coil is magnetised, it magnetises the core as well. This makes the magnetic field much stronger. When the current is switched off, the coil loses its magnetism, so the core does as well.

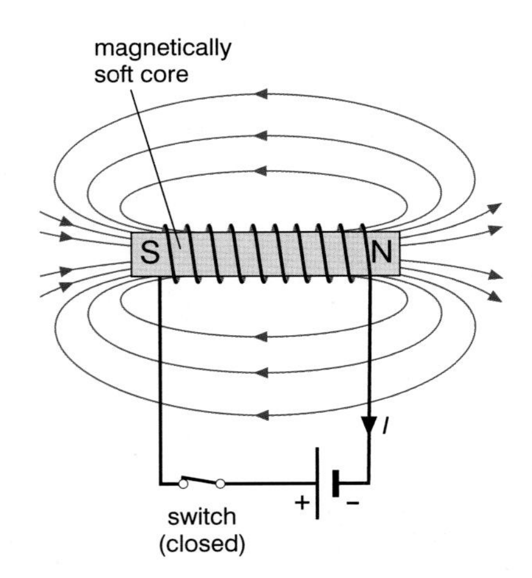

◁ Fig. 6.16 A solenoid produces a stronger magnetic field if it is wrapped round a soft iron core.

QUESTIONS

1. Describe how to produce a magnetic field using a battery, a switch and a coil of wire.
2. How can an electromagnet be made stronger?
3. Draw the magnetic field pattern for a solenoid.

Developing investigative skills

A student wants to investigate the factors that affect the strength of an electromagnet. He makes the electromagnet by winding a coil of wire around an iron nail. He then holds the electromagnet vertically in a clamp attached to a clamp stand.
He uses a battery to provide the current for the electromagnet, and connects it for a few seconds.

The student decides to investigate the effect of changing the current in the coil. To measure the strength of the electromagnet he finds out how many paper clips he can hang from the end of the electromagnet. His measurements are shown in the table.

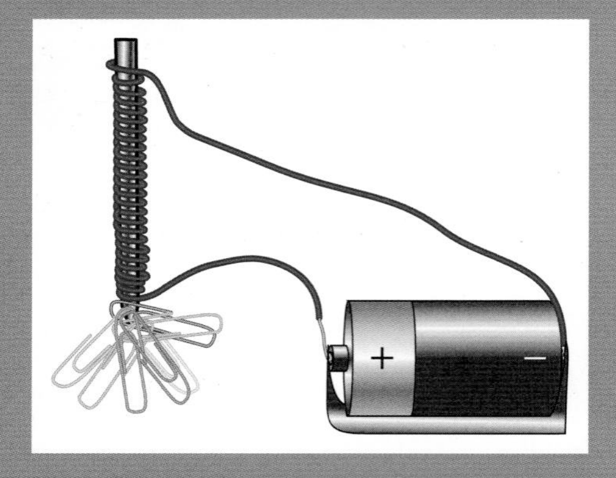

Δ Fig. 6.17 When the circuit is complete, paper clips hang off the nail.

Current (A)	Number of paper clips held
0	0
0.3	2
0.5	5
0.7	6
0.9	9
1.0	9

Devise and plan investigations

❶ Describe how the student can vary the current in the electromagnet coil.

❷ What factors should the student keep constant during the investigation? Why do these factors need to be controlled?

Analyse and interpret

❸ Draw a graph of the student's results.

❹ Describe the pattern (if any) shown by the graph.

Evaluate data and methods

❺ The student thought of three different ways to hang paper clips on the end of the nail.

a) all the paper clips hanging on the nail

b) the paper clips hanging in a line from the end of the nail with the paper clips interlocked

c) the paper clips hanging in a line from the end of the nail, but just held magnetically not joined together.

Describe advantages and disadvantages of each method.

❻ How could the student change their method so that they could achieve more precise measurements of the strength of the electromagnet?

❼ The student wants to continue making measurements with higher values of current. Suggest a difficulty the student will have as the current increases further.

FIELD PATTERNS FOR ELECTROMAGNETS

If a wire is carrying electric current, it generates a magnetic field around itself. The higher the current, the stronger the field. Some people believe that this field is a health hazard, particularly around high voltage transmission lines, but research into the topic has been unable to demonstrate any risk so far.

If the current is travelling along a long straight wire the field is as shown in Figure 6.18.

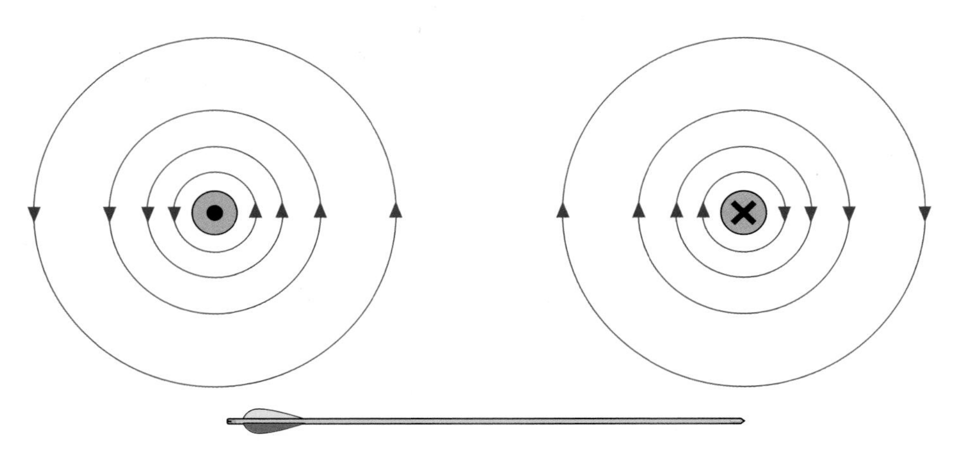

△ Fig. 6.18 The field pattern for a straight current-carrying conductor.

The dot in the centre of the wire indicates that the current is travelling directly towards you; the 'x' indicates that the current is travelling directly away. To remember this, think of an arrow. The dot is the tip of the arrow coming towards you, the 'x' is the flights on the tail of the arrow.

The field lines form continuous rings around the wire all along its length. The lines are shown closest together near to the wire, because the field is strongest there, and gets smaller further away from the wire.

If the current is travelling towards you, the magnetic field lines are going in an anticlockwise direction, and if away from you they are going clockwise. To remember this, think of a woodscrew or a corkscrew. In both cases, if the screw is travelling away from you it is going clockwise.

A note about the direction of the current

Electric current in the wires of a circuit is a flow of electrons around the circuit from the negative to the positive terminal of a battery. Unfortunately, early scientists guessed the direction of flow incorrectly. Consequently all diagrams were drawn showing the current direction from positive to negative. This way of showing the current has not changed, so the 'conventional current' that everyone uses gives the direction in which positive charges would flow.

Figure 6.19 shows the magnetic field pattern for a flat circular coil. Figure 6.20 shows the field pattern for a solenoid, which is a long coil of wire.

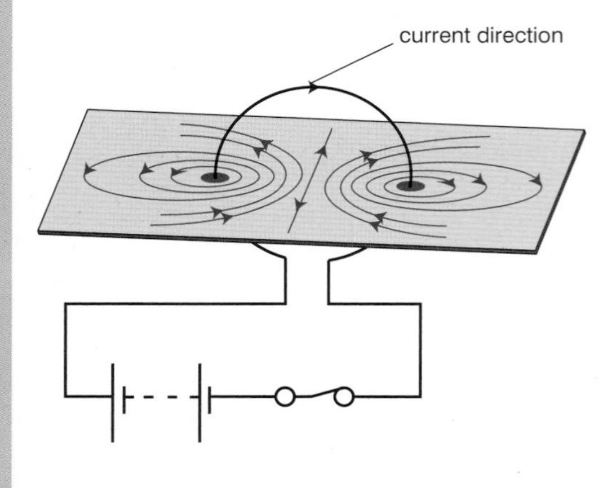

◁ Fig. 6.19 Magnetic field of a flat coil.

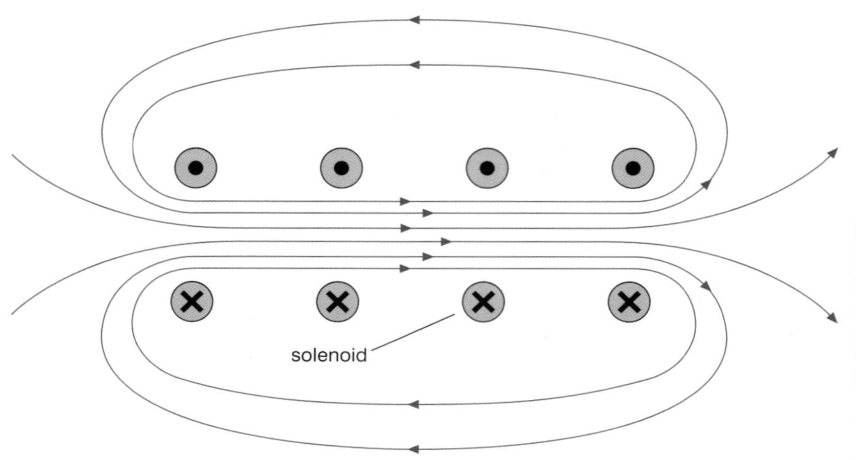

Δ Fig. 6.20 Magnetic field of a solenoid.

The magnetic field inside the solenoid is remarkably uniform, and this is used in MRI scanners to allow doctors to produce images of the inside of the body, like the one shown in Figure 6.21. To do this the whole body has to be placed inside the solenoid.

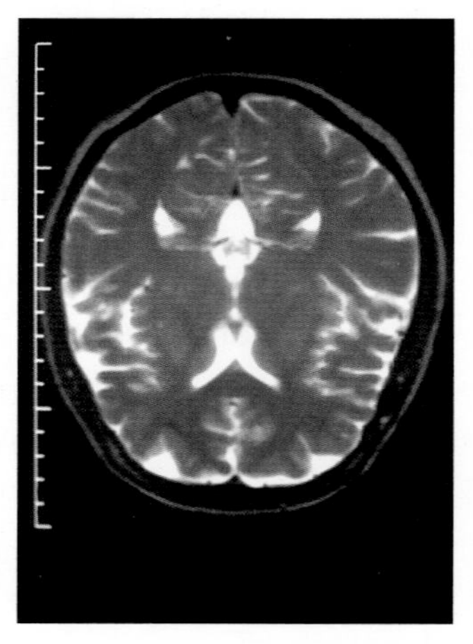

Δ Fig. 6.21 An MRI scan of the human brain.

QUESTIONS

1. Sketch the magnetic field produced by a solenoid. Describe how it is similar to the field of a bar magnet.
2. Describe how a coil can produce a uniform magnetic field. Which part of the field is uniform?
3. Sketch the field lines around a single wire where the current is a) into page b) out of page.

FORCE ON A CHARGED PARTICLE IN A MAGNETIC FIELD

If a charged particle, usually an electron, is stationary in a magnetic field, it does not experience any force from the field. So if a copper wire that is full of electrons is placed near a magnet, nothing happens. If you move the wire parallel to the magnetic field, nothing happens. However, if the charged particle starts to move through the field at an angle to the field, then it will experience a sideways force that will try to push it off its path.

FORCE ON A CURRENT-CARRYING WIRE

If a wire that is carrying an electric current is put in a magnetic field, then the moving electrons will be pushed sideways, and the whole wire will experience a sideways force. This called the motor effect. This effect is used to drive loudspeakers, and to make electric motors work.

In Figure 6.22a, a length of wire has been connected across a power supply and coiled around one pole of a permanent magnet. A magnetic field is induced in the coil when current is passed through the wire. The magnetic field of the permanent magnet repels that of the coil, causing the wire to jump off the magnet. You can see the wire jumping off the magnet in Figure 6.22b.

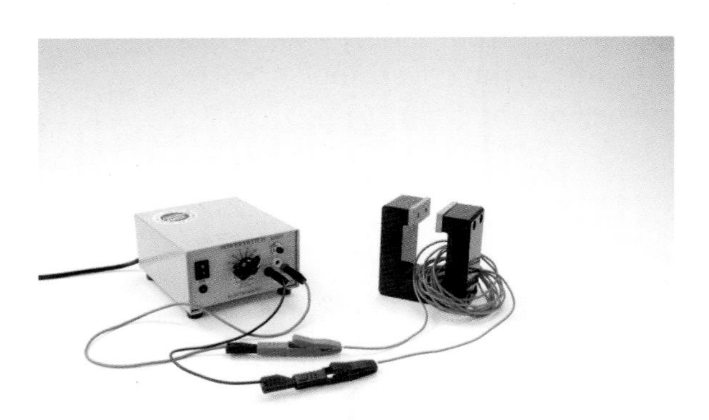

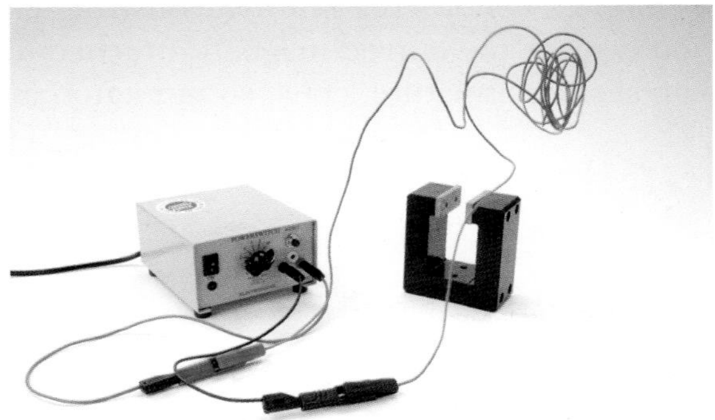

Δ Fig. 6.22 a) Coil around pole of permanent magnet. b) Coil jumps off pole of permanent magnet.

The motor

An electric motor is made from a coil of wire positioned between the poles of two permanent magnets (Figure 6.23). When a current passes through the coil of wire, it creates a magnetic field, which interacts with the magnetic field produced by the two permanent magnets. The two fields exert a force that makes the coil of wire move.

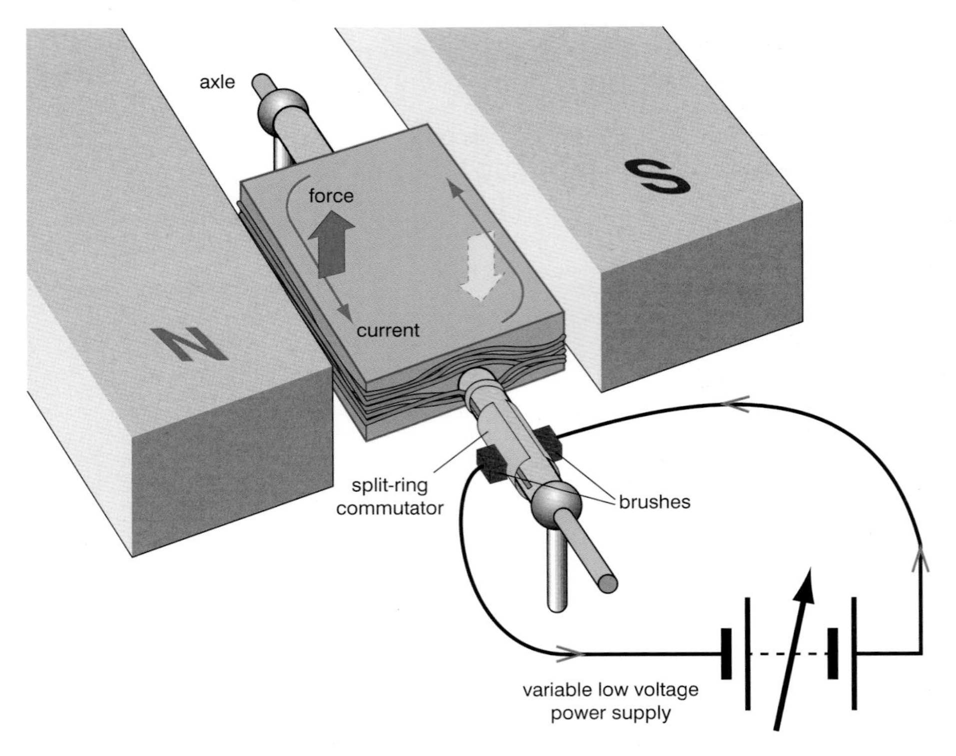

Δ Fig. 6.23 An electric motor.

Figure 6.24 shows the internal construction of a direct current motor. You can see the curved magnets attached to the inside of the cover. You can see the armature, coils, commutator and brushes on the right.

▷ Fig. 6.24 Inside a d.c. motor.

The loudspeaker

In a loudspeaker, the coil is attached to a paper cone. The changing current in the coil produces a changing magnetic field which interacts with the field from the permanent magnet. This creates a 'backwards and forwards' motion of the coil and paper cone. This makes the air vibrate – a sound wave.

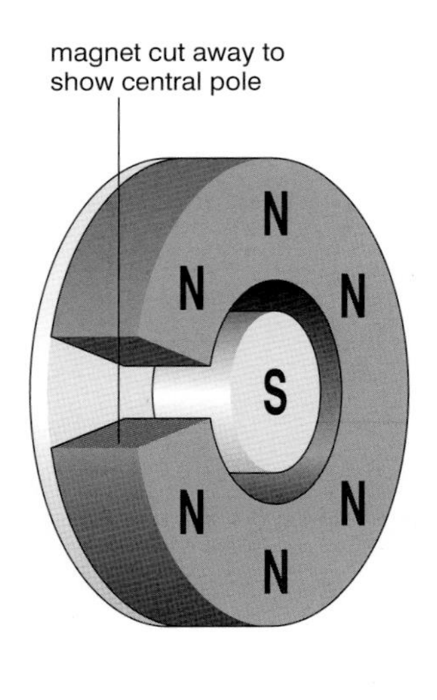

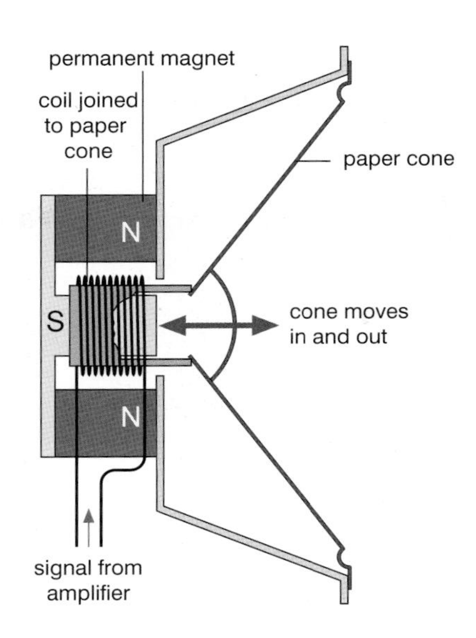

Δ Fig. 6.25 A loudspeaker.

QUESTIONS

1. Describe simply how a force is created on the coil of a motor.
2. Describe an experiment to demonstrate the motor effect.
3. Explain the basic operation of a loudspeaker.

PREDICTING THE DIRECTION OF A FORCE

If a wire carrying an electric current passes through a magnetic field, with the field at right angles to the wire, then the wire will experience a sideways force at right angles both to the wire and to the magnetic field. The size of the force depends on the magnitude of the current and the strength of the magnetic field. **Fleming's left-hand rule**, which is shown in Figure 6.26, predicts the direction of the force.

Remember that the current direction is that of the conventional current, and that the electrons are travelling the other way.

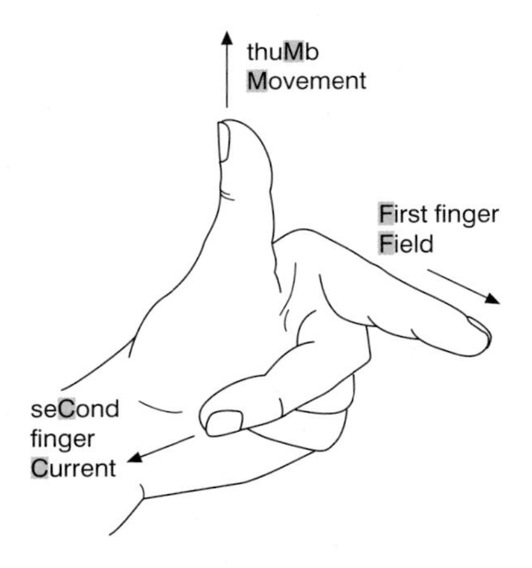

Δ Fig. 6.26 Fleming's left-hand rule predicts the direction of the force on a current-carrying wire.

If you experiment with Fleming's left-hand rule, you should be able to confirm that if you reverse either the magnetic field or the current then the force will be applied in the opposite direction, but that if you reverse *both* the field *and* the current then the force stays unchanged.

It is useful to look at the magnetic field lines for a wire in a uniform magnetic field, as shown in Figure 6.27.

Δ Fig. 6.27 How the magnetic field lines are changed by a wire in a uniform magnetic field.

The field lines from the magnet are pushed to one side by direction of the field lines from the wire. If you imagine that the lines are made of stretched elastic, then it is clear why the wire feels a sideways force.

Now if you consider again the electric motor in Figure 6.23, and apply Fleming's left hand rule, you will see that the force is upwards on the left hand side of the coil, but downwards on the right hand side of the coil, as the direction of the current is in the opposite direction. This is indicated by the arrows in parts 1 and 2 in Fig. 6.28.

The split-ring commutator (see Figure 6.23) ensures that the motor continues to spin. Without the commutator, the coil would rotate 90° from the position shown and then stop. This would not make a very useful motor. The commutator reverses the direction of the current (3 in Fig. 6.28) so that the forces on the coil are reversed and continue the rotating motion (4 in Fig. 6.28).

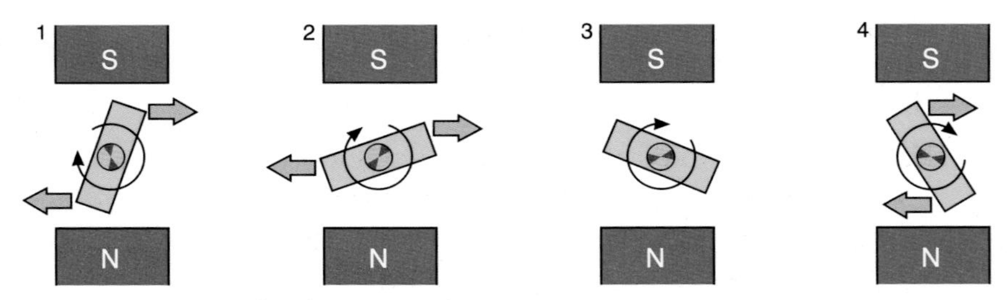

Δ Fig. 6.28 How an electric motor works.

Force increases with field and current

The motor can be used to show how the force on the current-carrying wires varies if you change the conditions. If you increase the current or if you increase the strength of the magnetic field, then the force is larger and the motor spins faster. If you change the direction of the current or if you reverse the magnetic field, the direction of the force is reversed and the motor spins in the opposite direction.

JAMES CLERK MAXWELL

The discovery that electricity and magnetism were linked was made in the 1820s by the Danish physicist Hans Christian Oersted. This discovery was built upon by others such as Michael Faraday, who showed that magnetism could be used to generate electricity – electromagnetic induction. Faraday developed the theory behind this connection, but he could not complete the work fully.

△ Fig. 6.29 James Clerk Maxwell.

The full picture of how magnetism and electricity are linked was developed by a Scottish physicist, James Clerk Maxwell. He showed that magnetism and electricity were simply different aspects of the same effect: electromagnetism. His four linked equations, now known as Maxwell's equations, were first presented to the Royal Society (in London) in 1864 and described the whole theory very elegantly.

Maxwell's theory went further. It predicted that magnetic and electric waves would travel very quickly, leading to the additional insight that light itself was an electromagnetic wave. Maxwell even predicted that there should be other forms of electromagnetic wave, invisible to the eye. His prediction was later found to be correct with the first production of radio waves by Heinrich Hertz. Now, of course, we know about the complete electromagnetic spectrum.

Maxwell's 'discovery' of electromagnetic waves also created a difficulty. His theory said that these waves would always travel at the same speed in a vacuum, no matter how this speed was measured. This did not match the theory of Newton, which said that any speed measurement you made would depend on how you were moving as well. Only when Albert Einstein developed his theory of relativity some 50 years later was the matter resolved – and Maxwell was right.

James Clerk Maxwell is not as widely known as some other physicists, but his achievements, particularly in electromagnetism, stand alongside the best.

QUESTIONS

1. Describe how to use Fleming's left hand rule to predict the direction of a resulting force when a wire carries a current perpendicular to a magnetic field.
2. Explain what the commutator does in a motor.
3. Describe three ways to increase the force created (and so increase the speed at which a motor spins).
4. Describe two ways to reverse the direction in which a motor spins.

End of topic checklist

An electric current in a **conductor** produces a **magnetic field** around it.

An **electromagnet** is a magnet made from a coil of wire. It may also have a core made of a soft magnetic material such as iron, which increases the strength of the magnetic field. The magnetic field is made when electric current passes through the coil.

Fleming's left-hand rule predicts the direction of the force on a wire that is carrying a current when the thumb, and first and second fingers are held at right angles. The first finger points in the direction of the field, the second finger in the direction of the current and the thumb gives the direction of the force (movement).

A **commutator** allows the coils in an electric motor to be connected to the opposite terminals each time the motor rotates through 180°.

The facts and ideas that you should understand by studying this topic:

- ❍ Know that an electric current in a conductor produces a magnetic field around it.
- ❍ Describe how to construct an electromagnet.
- ❍ Describe simple experiments to show the magnetic field pattern around electromagnets.
- ❍ Describe the magnetic field patterns produced by a straight wire, a flat circular coil and a solenoid.
- ❍ Know that a force is exerted on a moving charge in a magnetic field.
- ❍ When the current-carrying conductor is placed in a second magnetic field, a force is exerted on the conductor.
- ❍ In a motor, this effect is used to create forces in opposite directions on either side of a coil, so that a rotation is created and the motor coil spins.
- ❍ Understand how to use Fleming's left-hand rule to predict the direction of the force produced.

End of topic questions

1. Describe how to test to see which end of an electromagnet is a north pole. **(4 marks)**

2. The diagram below shows a simple electromagnet made by a student. Suggest two ways in which the electromagnet can be made to pick up more nails. **(2 marks)**

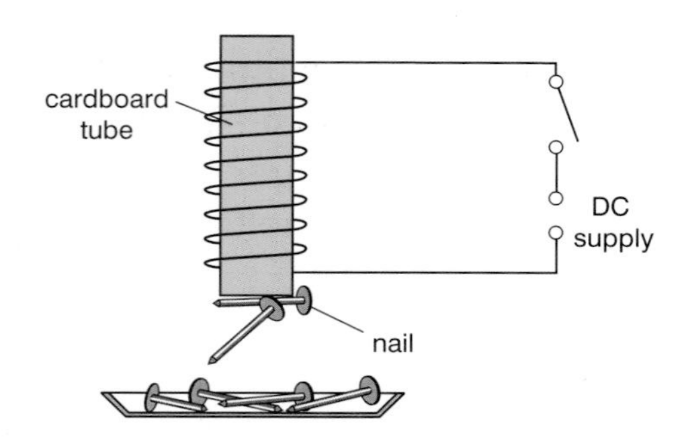

3. The diagram below shows an electric bell. Explain how the bell works when the switch is closed. **(4 marks)**

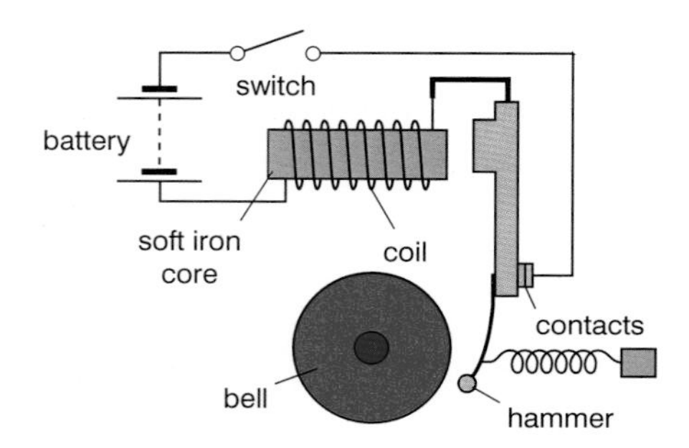

4. Describe two ways to increase the strength of an electromagnet. **(2 mark)**

5. A charged particle moves a) parallel to a magnetic field b) perpendicular to a magnetic field. Describe what happens to the particle in each case. **(2 marks)**

6. Explain why a motor with more turns of wire will turn faster. **(2 marks)**

7. Suggest why commercial motors often use several coils of wire, instead of one large coil. **(2 marks)**

8. Would it be possible to produce a uniform magnetic field between two electromagnets? Give a reason for your answer. **(2 marks)**

9. For the diagram of a simple loudspeaker write a detailed explanation of the way the speaker converts a varying signal into sound. **(6 marks)**

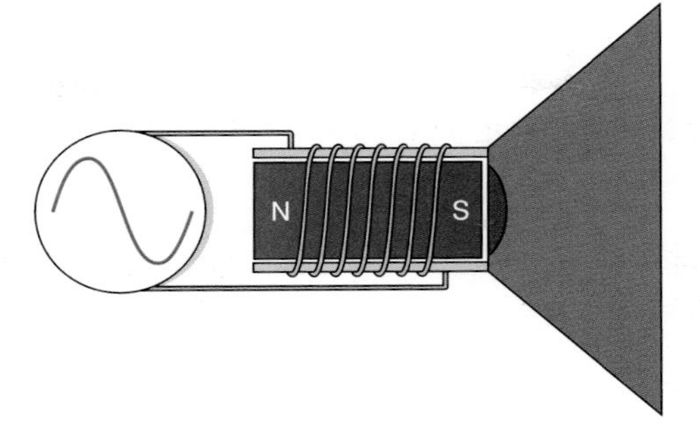

10. The diagrams show two views of a vertical wire carrying a current up through a horizontal card. Points P and Q are marked on the card. Copy both diagrams.

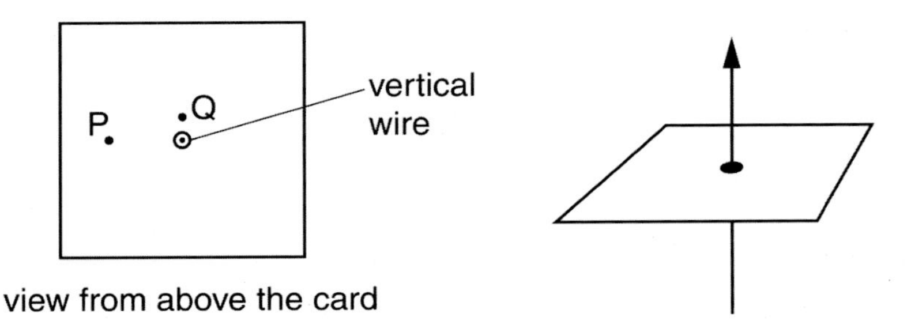

a) On the right hand diagram:

i) draw a complete magnetic field line (line of force) through P and indicate its direction with an arrow

ii) draw an arrow through Q to indicate the direction in which a compass placed at Q would point. **(3 marks)**

b) State the effect on the direction in which compass Q points of:

i) increasing the current in the wire

ii) reversing the direction of the current in the wire. **(2 marks)**

11. The diagram shows the view from above of a vertical wire carrying a current up through a horizontal card. A cm grid is marked on the card.
Point W is 1 cm horizontally away from the wire.

State the magnetic field strength at S, T and W in terms of the magnetic field strength at R. Use one of the alternatives, **weaker, same strength** or **stronger** for each answer. **(3 marks)**

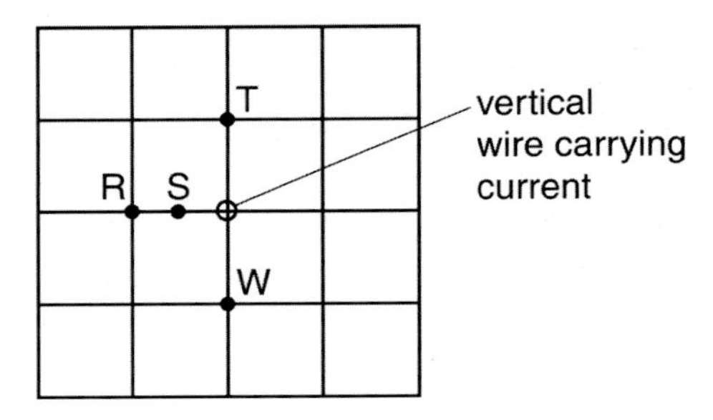

Δ Fig. 6.30 Generators in a hydro electric station.

Electromagnetic induction

INTRODUCTION

Generating electric current from a battery is fine if you want to light a torch, but if you want enough light for a street, or a city, you will need to rely on something else: electromagnetic induction. This is the process that power stations use to generate electricity. Whatever the source of the energy – coal, wind or nuclear sources – the key process in the generation of electricity makes use of electromagnetic induction. (Solar cells, which convert sunlight directly to electricity, are an exception.)

Michael Faraday was the first person to generate electricity from a magnetic field using electromagnetic induction. The large generators in power stations generate the electricity we need using this process.

KNOWLEDGE CHECK

✓ Know that an electric current in a wire is due to the motion of electrons.
✓ Know that mains electricity is supplied through high-voltage cables.
✓ Know that this high voltage is reduced at a local level.

LEARNING OBJECTIVES

✓ Be able to describe how a voltage is induced in a conductor or a coil.
✓ Be able to describe describe the generation of electricity by the rotation of a magnet within a coil of wire and of a coil of wire within a magnetic field, and describe the factors that affect the size of the induced voltage.
✓ Be able to describe the structure of a transformer, and understand that a transformer changes the size of an alternating voltage by having different numbers of turns on the input and output sides.
✓ Explain the use of step-up and step-down transformers in the large-scale generation and transmission of electrical energy.
✓ Know and use the relationship between input (primary) and output (secondary) voltages and the turns ratio for a transformer:
$\frac{\text{input (primary) voltage}}{\text{output (secondary) voltage}} = \frac{\text{primary turns}}{\text{secondary turns}}$.
✓ Know and use the relationship: input power = output power, $V_p I_p = V_s I_s$ for 100% efficiency.

INDUCED CURRENT

A potential difference is induced across the ends of a wire when:

- the wire is moved through a magnetic field ('cutting' the field lines), or
- the magnetic field is moved past the wire (again 'cutting' the field lines), or
- the magnetic field around the wire changes strength.

If the wire is part of a complete circuit, an induced current passes along the wire.

The faster these changes, the larger the induced voltage. So the size of the current depends on the field strength and the speed with which the wire is moved.

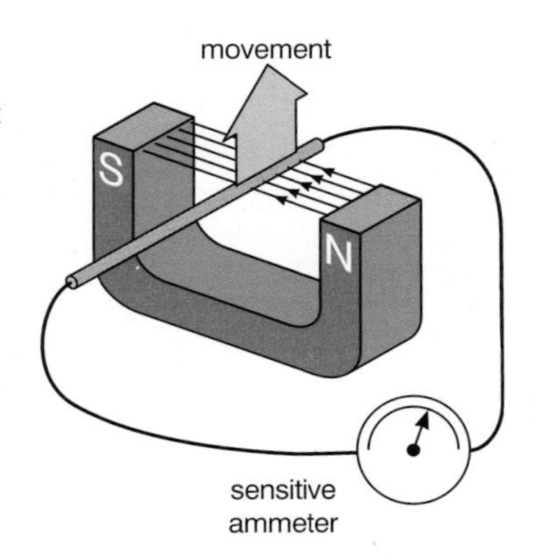

Δ Fig. 6.31 Electromagnetic induction.

QUESTIONS

1. Describe how an electric current can be induced in a wire.
2. If a wire is placed in a magnetic field, will a current be induced? Explain your answer.
3. How can the size of the induced current be changed?

GENERATING ELECTRICITY

In practice, the changes are induced in a coil of wire, because a larger current is created if more turns of wire cut through the magnetic field. In this case the voltage is created by the change in the number of field lines going through the coil as the coil or the magnet rotates. The magnitude of the induced voltage is set by the rate at which field lines are cut, and so it will depend on:

- the area of the coil (the larger the coil, the more field lines will be cut)
- the number of turns in the coil (each extra turn effectively increases the number of lines cut; but you cannot use too many turns because if you use a wire that is too thin it will overheat from the current carried)
- the strength of the magnetic field
- the speed of rotation.

QUESTIONS

1. Why is a coil of wire used to generate electricity?
2. What four factors affect the magnitude of the current?
3. Which will give a larger induced current: a coil of area 1 m^2 or a coil of area 2 m^2 if both have the same number of turns, the same speed of rotation and the same magnetic field strength?

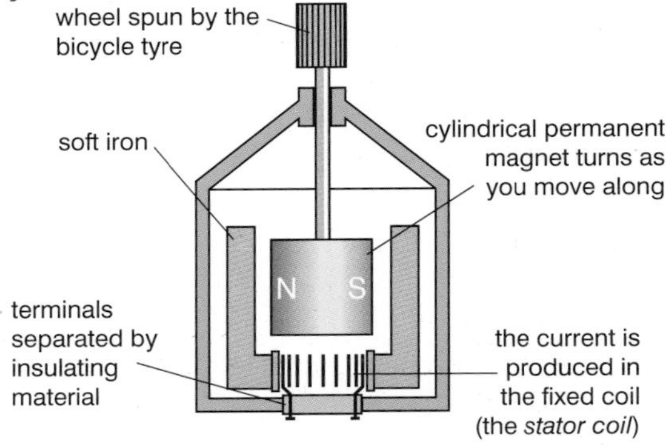

Δ Fig. 6.32 In a bicycle dynamo, the magnet rotates and the coil is fixed.

Dynamos

A **dynamo** is a simple current generator. It looks very much like an electric motor. Turning the permanent magnet inside the coil reverses its magnetic field every time the magnet is rotated by 180 degrees. The changes in the **magnetic field** through the coil induce an alternating current in the wires. The frequency of the electricity depends on how fast the dynamo is spinning.

Generators

Power station generators also produce alternating current (a.c.). Power stations use electromagnets rather than permanent magnets to create the magnetic field, and then pass the magnetic field through the rotating coils. The generator rotates at a fixed rate, producing a.c. at 50 Hz or 60 Hz, depending on the country.

Spinning a coil of wire in a magnetic field induces a continuously varying voltage much larger than that from a single wire. The current produced is removed via slip rings. The output is an alternating current.

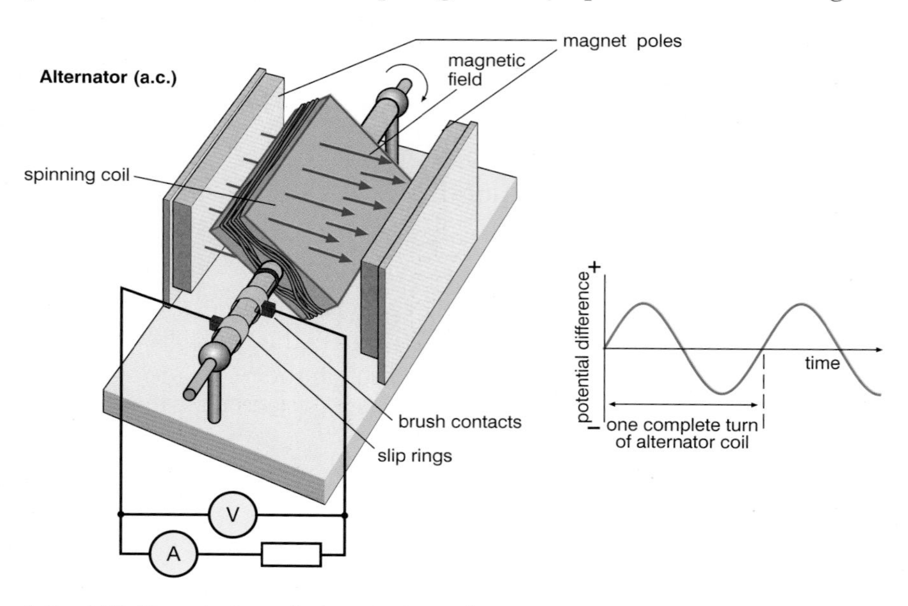

Δ Fig. 6.33 The spinning coil gives an alternating output.

QUESTIONS

1. Describe how a dynamo works.
2. What does the frequency of the electricity generated depend on?
3. Describe how to increase the voltage produced and explain why the output is an alternating current.

Developing investigative skills

A student wants to investigate the voltage induced in a coil by a moving magnet. To do this, they hang a magnet from a spring so that it can oscillate vertically. The student sets the magnet moving so that it oscillates in and out of a coil. To measure the voltage induced the student connects the coil to a cathode ray oscilloscope (CRO).

The student wants to vary the speed of the magnet's motion and measure the maximum voltage induced. However, the student cannot think of a way to measure the speed of the magnet and finds it difficult to judge the maximum voltage from the CRO screen, even with the timebase circuit switched off. The student adapts their method to a more qualitative approach, simply judging whether or not larger speeds give bigger voltages.

Devise and plan investigation

1. The student plans to vary the speed of the magnet by stretching the spring to different lengths before releasing it. Will this method work?
2. What difference will switching off the timebase circuit make to the student's observations on the CRO screen?
3. On the CRO screen, what difference will the student notice between the magnet moving into the coil and the magnet moving out of the coil?

Analyse and interpret

4. The student expects larger voltages to be induced when the magnet moves more quickly. How will the student be able to judge voltages from the CRO screen?
5. Sketch a graph to indicate how the induced voltage will vary over time after the student sets the magnet oscillating.

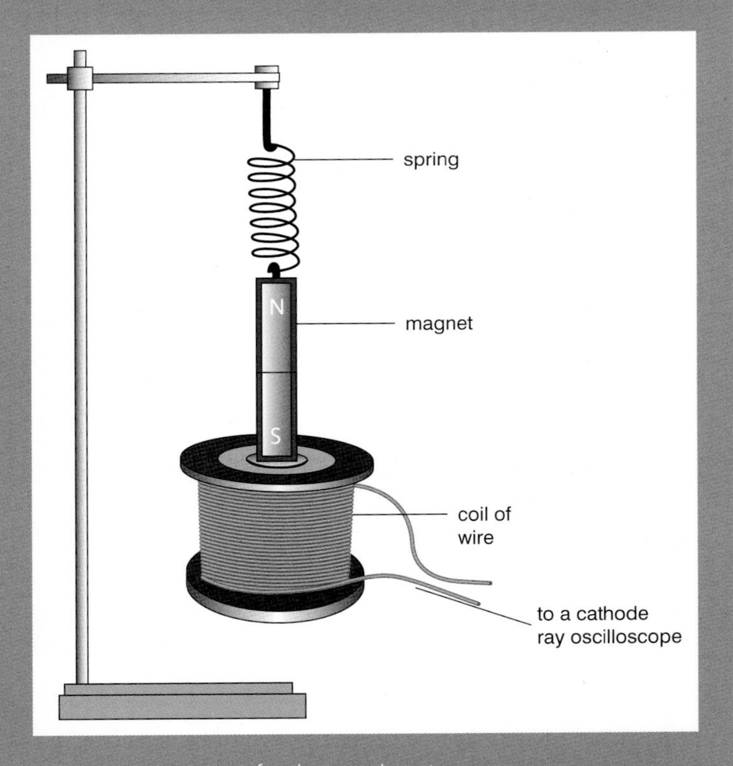

Δ Fig. 6.34 Apparatus for the experiment.

Evaluate data and methods

6. The student adapted their plan to a more qualitative approach. What does *qualitative* mean in this context?
7. Suggest how the student could adapt their method to record accurate measurements of the speed of the magnet and the induced voltage.

EXTENSION

Michael Faraday discovered how to generate electricity using this method in 1831.

Imagine that you are Michael Faraday on the day of your momentous discovery.

Write a letter to the Royal Society describing both the equipment needed and the different methods of controlling the voltage generated.

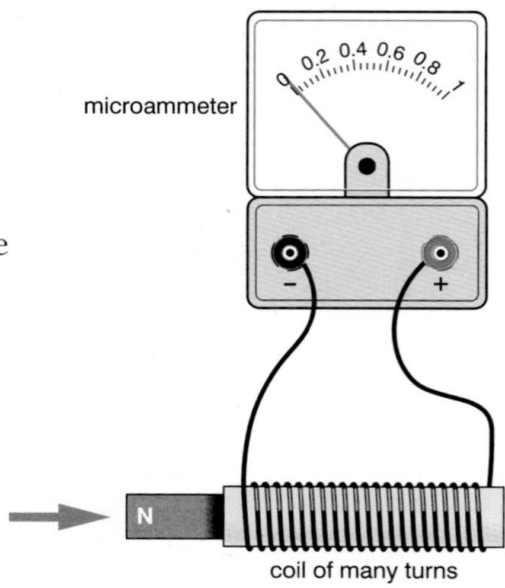

▷ Fig. 6.35 How Michael Faraday generated electricity.

Transformers

A **transformer** consists of two coils of insulated wire wound on a piece of iron. If an alternating voltage is applied to the first (primary) coil, the alternating current produces a changing magnetic field in the core. This changing magnetic field induces an alternating current in the second (the secondary) coil. The strength of the magnetic field is increased because the iron is soft and easily magnetised.

If there are more turns on the secondary coil than on the primary coil, then the voltage across the secondary coil will be greater than the voltage across the primary coil. The exact relationship between turns and voltage is:

$$\frac{\text{primary coil voltage } (V_p)}{\text{secondary coil voltage } (V_s)} = \frac{\text{number of primary turns } (n_p)}{\text{number of secondary turns } (n_s)}$$

When the secondary coil has more turns than the primary coil, the voltage increases in the same proportion. This is a step-up transformer.

A transformer with fewer turns on the secondary coil than on the primary coil is a step-down transformer, which produces a smaller voltage across the secondary coil.

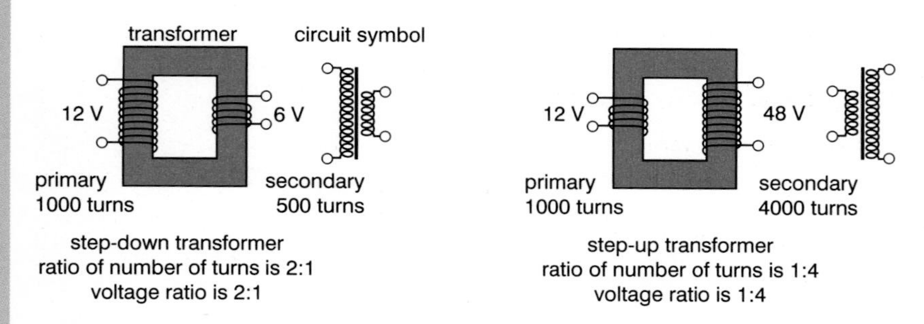

Δ Fig. 6.36 Transformers are widely used to change voltages. They are frequently used in the home to step down the mains voltage of 230 V to 6 V or 12 V.

WORKED EXAMPLE

Calculate the output voltage from a transformer when the input voltage is 230 V and the number of turns on the primary coil is 2000 and the number of turns on the secondary coil is 100.

Write down the formula: $\frac{V_p}{V_s} = \frac{n_p}{n_s}$

Substitute the known values: $\frac{230}{V_s} = \frac{2000}{100} = 20$

Rewrite this so that V_s is the subject: $V_s = \frac{230}{20}$

Work out the answer and write down the unit: $V_s = 11.5$ V

The current used by the transformer must change as well. No transformer is 100 per cent efficient, because all transformers produce some heat when they are working. But if it were 100 per cent efficient, then the electrical power going in would equal the electrical power going out. That is to say:

primary coil voltage (V_p) × primary coil current (I_p)
= secondary coil voltage (V_s) × secondary coil current (I_s)

For example, if the output is 12 V, 10 A, that is 120 watts of power going out of the transformer. If you know that the input voltage is 240 V, then the input current will be 0.5 A.

EXTENSION

A bathroom shaver socket contains an isolating transformer. The socket has an output voltage that is the same as the input voltage. It sometimes has an alternative output voltage. This means that people who travel around the world can use their electric shavers in their hotel rooms.

The mains supply is hidden behind the socket face. The only 'bare' terminals exposed are the two holes that the shaver plugs into. These holes are not 'live', so there is no risk of being electrocuted if you touch them with damp hands.

Research how shaver sockets work in a way that makes them safe to use in a bathroom.

Prepare revision notes to share with your classmates.

Use plenty of diagrams and colour to help them remember your points.

Δ Fig. 6.37 A shaver power socket.

QUESTIONS

1. What is the purpose of the iron core in a transformer?
2. Describe the difference between a step-up transformer and a step-down transformer.
3. A transformer has input voltage 2 V. There are 20 turns on the primary and 200 turns on the secondary. What is the output voltage?
4. A transformer has input voltage 1.5 V. There are 60 turns on the primary and 240 turns on the secondary. What is the output voltage?
5. A transformer has an input voltage of 2 V. There are 50 turns on the primary coil. The secondary coil has 600 turns.
   a) What is the output voltage?
   b) The secondary coil has a resistance of 12 Ω. What is the secondary current?
   c) What is the primary current?

TRANSMITTING ELECTRICITY

Most power stations burn fuel to heat water into high-pressure steam, which is then used to drive a **turbine**. The turbine turns an a.c. generator, which produces the electricity.

To minimise the power loss in transmitting electricity, the current must be kept as low as possible. The higher the current, the more the transmission wires will be heated by the current and the more energy is wasted as heat.

This is where transformers are useful. This is also the reason that mains electricity is generated as alternating current. When a transformer steps up a voltage, it also steps down the current and vice versa. Power stations generate electricity with a voltage of 25 000 V. Before this is transmitted, it is converted by a step-up transformer to 400 000 V. This is then reduced by a series of step-down transformers to 230 V before it is supplied to homes.

Δ Fig. 6.38 This turbine wheel is from a hydroelectric power station. It is turned by water taken from a reservoir behind a dam.

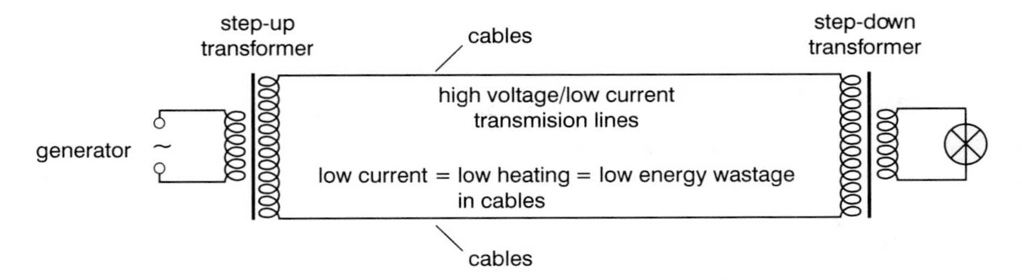

Δ Fig. 6.39 Transformers used to distribute electricity from the power station to consumers.

Energy losses in cables

Except for some lengths of superconducting cable (which has zero resistance but needs to be kept at a temperature below –200 °C), the distribution cables used by electricity companies do not have zero resistance. A typical cable with a length of 100 km may have a resistance of 4 ohms. Now consider the problem facing the company when they want to send 4 MW of power to a town 100 km away. They could send either 10 A at 400 000 V, or 160 A at 25 000 V, or 17 400 A at 230 V.

The 230 V solution is useless. To send 17 400 A through a resistor of 4 ohms requires a potential difference (p.d.) across the wire of 68 000 V. More power would be lost from the cables than was put into them by the power station. This would be impossible.

At 25 000 V, the p.d. across the cable would be:

$$\begin{aligned} V &= I \times R \\ &= 160\text{ A} \times 4\ \Omega \\ &= 640\text{ V} \end{aligned}$$

The power lost in the cables would be:

$$\begin{aligned} P &= V \times I \\ &= 640\text{ V} \times 160\text{ A} \\ &= 102\ 000\text{ W} \end{aligned}$$

Of the 4 000 000 W being sent, this is 2.6 per cent. This is not too bad, as electricity supply companies expect to lose a total of 5–10 per cent of generated power between the power station and the customer.

At 400 000 V, 10 A, the power lost in the cables is just 400 W, which is 0.01 per cent of the power being sent. The 400 000 V solution costs more, and so the electricity company will have to work out which high voltage solution is best.

▷ Fig. 6.40 This 400 000 V distribution power line absorbs very little of the power that it carries, but it cost approximately US$ 500 000 per km to build.

THE INDUCTION COOKER

How can a cooker heat food without the cooker getting hot? The answer lies in electromagnetic induction.

In an induction cooker, a copper coil is placed under the saucepan. An alternating current in the copper coil creates a magnetic field around it. Because it is an alternating current, it produces a changing magnetic field. This changing magnetic field cuts through the metal of the pan and generates a an alternating current directly in the pan – there is a complete circuit around the pan (think of it as a transformer where the secondary coil has a single, very wide turn).

The current in the pan experiences a resistance and so it heats up, heating the food inside.

The process is controlled more quickly than a hot plate. When you switch off the current, the magnetic field collapses straight away, so you do not waste heat heating the top plate of the cooker. But is it safe? Will you get an electric shock from touching the pan? No, because although there is a large current in the pan (to get lots of resistive heating) there is only a small voltage – too small to drive a current through the cook's body.

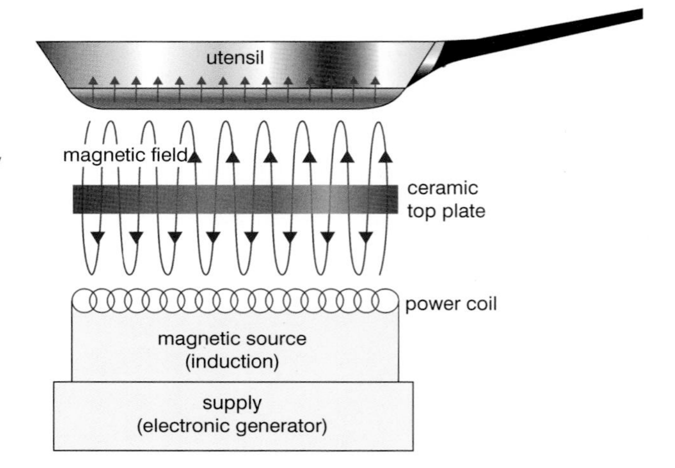

Δ Fig. 6.41 How an induction cooker works.

The induction cooker is often described as the cooker that 'doesn't get hot'. This is true in one sense, but of course some heat will conduct back to the cooker from the pan because they are in contact. Also, it does not work for every type of pan. If you think of the effect alongside the transformer, you will see that the iron core of the transformer (which increases the effect) is missing. The pan needs to have some ferrous content to improve its operation.

QUESTIONS

1. Explain why electricity is transmitted at high voltages if the distances involved are large.
2. Describe how transformers help in the transmission of high-voltage electricity.
3. How much power would be lost if the electricity were sent at 50 000 V with a resistance of 4 Ω and a current of 100 A?
4. What percentage of the 4 MW that the company want to send is this?

End of topic checklist

A voltage is **induced** in a conductor or coil when there is relative motion between the conductor and a magnetic field.

Transformers change voltages – the ratio of the number of turns in the coils is the same as the ratio of the voltages across the coils.

A **step-up transformer** increases the voltage. A **step-down transformer** decreases the voltage.

A **dynamo** is a simple current generator.

The facts and ideas that you should understand by studying this topic:

- ❍ Know that a voltage is induced across a conductor when the conductor moves through a magnetic field or when a magnetic field moves past the conductor.
- ❍ Describe the factors that affect the size of the induced voltage.
- ❍ Know that a current is induced if there is a complete circuit.
- ❍ Describe the structure of a transformer.
- ❍ Describe the action of a transformer and the factors that control the values of the voltage produced.
- ❍ Explain the use of step-up and step-down transformers in the large-scale generation and transmission of electricity.
- ❍ Transformers are used to produce high voltages for electricity transmission. This reduces the energy losses across the cables.
- ❍ Know and use the relationship between input (primary) voltage and output (secondary) voltage and the turns ratio for a transformer.
- ❍ Know and use the relationship linking the input power to the output power (for 100% efficiency).

End of topic questions

1. State the conditions for there to be an induced potential difference and an induced current. **(3 marks)**

2. Explain in your own words why an alternating current in the left-hand coil of the diagram below induces a current in the other coil, but a direct current in the left-hand coil does not. **(4 marks)**

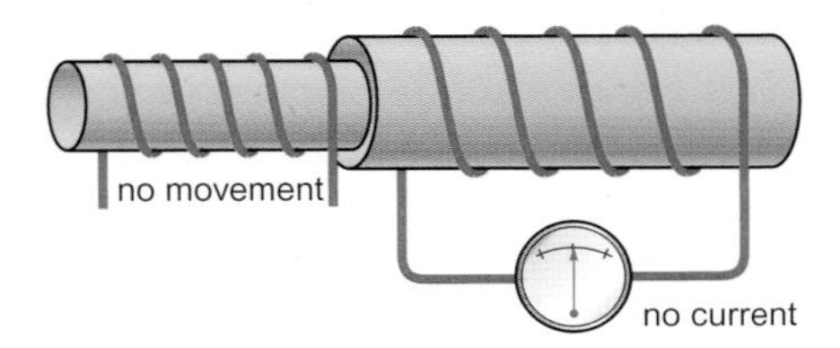

3. The diagram below shows a transformer.

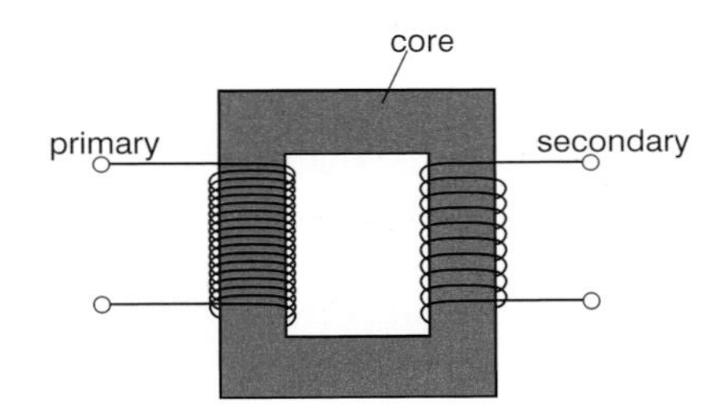

   a) What material is used for the transformer core? **(1 mark)**

   b) What happens in the core when a current is switched on in the primary coil? **(3 marks)**

   c) What happens in the secondary coil when the primary current is switched on? **(3 marks)**

   d) If the primary coil has 12 turns and the secondary coil has 7 turns, what will the primary voltage be if the secondary voltage is 14 V? **(3 marks)**

4. A computer that plugs into the mains needs to operate on a potential difference of 12 V. Explain whether you would use a step-up or a step-down transformer. Give a reason for your answer. **(3 marks)**

5. Describe how there is a potential difference across the secondary coil of a transformer, when the two coils are not connected to each other. **(4 marks)**

6. 'A transformer works with alternating or direct current.' Is this true or false? Explain your answer. **(3 marks)**

End of topic questions continued

7. A transformer has 150 turns of wire on the primary coil and 600 turns of wire on the secondary coil.

a) Is it a step-up or a step-down transformer? **(1 mark)**

b) If the output potential difference across the secondary coil is 800 V, what is the input PD across the primary coil? **(3 marks)**

8. A transformer is required to change an input potential difference of 25 000 V to an output PD of 40 000 V.

a) Use ratios to suggest several possible pairs of values for the number of turns on primary and secondary coils. **(3 marks)**

b) If the secondary coil had 1600 turns, how many turns would the primary coil have? **(3 marks)**

9. Explain in your own words where transformers are used in the distribution of electricity. **(6 marks)**

10. Explain in your own words why the distribution of electricity uses such high values of potential difference. **(6 marks)**

11. The primary coil of a transformer has a potential difference of 240 V, and a current of 20 mA. If the PD across the secondary coil is 12 V, calculate the current in this coil. **(3 marks)**

Exam-style questions

Sample student answers

Question 1

A student uses a 12 V low-voltage power supply in the classroom.

It plugs into the 230 V mains supply.

a) The student realises that the power supply must contain a step-down transformer.

i) Describe the operation of a transformer.

A current in the primary coil makes a magnetic field which goes into the iron core. ✓✓ *The magnetic current goes into the second coil and makes an induced voltage. The transformer is there to make different voltages.* ✗✗ (2) **(4)**

ii) Why does the power supply contain a *step-down* transformer?

To make the voltage smaller (1) **(1)**

b) The student wants to use the power supply to provide a current of 3 A.

i) Write down the equation linking input power and output power of a transformer, assuming 100% efficiency.

Power = V x I ✓ (1) **(1)**

ii) Calculate the input current from the 230 V mains supply if the student is to use an output current of 3 A at 12 V.

230 x I = 3 x 12 ✓ (1)

I = 36/230 ✓ (1)

= 0.16 A ✓ (1) **(3)**

EXAMINER'S COMMENTS

a) i) The candidate has a rather confused idea about how a transformer operates. Mentioning electric currents and magnetic currents in the same answer is a common mistake and suggests a lack of thorough learning. Questions that require an answer in a sequence like this need to be learned carefully.

A better answer would be that the changing magnetic field goes into the second coil and a voltage is induced. Transformers change the size of the voltage.

ii) Correct.

b) This part of the question has been handled very well – full marks.

Exam-style questions continued

c) Step-up and step-down transformers are also used in the large-scale transmission of electrical energy.

i) Explain the benefits of using transformers in this way.

It means the electricity can be sent using high voltages which wastes less energy. ✓✓✗ (2)

(4)

ii) Describe a potential hazard in transmitting electrical energy this way and describe how this risk can be reduced.

It is more dangerous to use high voltages, ✗ *so the companies put warning signs near to high voltage cables.* ✓ (1)

(2)

(Total 15 marks)

c) i) The candidate cannot score full marks as they have not given four points – always make sure you have given enough points to cover the mark allocation. What is given is fine, but the candidate has missed the idea that high voltages lead to smaller currents (1 mark) and that wasting less energy saves money or makes the process more efficient (1 mark).

ii) The candidate loses one mark for simply saying that high voltages are dangerous – that was given in the question – and not going on to say what the danger is (risk of shock or electrocution).

Exam-style questions continued

Question 2

The diagram shows two bar magnets.

a) Copy the diagram. Draw on the magnetic field pattern on your diagram. **(3)**

b) Describe how you could test the area near a bar magnet to find the magnetic field pattern. **(3)**

c) The diagram shows a simple electric motor.

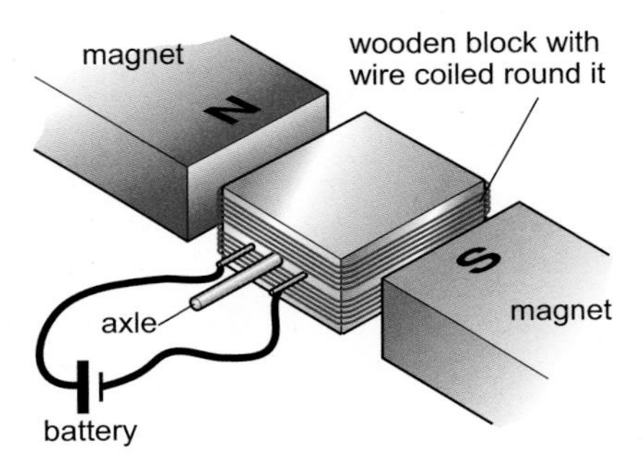

i) Explain why the coil spins when an electric current is supplied to it. **(4)**

ii) The motor spins faster when the current is increased.
Explain why. **(2)**

iii) When the current is reversed, the motor spins in the opposite direction.
Explain why. **(2)**

(Total 14 marks)

What do these things have in common: treatment for cancer, finding leaks in underground pipes, and dating archaeological specimens? The answer is that they all use radioactivity in some way. Cancer is treated by targeting the tumour with radioactive substances which destroy the tumour cells but do minimal damage to the surrounding tissues. Leaks in underground pipes can be detected by adding a radioactive substance to the fluid flowing in the pipe and then using a detector above the ground to trace the amount of radioactivity emitted. Archaeological specimens can be dated by carbon atoms, because of the fact that everything that lives, or once lived, contains carbon.

STARTING POINTS

1. What is an atom?
2. What happens when charged particles are placed in a magnetic field?
3. What is meant by **a)** atomic number **b)** mass number of an element?
4. What do you have to do to balance any equation?
5. What do you understand by the term 'radioactivity'?
6. What happens in a nuclear power station?

CONTENTS

7 Radioactivity, fission and fusion

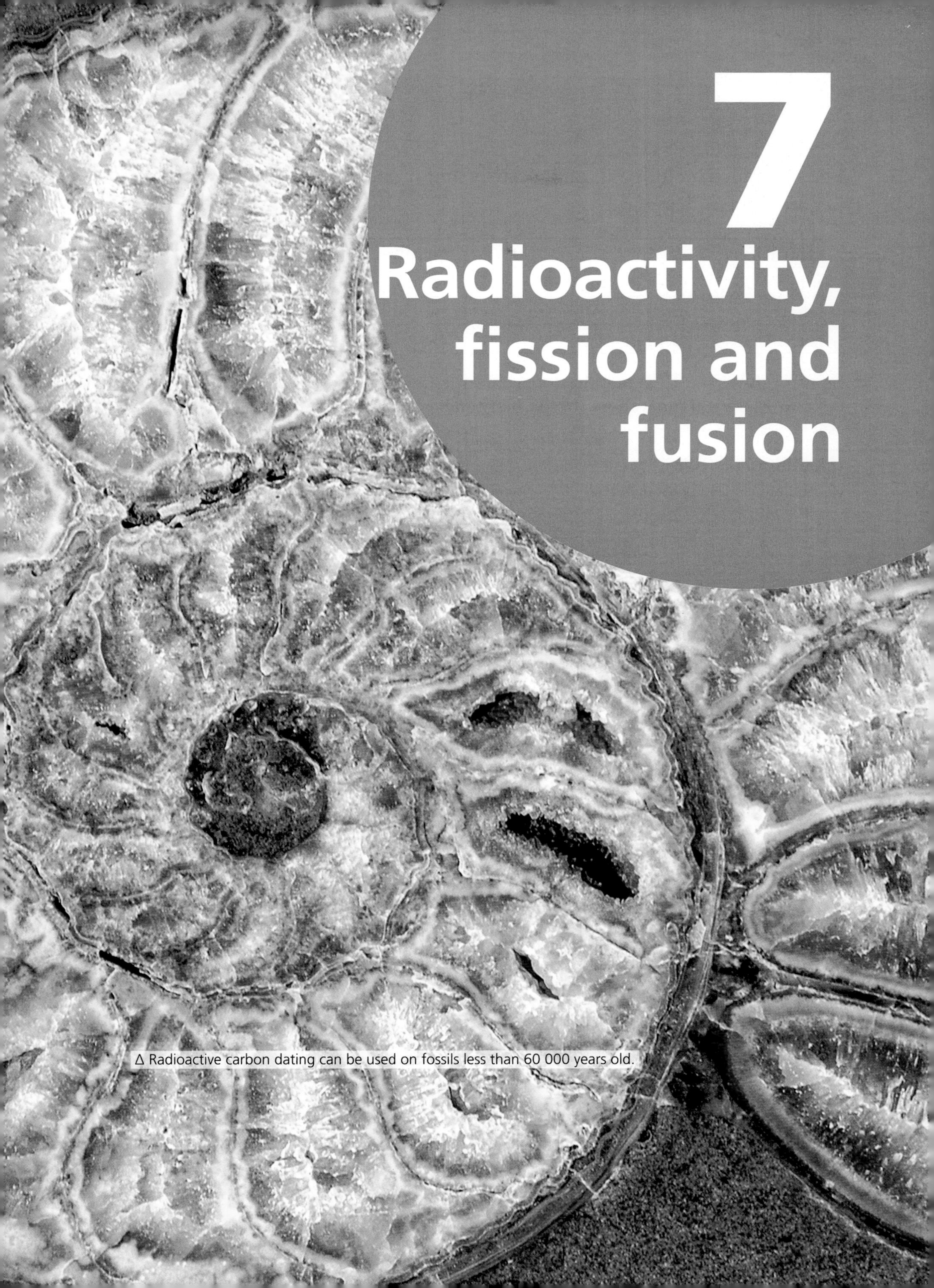

Δ Radioactive carbon dating can be used on fossils less than 60 000 years old.

Units

INTRODUCTION

For the topics included in radioactivity and particles, you will need to be familiar with:

Quantity	Unit	Symbol
time	hour, minute, second	h, min, s
length	metre, centimetre	m, cm
activity	becquerel	Bq

Some of these you have met previously but the new units will be defined when you meet them for the first time.

LEARNING OBJECTIVES

✓ Use the following units: becquerel (Bq), centimetre (cm), hour (h), minute (min) and second (s).

Radioactivity

INTRODUCTION

The discovery of radioactivity in the late 1800s came at a time of great development in our knowledge of the atom. Far from being the 'fundamental building blocks of nature' as had first been thought, atoms were revealed as collections of particles in many different combinations.

△ Fig. 7.1 Painting luminous watch dials with uranium salts in the early 20th century.

Piecing the whole puzzle together took many years as the properties of unstable atoms were studied and new atoms were discovered. Unfortunately, the dangers of radiation were not known at the time, so many of the early experimenters were exposed to their harmful effects.

Now that the danger is understood, radioactive materials are safely used in a number of everyday situations such as smoke alarms.

KNOWLEDGE CHECK

✓ Know the basic structure of an atom in terms of protons, neutrons and electrons.
✓ Have some background knowledge of some issues relating to radioactivity in everyday life – waste from nuclear power stations, and so on.

LEARNING OBJECTIVES

✓ Be able to describe the structure of an atom in terms of protons, neutrons and electrons and use symbols such as p, e and n to describe particular nuclei.
✓ Know the terms atomic (proton) number, mass (nucleon) number and isotope.
✓ Know that alpha (α) particles, beta (β^-) particles, and gamma (γ) rays are ionising radiations emitted from unstable nuclei in a random process.
✓ Be able to describe the nature of alpha (α) particles, beta (β^-) particles, and gamma (γ) rays, and recall that they may be distinguished in terms of penetrating power and ability to ionise.
✓ Be able to investigate the penetration powers of different types of radiation using either radioactive sources or simulations.
✓ Be able to describe the effects on the atomic and mass numbers of a nucleus of the emission of each of the four main types of radiation (alpha, beta, gamma and neutron radiation).
✓ Understand how to balance nuclear equations in terms of mass and charge.
✓ know that photographic film or a Geiger–Müller detector can detect ionising radiations.
✓ Be able to list the sources of background (ionising) radiation from Earth and space.
✓ Know that the activity of a radioactive source decreases over a period of time and is measured in becquerels.

✓ Know the definition of the term half-life and understand that it is different for different radioactive isotopes.
✓ Be able to use the concept of the half-life to carry out simple calculations on activity, including graphical methods.
✓ Be able to describe uses of radioactivity in industry and medicine.
✓ Be able to describe the difference between contamination and irradiation.
✓ Be able to describe the dangers of ionising radiations, including:
- that radiation can cause mutations in living organisms
- that radiation can damage cells and tissue
- the problems arising from the disposal of radioactive waste
- how the associated risks can be reduced.

THE STRUCTURE OF AN ATOM

All **elements** are made up of atoms, consisting of **protons**, **neutrons** and **electrons**. The protons and electrons have electrical charges that are equal but opposite. Because atoms, in general, do not have an electric charge, they usually contain the same number of protons and electrons.

The **nucleus** is made of protons and neutrons, bound together by an extremely strong force, far stronger than gravity, magnetism or electricity, and completely different from any of them. The electrons form a loose cloud on the outside of the atom with the nucleus in the middle. The relative masses and charges are shown in Table 7.1

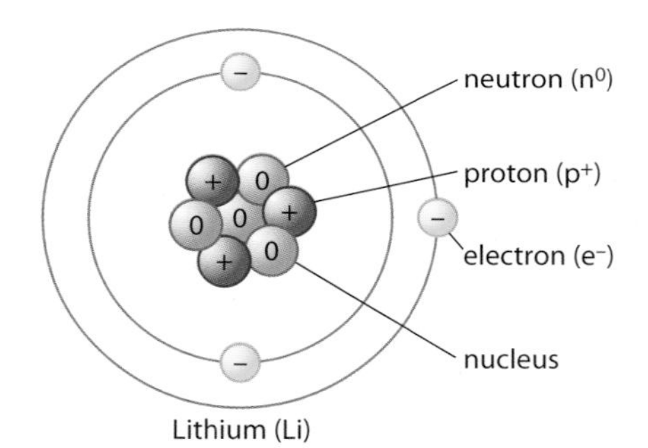

◁ Fig. 7.2 A lithium atom.

Particle	Proton	Neutron	Electron
Relative mass	1.0	1.001	$\frac{1}{1838}$
Relative charge	+1	0	–1

Δ Table 7.1 Atomic particles.

ATOMIC NUMBER, MASS NUMBER AND ISOTOPES

The behaviour of the atom is fixed by its nucleus. Each nucleus is represented by its chemical symbol with two extra numbers written before it. Here is the symbol for radium-226:

$^{226}_{88}\text{Ra}$

The top number is the **mass number** (the total number of protons and neutrons).

The bottom number is the **atomic number** (the number of protons).

It is common for several different **nuclides** (nuclei) to have the same number of protons but different numbers of neutrons. For example, there are two types of copper nuclei, $^{63}_{29}Cu$ and $^{65}_{29}Cu$. The first type contain 29 protons and 63 nucleons (particles in a nucleus: protons and neutrons), hence $(63 - 29) = 34$ neutrons. The second type is the same, except that it contains 36 neutrons.

In natural copper, just under 70 per cent of the nuclei are $^{63}_{29}Cu$, and just over 30 per cent are $^{65}_{29}Cu$. These are the two stable **isotopes** of copper. Because they both contain 29 protons, they are surrounded by 29 electrons in the same pattern. It is the structure of the electrons around the nucleus that fix how the chemistry will work, so these two isotopes have the same chemistry.

△ Fig. 7.3 The copper on the roof of this building contains 69 per cent $^{63}_{29}Cu$ nuclei and 31 per cent $^{65}_{29}Cu$ nuclei.

In addition there are various **radioactive** isotopes of copper. These have different numbers of neutrons so their nuclei are unstable. There are nine radioactive isotopes, with mass numbers that vary from $^{59}_{29}Cu$ to $^{69}_{29}Cu$. These two extremes are very unstable with **half-lives** (see page 357) of a few minutes. $^{64}_{29}Cu$ has a half-life of 12 hours. So all of the radioactive isotopes of copper are extremely radioactive. Some radioactive isotopes of other elements are much less radioactive, and have half-lives measured in years, or even thousands of years.

QUESTIONS

1. What is in an atom?

2. Why are atoms electrically neutral?

3. Explain the meaning of the terms atomic number, mass number and isotope.

4. Why are different isotopes of an element identical chemically?

PARTICLE PHYSICS AND THE LARGE HADRON COLLIDER

The Large Hadron Collider, at CERN in Switzerland, is the world's largest particle accelerator. It is designed to attempt to answer some of the most fundamental questions of physics of our time. It is the result of many years of research into atomic and subatomic particles. A hadron is a particle that consists of quarks (subatomic particles) – protons and neutrons are hadrons, for example.

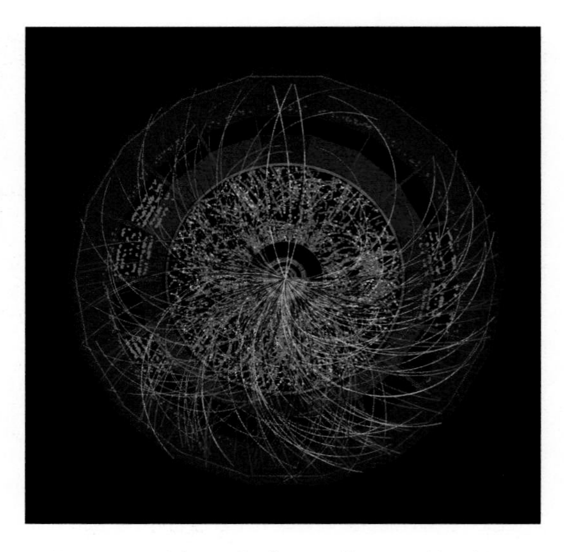

∆ Fig. 7.4 Particle tracks from collisions of lead ions in the Large Hadron Collider.

For 20 years after the discovery of the nucleus in Rutherford's scattering experiment in 1909, further investigations continued to reveal the existence of the proton and the neutron. At this point, the model of the atom consisted of a central nucleus, containing the proton and neutron, with electrons in a cloud around it. All the atoms of the Periodic Table are various combinations of these particles. Information about atomic structure also gave insights into radioactivity, and if the electrons were only in particular states, it also accounted for the emission and absorption of light. The electron patterns allowed chemical behaviour to be included.

However, experiment is always the test of any theory, and there were still some experimental observations that could not be fitted into the atomic model. In particular, the problem of cosmic rays had to be accounted for.

Cosmic rays are high-energy particles that arrive from space (the cosmos). As they hit the atmosphere, various interactions occur and we can often detect particles on the ground that are made in these collisions.

The evidence from these observations indicated the existence of other 'sub-atomic' particles. To avoid the rather random pattern of waiting for evidence to arrive from space, scientists built particle accelerators and continued to look at the effects of high-energy particles colliding with each other.

More than 200 separate sub-atomic particles were identified. The previously clear picture of fundamental particles – proton, neutron, electron – became very muddled. The results suggested that there is a hidden level of deep atomic structure. This led to the development of the Standard Model. This is a model which links the electromagnetic, weak and strong nuclear forces, which are involved in the dynamics of the known sub-atomic particles. This model is our best understanding at this time of what the fundamental particles of nature are.

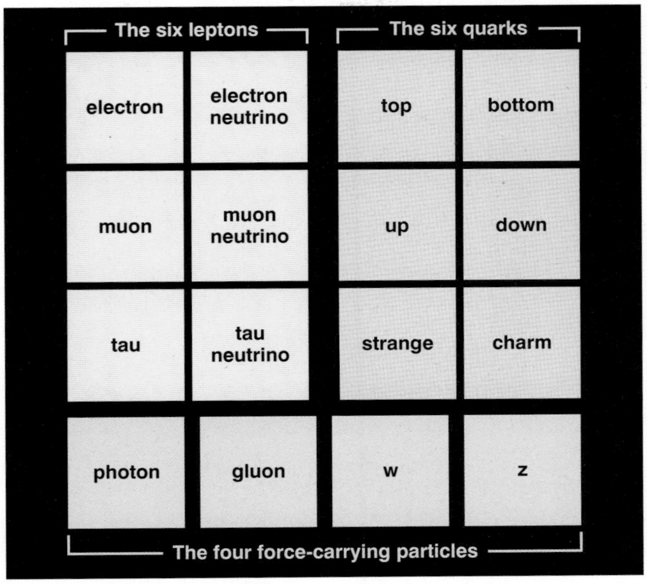

The six leptons		The six quarks	
electron	electron neutrino	top	bottom
muon	muon neutrino	up	down
tau	tau neutrino	strange	charm
photon	gluon	w	z
The four force-carrying particles			

Δ Fig. 7.5 The main sub-atomic particles.

To reveal deeper atomic structure, experiments are needed with particles at higher and higher energies. This has led to the construction of the Large Hadron Collider near Geneva. Here, teams of scientists from many countries hope to find more evidence about how nature works at the smallest scale.

TYPES OF RADIOACTIVE DECAY

Inside the atom the central nucleus of positively charged protons and neutral neutrons is surrounded by shells, or orbits, of electrons. Most nuclei are very stable, but some 'decay' and break apart into more stable nuclei. This breaking apart is called **radioactive decay**. Atoms whose nuclei do this are radioactive.

Radioactive decay is random: it is not possible to predict when a nucleus will decay. When a radioactive nucleus decays it may emit one or more of the following:

- alpha (α) particles
- beta (β) particles
- gamma (γ) rays.

A stream of these particles or rays is referred to as **ionising radiation** (often called nuclear radiation, or just 'radiation' for short). Ionising radiation is radiation that has enough energy to cause other atoms to lose electrons and form positively charged ions. Ionisation occurs when the radiation from radioactive decay knocks one or more outer electrons out of the atoms the radiation passes through.

Δ Fig. 7.6 A device for measuring radiation, a radioactive source, detector and meter.

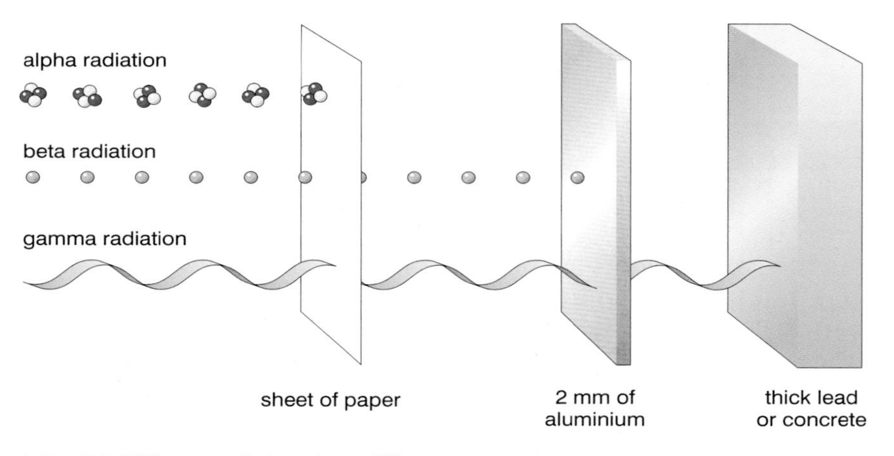

Δ Fig. 7.7 Different radiations have different penetrating power.

	Alpha (α)	Beta (β)	Gamma (γ)
Description	A positively charged particle, identical to a helium nucleus (two protons and two neutrons)	A negatively charged particle, identical to an electron	Short wavelength electromagnetic radiation Uncharged
Penetration	4–10 cm of air. Stopped by a sheet of paper	About 1 m of air. Stopped by a few mm of aluminium	Almost no limit in air. Intensity greatly reduced by several cm of lead or several metres of concrete
Ability to ionise	High	Medium	Low

Δ Table 7.2 Radioactive particles.

QUESTIONS

1. What is radioactive decay?
2. Is a radioactive isotope stable or unstable before it decays?
3. Use the descriptions of alpha, beta and gamma radiation given in the table to explain why they have different penetrating powers.

EXTENSION

Radioisotopes that are beta emitters can be used to monitor the thickness of paper as it is manufactured, using a set-up like the one shown here.

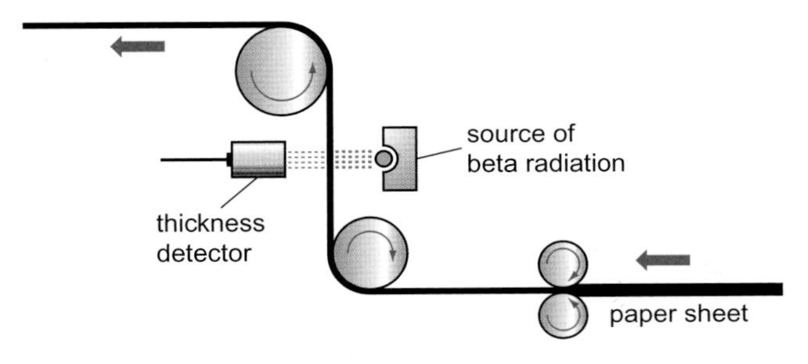

△ Fig. 7.8 A beta emitter for checking the thickness of paper.

Imagine you are the managing director of a local paper mill. You are giving a presentation to the local community, during which you will explain why beta emitters are ideal for monitoring the thickness of paper, and how you will minimise the risk to the local environment from your beta sources. Think about the penetrating power of beta particles, and how they could be shielded.

THE EFFECTS OF RADIOACTIVE DECAY ON A NUCLEUS

An alpha particle contains two protons and two neutrons. So, with four nucleons and two protons, it is written as $^4_2\alpha$. It is the same particle as the nucleus of a helium atom, which is written $^4_2$He.

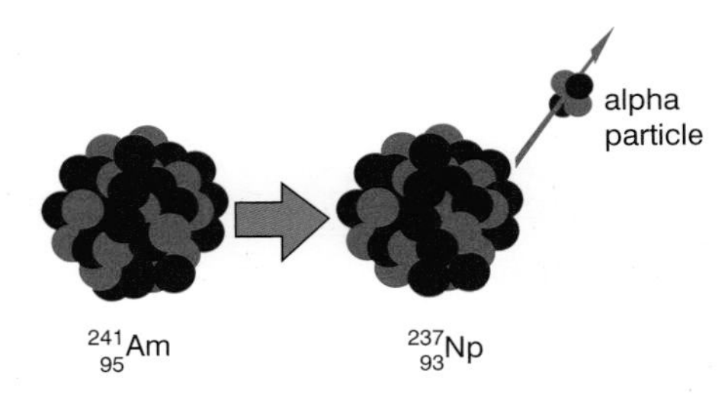

△ Fig. 7.9 Alpha decay.

Since an alpha particle contains two protons and two neutrons, when a nucleus emits an alpha particle it loses four **nucleons**, and its mass number decreases by 4. It loses two protons and its atomic number decreases by 2.

A beta particle is written $^{0}_{-1}\beta$ to show that it is not a nucleon and has a mass so small that it can be ignored. It is of exactly the opposite charge to a proton. A beta particle is the same as an electron but is emitted from the nucleus, and could be written $^{0}_{-1}$e.

When a nucleus emits a beta particle, a neutron inside the nucleus has changed into a proton plus an electron. The total charge is unchanged, but the nucleus has lost a neutron and gained a proton. The electron, of course, is emitted as the beta particle.

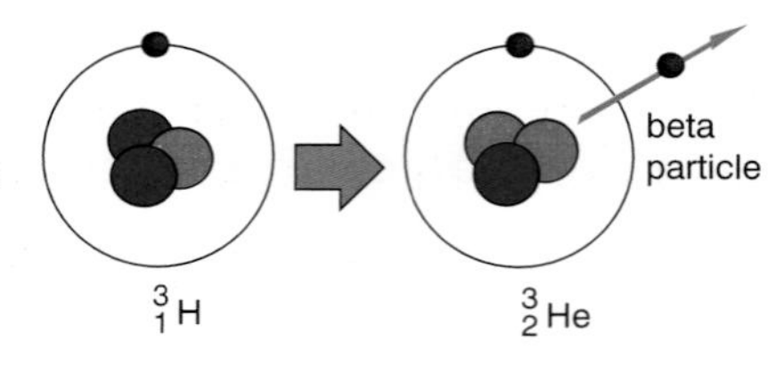

Δ Fig. 7.10 Beta decay.

Radioactive decay occurs in atoms that have an unstable nucleus. This can cause the forces within a nucleus to be unbalanced. Some atoms contain an excess of neutrons and this can lead to neutron emission (**neutron radiation**) which is another type of radioactive decay. When this happens, a neutron is emitted from the nucleus which then leaves the nucleus more stable.

Helium-5 is an unstable nucleus with an excess of neutrons. It decays by losing a neutron:

$$^5_2He \rightarrow ^4_2He + ^1_0n$$

The atomic number of the parent nucleus does not change, as only a neutron is emitted. It simply forms a different isotope of helium.

As we shall see, neutrons are also emitted in a nuclear fission reaction. In both cases, neutron radiation is more penetrating than alpha or beta radiation and passes through most materials. However, the emitted neutrons will collide with other nuclei and may cause them to become a different isotope by capturing a neutron. In this way, neutrons can cause stable atoms to become radioactive.

When a nucleus emits a gamma ray, it is a wave that is emitted as the unstable nucleus (which has excess energy) re-organises itself inside. No particles are emitted, and so the mass number and the atomic number do not change. Table 7.3 summarises these changes.

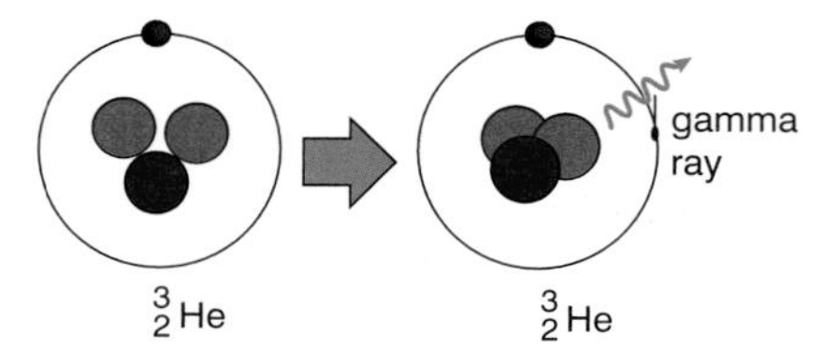

Δ Fig. 7.11 Gamma ray emission.

Nucleus emits	Mass number	Atomic number
Alpha particle	Decreases by 4	Decreases by 2
Beta particle	Does not change	Increases by 1
Gamma ray	Does not change	Does not change

Δ Table 7.3 Emission of different particles.

NUCLEAR EQUATIONS

You can write down nuclear changes as **nuclear equations**. The rules are as follows:

- The numbers in the top row must be balanced on each side of the equation, because the number of nucleons will not change. (For example, you could get 231 on the left and 227 + 4 for an alpha emission, and 231 on the left and 231 + 0 on the right for beta emission.)
- The numbers in the lower row must be balanced on each side of the equation, because the total electrical charge will stay unchanged.

On the right are two examples. Note particularly what happens when a beta particle is emitted. Because a neutron has changed into a proton, the atomic number has increased by 1!

Alpha decay – the nucleus emits an α-particle (2 protons and 2 neutrons)

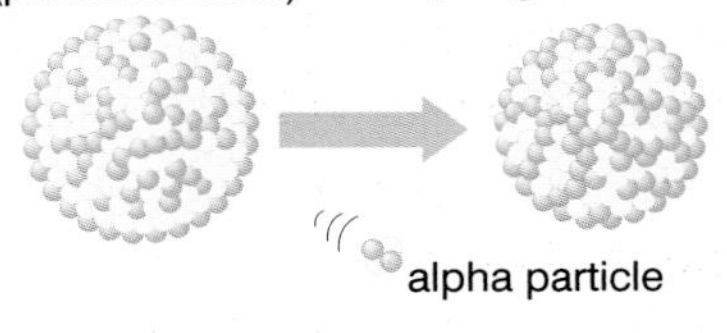

$${}^{226}_{88}\text{Ra} \rightarrow {}^{222}_{86}\text{Rn} + {}^{4}_{2}\alpha$$

Beta decay – a neutron changes into a proton in the the nucleus

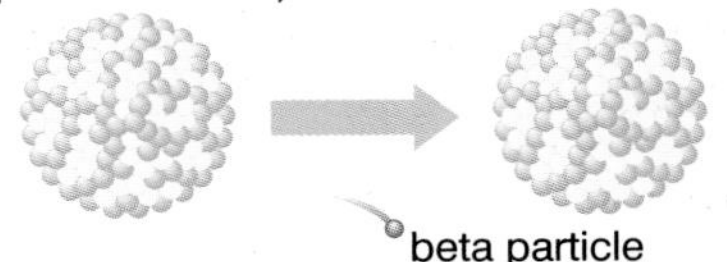

$${}^{218}_{84}\text{Po} \rightarrow {}^{218}_{85}\text{At} + {}^{0}_{-1}\beta$$

Gamma decay – the nucleus emits a γ-ray to get rid of excess energy

Δ Fig. 7.12 Examples of radioactive decay.

QUESTIONS

1. How does alpha decay affect a nucleus?
2. How does beta decay affect a nucleus?
3. Describe how emitting gamma radiation affects the nucleus.
4. After emitting alpha or beta radiation, the atom has changed to be a different element. Explain why.

DETECTING RADIOACTIVITY

All ionising radiation is invisible to the naked eye, but it affects photographic plates. Individual ionising events can be detected using a **Geiger–Müller (G–M) tube** as shown in Figure 7.13. The count rate is the number of ionising events detected per second or per minute.

There is *always* ionising radiation present. This is called **background radiation**. Background radiation is caused by radioactivity in soil, rocks and materials like concrete, radioactive gases in the atmosphere and cosmic rays, high-energy particles which come from somewhere in outer space, though we are still not sure exactly where.

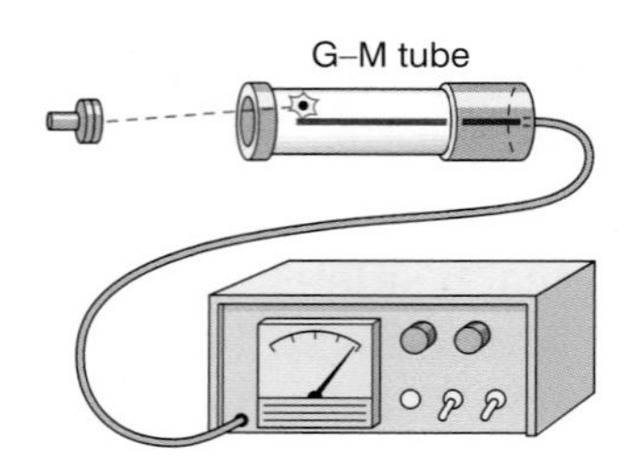

Δ Fig. 7.13 A Geiger-Müller tube.

EXTENSION

Background radiation comes from a variety of natural sources: traces of radioactive substances in the body (which have mainly come from natural radioisotopes in food and in the air), radioactive gas (such as radon) from the Earth, cosmic rays and radioactive rocks. The table shows the percentage of a typical person's annual radiation dose that comes from various sources.

Source	Percentage of annual dose
cosmic rays	5.0
radiation from rocks	3.1
radon gas	37.0
radioactive isotopes in the body	2.5
medical sources	48.4
consumer products	2.1
others (including air travel and occupational hazards)	1.9

Δ Table 7.4 Sources of the average person's annual exposure to radiation.

These sources can be divided into natural sources and those made by humans.

1. Which sources in the table can be described as natural?

2. Which sources in the table are made by humans?

3. Which category (natural or made by humans) contributes most to the average person's annual dose?

4. Which types of radiation might a person be exposed to that would come under the 'medical' category?

5. Name another source of background radiation made by humans.

Developing investigative skills

A teacher investigates four different radioactive sources U, V, W and X. He has paper, aluminium and lead to use as absorbers.

He measures the count rate at a fixed distance from each source, with and without different absorbers placed in front of the source.

The table shows his results.

Radioactive source	Count rate (counts per minute)			
	A few mm of aluminium	A few cm of lead	No absorber	A few sheets of paper
U	201	21	202	203
V	19	22	301	21
W	21	23	102	77
X	139	21	250	141

Devise and plan investigations

1. What measuring instrument would the teacher need for this investigation?
2. Describe the safety precautions the teacher should take.

Analyse and interpret results

3. Use the data above to identify which radiations are emitted by each source. Justify your answers with explanations.

Evaluate data and methods

4. For source U more radiation was detected with paper than without paper. Suggest why.

DECREASING ACTIVITY OVER TIME

The activity of a radioactive source is the number of ionising particles it emits each second. Over time, fewer nuclei are left in the source to decay, so the activity drops. The time taken for half the radioactive nuclei to decay is called the **half-life**. Note that this time is the same whenever you start measuring the source of radioactivity, but is different for different radioactive isotopes.

The level of radioactivity or activity is measured in **becquerels** (Bq). A source that is one becquerel (written as 1 Bq) has one nucleus decay and emit radiation per second. A source in a laboratory may be a few hundred Bq, but industrial sources can be several kBq, if not MBq or GBq. Activity can also be compared with a G-M tube as a count rate.

Half-life

Starting with a pure sample of radioactive nuclei, half the nuclei will have decayed after one half-life. The remaining undecayed nuclei still have the same chance of decaying as before, so after a second half-life, half of the remaining nuclei will have decayed. After two half-lives, a quarter of the nuclei will remain undecayed.

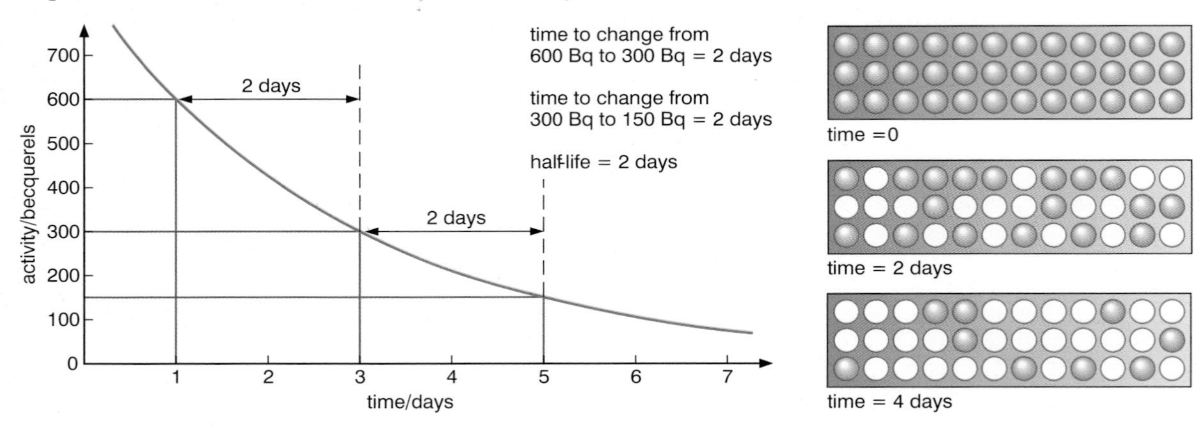

Δ Fig. 7.14 The half-life is 2 days. Half the number of radioactive nuclei decays in 2 days.

WORKED EXAMPLE

A radioactive element is detected by a Geiger–Müller tube and counter as having an activity of 400 counts per minute. Three hours later the count is 50 counts per minute. What is the half-life of the radioactive element?

Write down the activity and progressively halve it. Each halving of the activity is one half-life:

0	400 counts per minute
1 half-life	200 counts per minute
2 half-lives	100 counts per minute
3 half-lives	50 counts per minute

Therefore 3 hours corresponds to 3 half-lives and 1 hour therefore corresponds to 1 half-life.

Developing investigative skills

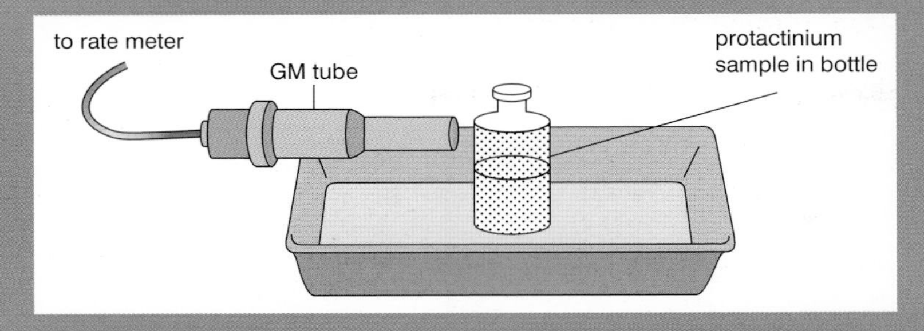

Δ Fig. 7.15 Apparatus for experiment.

A teacher demonstrates the radioactive decay of protactinium. Before starting the investigation, the teacher makes measurements to allow for background radiation. Then she sets a G–M tube close to a bottle containing the protactinium source. The G–M tube is connected to a ratemeter. The teacher starts a stopclock and measures the activity of the source every 30 seconds for 5 minutes. The data is shown in the table.

Time (minutes)	Count rate, corrected for background radiation (counts per second)
0	80
0.5	60
1.0	44
1.5	35
2.0	28
2.5	21
3.0	16
3.5	14
4.0	10
4.5	7
5.0	6

Δ Table 7.5 Measuring radioactive decay of protactinium.

Devise and plan investigation

❶ Why does the teacher carry out this experiment, rather than a student?

❷ What measurements should the teacher make to correct for background radiation?

❸ What was the independent variable in this investigation? What was the dependent variable?

Analyse and interpret

➍ Draw a graph of the results.

➎ Use your graph to find the half-life of the sample.

Evaluate data and methods

➏ Radioactivity is a random process. What does this mean and is there any evidence from the graph to support the idea?

➐ Why was it important to correct for background radiation?

➑ Would repeating the experiment and taking an average give a more accurate value of the half-life? Explain your answer.

QUESTIONS

1. Give some sources of background radiation.
2. What does the term *half-life* mean?
3. A radioactive element has an activity of 1200 counts per minute. Four hours later the count rate is 300 counts per minute. What is the half-life of the radioactive element?
4. A radioactive element has an activity of 1000 counts per minute. Twelve hours later the count rate is 125 counts per minute. What is the half-life of the radioactive element?
5. A radioactive sample has a half-life of 8 hours. If the activity is 800 Bq now, after how long will the activity be 100 Bq?

USES OF RADIOACTIVITY

Gamma rays can be used to kill bacteria. This is used in sterilising medical equipment and in preserving food. The food can be treated after it has been packaged. The food is **irradiated** with gamma rays but this does *not* make the food itself radioactive.

A smoke alarm includes a small radioactive source that emits alpha radiation. The radiation produces ions in the air which conduct a small electric current. If a smoke particle absorbs the alpha particles, it reduces the number of ions in the air, and the current drops. This sets off the alarm.

Beta particles are used to monitor the thickness of paper or metal. The number of beta particles passing through the material is related to the thickness of the material.

A gamma source is placed on one side of a weld and a photographic plate on the other side. Weaknesses in the weld will show up on the photographic plate.

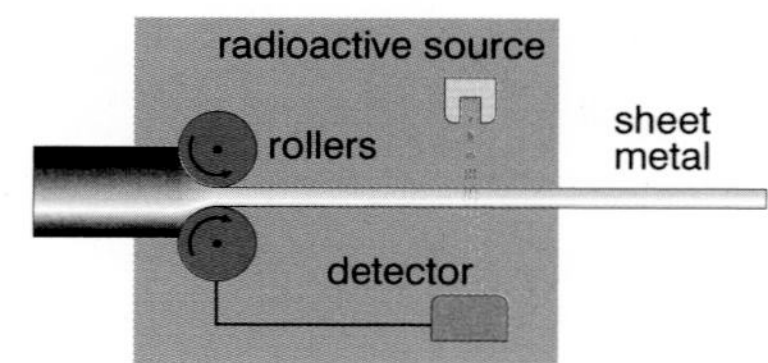

Δ Fig. 7.16 Sheet thickness control.

In **radiotherapy** high doses of radiation are fired at cancer cells. This irradiates the cancer cells and kills them. It is important to understand that when irradiated the cancer cells do not become radioactive themselves. Here, as in the case of X-rays, the radiation that can cause cancer is also an important tool in treating it.

Tracers are radioactive substances with half-lives and radiation types that suit the job they are used for. The half-life must be long enough for the tracer to spread out and be detected, but not so long that it stays in the system and causes damage.

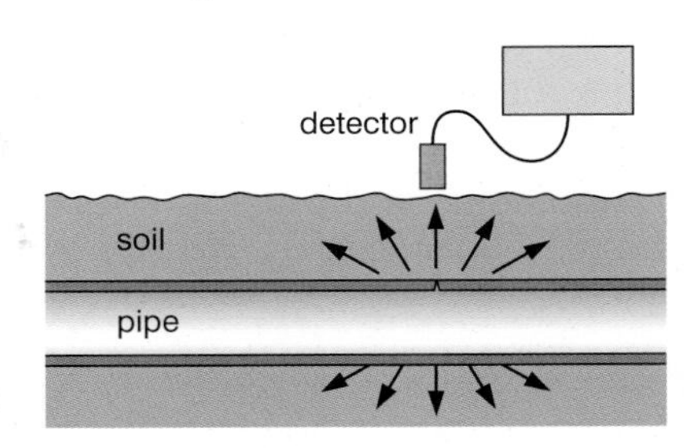

Δ Fig. 7.17 Tracers detect leaks.

- Medical tracers are used to detect blockages in vital organs. A gamma camera is used to monitor the passage of the tracer through the body.
- Agricultural tracers monitor the flow of nutrients through a plant.
- Industrial tracers can measure the flow of liquid and gases through pipes to identify leakages.

Radioactive dating

Igneous rock, which is formed through the cooling and solidification of magma and lava, contains small quantities of uranium-238, a type of uranium that decays with a half-life of 4500 million years, eventually forming lead. The ratio of lead to uranium in a rock sample can be used to calculate the age of the rock. For example, a piece of rock with equal numbers of uranium and lead atoms in it must be 4500 million years old – but this would be unlikely, as the Earth itself is 4500 million years old.

Δ Fig. 7.18 This man's body was preserved in a bog in Denmark. Radioactive carbon dating showed that he had been there for over 2000 years.

Carbon in living material contains a constant, small amount of the radioactive isotope carbon-14, which has a half-life of 5700 years. When the living material dies the carbon-14 nuclei slowly decay. The ratio of carbon-14 nuclei to the non-radioactive carbon-12 atoms can be used to calculate the age of the plant or animal material. This method is called **radioactive carbon dating**.

Other uses of isotopes

Radioisotopes have the same chemistry as non-radioactive isotopes of the same element. This can be very valuable in research as well as medicine. A well-known example is the use of radioactive iodine in treatment of cancer of the thyroid. The thyroid gland can become cancerous and the cancer then spreads through the body. Because the cells of the thyroid absorb far more iodine than other parts of the body, the cancer can be targeted by injecting the body

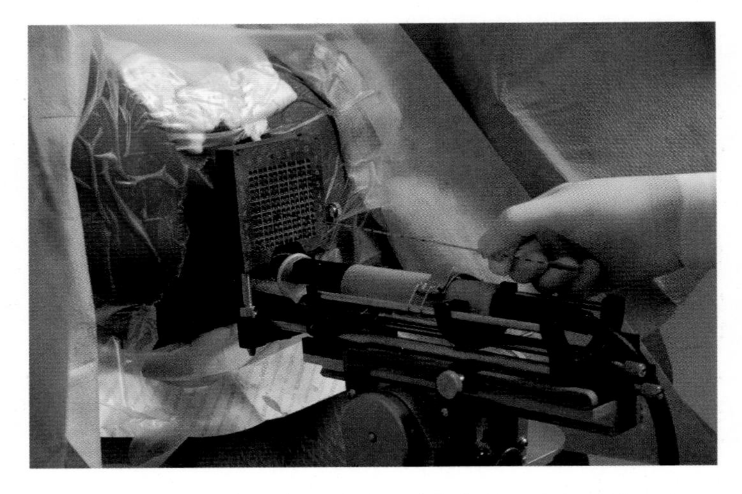
Δ Fig. 7.19 Radioactive iodine is injected during a treatment for cancer.

with radioactive iodine. The iodine is absorbed by the cancerous cells wherever they are in the body, after which it kills them. Another example is carbon-11, which is used in a technique called positron emission tomography (PET) scanning, which builds up images of the organs inside a human body. Phosphorus-32 is often used to label label amino acids and phosphoproteins in biochemical studies.

DANGERS OF IONISING RADIATION

Alpha, beta and gamma radiation can all damage living cells. Alpha particles, due to their strong ability to ionise other particles, are particularly dangerous to human tissue. Gamma radiation is dangerous because of its high penetrating power. However, the cell has repair mechanisms that make ordinary levels of radiation relatively harmless.

Nevertheless, radiation can be very useful – it just needs to be used *safely*.

Safety precautions for handling radioactive materials include:

- Use forceps to hold radioactive sources – don't hold them directly.
- Do not point radioactive sources at living tissue.
- Store radioactive materials in lead-lined containers – and lock the containers away securely.
- Check the surrounding area for radiation levels above normal background levels.
- Reduce time spent with radioactive materials to reduce the total exposure.

High levels of radiation are extremely hazardous. People who handle highly radioactive materials must wear special film badges (containing photographic film) that monitor the dose that they are receiving. They may need to wear protective clothing, perhaps containing sheets of lead, and they need to shower and check for radioactivity on their bodies at the end of each shift.

High levels of radiation can cause **mutations** in living organisms. A mutation is a change in the genetic material in the nucleus of a cell. This can make the cell behave in a different way. Sometimes it can cause the cells to be cancerous.

Many people think that if you expose materials to alpha, beta or gamma radiation then the materials themselves can become radioactive. However, this is not the case. Dried herbs and spices used for cooking are **irradiated** by having alpha, beta or gamma radiation fired at them. This kills any bacteria or mould and so helps keep the herbs and spices keep longer. The spices do not themselves contain a source of ionising radiation – they are **not contaminated**. Cotton swabs used for first aid are often sterilised by irradiation during manufacture. They are not contaminated with radioactive material, so do not emit ionising radiation themselves.

The use of radium in luminous paint for watches and aircraft instruments was common in the 1920s and 1930s. Radium emits alpha particles which caused pigments in the paint to glow in the dark. Unfortunately, many of the workers who used the paint became contaminated with the radioactive radium through licking and handling the brushes. This meant that alpha radiation was emitted from within their bodies, causing cancers to develop.

The problem of radioactive waste

As we shall see in the next topic, Fission and fusion, some isotopes of uranium are used in nuclear power stations to generate electricity. Uranium is mined from rocks underground. The dust caused by mining is dangerous as it contains radioactive uranium, which can be breathed in or absorbed through the skin. As it decays inside the body, the uranium emits ionising radiation which will damage cells and tissue.

Used fuel and other materials at nuclear power stations can be highly radioactive. This radioactive waste has to be stored safely to reduce the risk of harm for workers and to prevent leaks that would contaminate the surrounding air, ground or water. Highly radioactive waste can develop high temperatures as it decays, and needs cooling. Some isotopes in radioactive waste have very long half-lives and the waste will remain harmful to living things for millions of years. This material may be encased in glass or metal containers and buried in managed areas deep underground.

QUESTIONS

1. Give two uses of gamma rays for sterilising.
2. What are tracers? Give three different areas where they are used.
3. Describe two medical and two non-medical uses of radioactive isotopes.
4. Explain why safety precautions need to be taken when handling radioactive materials.
5. Explain the difference between irradiation and contamination.

HOW OLD IS THE EARTH?

How old is the Earth? About 4500 million years old. But how do we know that?

During the 19th century, the science of geology was developing. Rock samples and fossils were collected from all over the world and organised into different categories. Their similarities were noted and a timeline for the different rock samples was worked out. However, the timeline was only relative – it was only possible to say that one rock was older than another; there was no way to put an actual figure on it in years.

Some observers looked at the way rivers and seas washed fragments away – evidence of the land being broken down is everywhere. Some people believed that the Earth could not be very old because if it was old, the rivers would have smoothed it all out by now. Others said that was not important as long as there was a way for the land to be built back up again.

The theory of plate tectonics, which describes huge land masses pushing against each other, was not developed until well into the 20th century; before then, there was no way to explain how mountains were created.

Δ Fig. 7.20 The boundary of two tectonic plates in Iceland.

Also during the 19th century, the British naturalist Charles Darwin had published his theory of evolution. This argued that changes to species to happen only gradually over long periods of time. Supporters of this theory therefore believed the Earth to be ancient.

Another approach was needed. It seemed sensible that even if it was impossible to work out the age of the Earth, it might be possible to work out the age of the Sun. Whatever that turned out to be, the Earth must be younger.

Surprisingly, working out the age of the Sun was easier to do – or at least it seemed that way. There is a very simple experiment for measuring how much energy the Sun produces. It only requires a pan of water, a thermometer and a clock. The pan of water is put into the sunlight, so that the increase in temperature in a fixed amount of time can be measured. Because we know how much energy it takes to raise the temperature of water, we can work out how much energy arrived at our pan from the Sun in that time. Then we just need to work out what fraction of the Sun's total output that is – you should be able to figure out how to do that!

Using the total mass of the Sun, it is easy to work out how long the Sun will take to burn up its fuel. The problem was that in the 19th century, the process of nuclear fusion (which makes the Sun 'burn') was not known. Scientists assumed that the Sun burned like a fire, using a combustion reaction. Calculated in this way, the Sun's age would be only a few thousand years – indicating that the Sun had to be much younger than the Earth.

Rocks could only be dated accurately when radioactivity and the idea of half-life had been understood in the 20th century. The proportion of radioactive uranium in rocks now compared to that when rocks are formed allowed convincing evidence to be found for the true age of the Earth. The discovery of nuclear fusion allowed a true age for the Sun to be calculated also. It is now calculated at about 4.5 billion years.

End of topic checklist

A substance is **radioactive** when it has nuclei that are not stable. When the particle or energy is emitted from the unstable nucleus, **radioactive decay** takes place. Radioactivity is a random process.

The activity of a radioactive substance is measured in **becquerels** (Bq).

Alpha, **beta** and **gamma** radiation are types of **ionising radiation**. They have different properties in terms of their ionising ability, **penetrating power**.

Neutron radiation is the release of a high-speed neutron from the nucleus.

Background radiation is low level nuclear radiation that exists everywhere, from rocks and other environmental sources.

Half-life is the time taken for half of the radioactive nuclei in a sample to decay.

Isotopes are atoms of the same element that contain different numbers of neutrons. Isotopes have the same **atomic number** but different **mass numbers**.

The facts and ideas that you should understand by studying this topic:

- ❍ Describe the structure of an atom and use symbols to describe particular nuclei.
- ❍ Understand the terms atomic (proton) number, mass (nucleon) number and isotope.
- ❍ Understand that alpha, beta and gamma radiations are ionising radiations emitted from the nuclei of unstable atoms.
- ❍ Understand how these radiations can be detected.
- ❍ Describe the nature and properties of alpha, beta, gamma and neutron radiations.
- ❍ Describe the effect on the nucleus of emitting alpha, beta or gamma radiation.
- ❍ Understand how to balance the masses and charges in nuclear equations.
- ❍ Explain the sources of background radiation.
- ❍ Understand that that activity of a sample decreases over time.
- ❍ Understand the term half-life and use the concept to carry out simple calculations on activity.
- ❍ Describe the uses of radioactivity such as in medical and non-medical tracers, in radiotherapy and in radioactive dating.
- ❍ Describe the dangers of ionising radiations from contamination and irradiation, including damage to cells and tissues, mutations in living organisms and problems arising from the disposal of radioactive waste, and describe how the risks can be reduced.

End of topic questions

1. Copy and complete this table to show the particles in these atoms. **(12 marks)**

Atom	Symbol	Number of protons	Number of neutrons	Number of electrons
Hydrogen	${}^{1}_{1}\mathrm{H}$			
Carbon	${}^{12}_{6}\mathrm{C}$			
Calcium	${}^{40}_{20}\mathrm{Ca}$			
Uranium	${}^{238}_{92}\mathrm{U}$			

2. The graph shows how the activity of a sample of sodium-24 changes with time. Activity is measured in becquerels (Bq).

a) Sodium-24 has an atomic number of 11 and a mass number of 24. What is the composition of the nucleus of a sodium-24 atom? **(3 marks)**

b) Use the graph to work out the half-life of sodium-24. **(3 marks)**

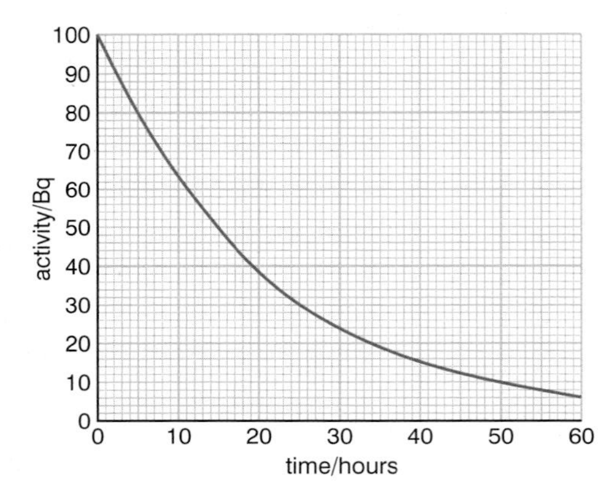

3. The following equation shows what happens when a nucleus of sodium-24 decays.

$${}^{24}_{11}\mathrm{Na} \rightarrow {}^{x}_{y}\mathrm{Mg} + {}^{0}_{-1}\beta$$

a) What type of nuclear radiation is produced? **(1 mark)**

b) What are the numerical values of *x* and *y*? **(2 marks)**

4. What is the difference between two or more isotopes of an element? **(2 marks)**

5. Describe the properties of beta radiation. **(3 marks)**

6. What do you understand by ionisation? **(2 marks)**

7. Radon-220 is radioactive. It decays to polonium-212. Write the nuclear equation for its decay by alpha radiation. The atomic number of radon is 86 and the atomic number of polonium is 84. **(3 marks)**

8. Carbon-14 decays by beta radiation. Write the nuclear equation. **(3 marks)**

9. A laboratory technician has three radioactive sources: cobalt-60, which is a source of gamma rays; strontium-90, which is a source of beta particles; and americium-241, which is a source of alpha particles. Suggest how the technician can tell which is which. **(6 marks)**

10. During experimental work, why should you take background radiation into account? How would you do this? **(3 marks)**

11. A sample of radioactive material has 800 atoms.

    a) How many atoms are left after two half-lives? **(2 marks)**

    b) How many half-lives does it take to reduce it to 50 atoms? **(2 marks)**

    c) If the material has a half-life of 30 minutes, how long does it take for the number of atoms to reduce to 100? **(2 marks)**

12. Iodine-131 has a half-life of 8 days. A sample has a count rate of 128 counts/minute.

    a) What will the count rate be after 3 half-lives? **(2 marks)**

    b) Calculate the time it will take for the activity rate to drop to 4 counts/minute. **(2 marks)**

13. The graph shows the decay curve for strontium-93. Determine its half-life. **(3 marks)**

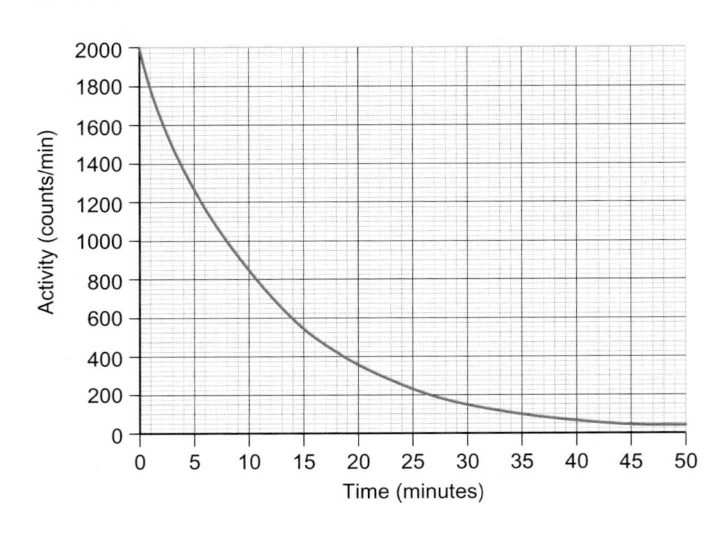

End of topic questions continued

14. In March 2011, traces of the radioisotope iodine-131 were found in air-monitoring stations in the UK following an earthquake that damaged the Fukushima nuclear plant in Japan. The graph shows how the count rate from the traces varied from the initial level, 80, over time.

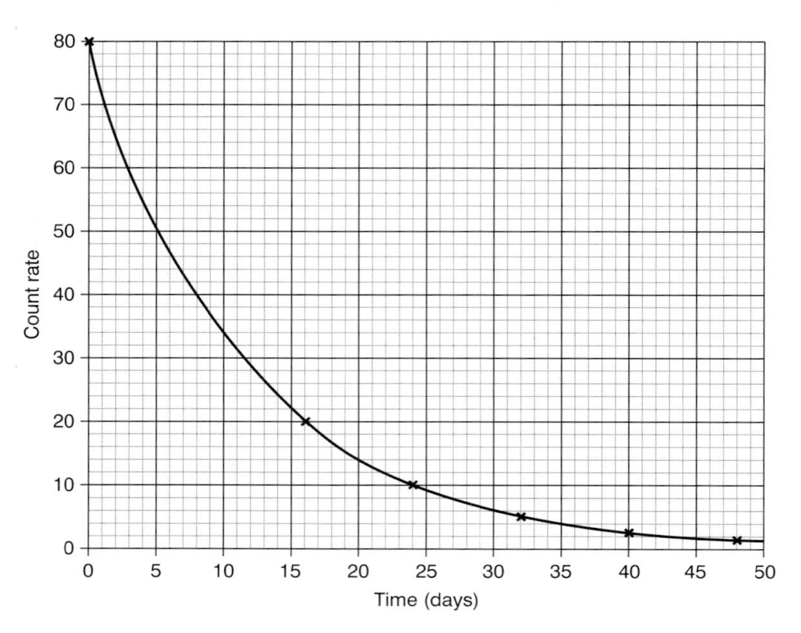

a) What is the half-life of the isotope? **(1 mark)**

b) If the initial count rate had been 2000, how many days would it have taken for the count rate to fall below 50? **(1 mark)**

c) If the initial count rate had been 2000, how many days would it have taken for the count rate to fall below 20? **(1 mark)**

d) If the initial count rate had been 2000, how many days would it have taken for the count rate to fall below 5? **(1 mark)**

15. A student has three sources of radiation: **A**, **B** and **C**. One emits alpha radiation, another emits beta radiation and the other emits gamma radiation. Describe an experiment you could do to identify the three sources. **(8 marks)**

16. This question is about tracers.

a) What is a tracer? **(1 mark)**

b) The table shows the half-life of some radioactive isotopes.

Radioactive isotope	Half-life
lawrencium-257	8 seconds
sodium-24	15 hours
sulphur-35	87 days
carbon-14	5700 years

Using only the information in the table, state which one of the isotopes is most suitable to be used as a tracer in medicine. Give a reason for your choice. **(4 marks)**

17. Beta radiation is used to test the thickness of aluminium foil in a factory.

a) Gamma rays and alpha radiation are not used for thickness testing for aluminium. Explain why. **(4 marks)**

b) Explain how the beta radiation is used to measure and control the thickness of aluminium. **(4 marks)**

c) The half-life of the beta source is 4 years.

The thickness control unit needs resetting every few months. Explain why and what may happen if it is not re-set. **(4 marks)**

Δ Fig. 7.21 This submarine gets its energy from a small nuclear reactor inside it.

Fission and fusion

INTRODUCTION

By splitting nuclei of uranium-235, electricity can be generated from the thermal energy produced from nuclear fission. Nuclear fusion where small nuclei are forced together to produce larger nuclei can produce massive amounts of energy. It happens in the stars but how practical is it to do this on Earth? In this section you will learn about both these developments. In particular, you will learn how splitting uranium-235 nuclei can be controlled well enough to be useful and generate electricity while minimising the dangers of releasing so much energy.

KNOWLEDGE CHECK

✓ Know the basic structure of the atom in terms of protons, neutrons and electrons.
✓ Know that radioactive materials have unstable nuclei.
✓ Know that to generate electricity, several processes rely on heating water.

LEARNING OBJECTIVES

✓ Know that nuclear reactions, including fission, fusion and radioactive decay, can be a source of energy.
✓ Know that the process of nuclear fission releases energy as kinetic energy of the fission products.
✓ Understand how fission of a nucleus of U-235, by collision with a neutron, produces two radioactive daughter nuclei and a small number of neutrons.
✓ Describe how a chain reaction can be set up if the neutrons produced by one fission strike other U-235 nuclei.
✓ Describe the role played by the control rods and moderator in the fission process and the role of shielding around a nuclear reactor.
✓ Be able to explain the difference between nuclear fusion and nuclear fission.
✓ Be able to describe nuclear fusion as the creation of larger nuclei resulting in a loss of mass from smaller nuclei, accompanied by a release of energy.
✓ Know that fusion is the energy source for stars.
✓ Be able to explain why nuclear fusion does not happen at low temperatures and pressures, due to electrostatic repulsion of protons.

NUCLEAR FISSION

A few nuclides have the strange property of splitting into two nuclei if they are hit by a slow-moving neutron. This property is called **nuclear fission.** What makes it of great practical importance is that when the nuclide splits:

- It gives out a lot of energy in the form of kinetic energy of the two smaller nuclei produced and the release of high-speed neutrons.
- It gives out two or three more neutrons that can cause the fission of more nuclei.

One of just a few nuclides that can be split by nuclear fission is the uranium isotope U-235. The equation for a typical reaction is:

$$^{235}_{92}\mathrm{U} + {}^{1}_{0}\mathrm{n} = {}^{137}_{56}\mathrm{Ba} + {}^{97}_{36}\mathrm{Kr} + {}^{1}_{0}\mathrm{n} + {}^{1}_{0}\mathrm{n} + Q$$

You will notice that this equation is balanced by the rules given above. Note also that the neutron has a mass of 1 and no charge.

The starting point is uranium metal. After the split, the result is barium metal and krypton gas. In this reaction, two neutrons are emitted, along with a lot of kinetic energy (represented by the letter Q).

The U-235 will split in one of many possible alternative ways. Instead of ending up as barium and krypton, it could turn into lanthanum and bromine, or cesium and rubidium. But generally both products (the **daughter nuclei**) will be radioactive, and two or three neutrons are emitted (**neutron radiation**).

These products of fission in a nuclear reactor are known as radioactive waste. Other materials, such as the control rods (see page 365) or the metal jackets that surround the fuel rods, may become radioactive due to neutron radiation. These materials may remain radioactive for many hundreds of years as the new isotopes produced often have very long half-lives.

The reactor itself is therefore highly radioactive and it must be shielded from the surroundings. Typically, the gamma radiation is stopped by dense materials. Lead, steel and concrete are often used. Sometimes water is used but this layer would have to be at least ten times thicker than lead shielding. Alpha and beta radiation are easily stopped by all these materials. The neutron radiation produced in the fission reaction is highly penetrating and potentially very harmful. The neutrons are absorbed by boron-10, which can be in boric acid dissolved in water.

It is tempting to think that as alpha and beta radiation are easy to absorb they are not dangerous. However, humans could be contaminated by breathing in dust or gases containing an alpha source, which could then be stored in the body for years. During that time, it would irradiate cells inside the body causing severe localised damage.

A chain reaction

If the neutrons emitted by fission of a nucleus strike other atoms and cause them to split, then a **chain reaction** will occur. Consider the case where each nucleus causes two more nuclei to split. When these two nuclei split, they will cause four nuclei to split, and then 8, 16, 32, 64, 128, and so on. More and more heat will be generated by the reaction. This uncontrolled chain reaction will cause an explosion, which is not something you want to happen in a nuclear power station!

Instead, it would be better to have a chain reaction in which every nucleus causes just one further nucleus to split. This would provide the steady heat that can generate electricity.

You can set up a row of dominoes so that each domino knocks down one further domino. This is a controlled chain reaction. This is shown in Figure 7.22.

You can also try setting up the dominoes so that one domino knocks down two dominoes, which then knock down four, etc. If you succeed you will get something more like an uncontrolled chain reaction.

Δ Fig. 7.22 A model for a controlled chain reaction.

QUESTIONS

1. What is nuclear fission?
2. What happens when a nuclide splits?
3. Describe how fission can lead to a chain reaction.

THE NUCLEAR REACTOR

Nuclear fission releases energy as kinetic energy of the fission products. The thermal energy from a nuclear chain reaction is used in a nuclear power station to heat water to produce steam. The steam drives a turbine to generate electricity.

Inside a reactor are the following components:

- Fuel rods: The uranium fuel (a proportion of which is uranium-235) is assembled into rods. The rods can easily be inserted and removed to load and unload the reactor. Each rod is used for a few months until the amount of U-235 is greatly reduced.

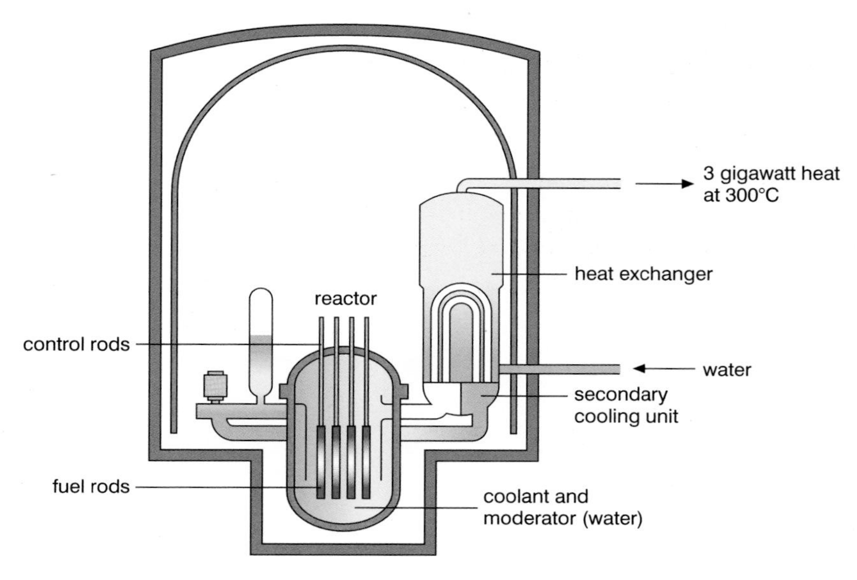

Δ Fig. 7.23 The components of a nuclear power station.

- **Control rods**: These rods are made of a metal that absorbs neutrons, such as cadmium or boron. If the rods are fully inserted, the chain reaction stops. If the rods are moved out, the reaction starts going faster and faster.
- **Moderator**: The gaps between the fuel rods and control rods are filled with a moderator such as carbon or **heavy water** (this is water that has an isotope of hydrogen, deuterium, combined with oxygen rather than the normal hydrogen). The job of the moderator is to slow down the neutrons emitted by the U-235. Without it, the neutrons will be going too fast and will escape from the reactor without creating fission. It may seem strange that a neutron can be going *too fast* to split an atom of U-235, but it is possible.

Cooling is also needed to prevent the reactor over-heating. This is done by circulating pressurised water or another fluid through the reactor. In a power station, the exhaust from the reactor is used to boil further water for the turbines.

Nuclear power stations will continue to be controversial, but the fact remains that they are extremely important. They generate 17 per cent of the world's electricity, and they have prevented the release of millions of tonnes of carbon dioxide and sulfur dioxide from coal and oil into the Earth's atmosphere.

◁ Fig. 7.24 Controlled fission is used in a nuclear power station.

QUESTIONS

1. Describe the purpose of the fuel rods, control rods and moderator in a nuclear fission reactor.
2. How is the reactor cooled?
3. The waste from a nuclear reactor must be handled very carefully. Explain why.

EXTENSION

Figure 7.25 shows a chain reaction. If the energy released in a chain reaction is to be used to produce power, it needs to be controlled. This is done using control rods, which are made from materials that absorb neutrons, such as boron.

1. Explain how control rods control the reaction.
2. What would you do to increase the rate of reaction in a nuclear reactor?
3. What would you do to reduce the rate of reaction in a nuclear reactor?

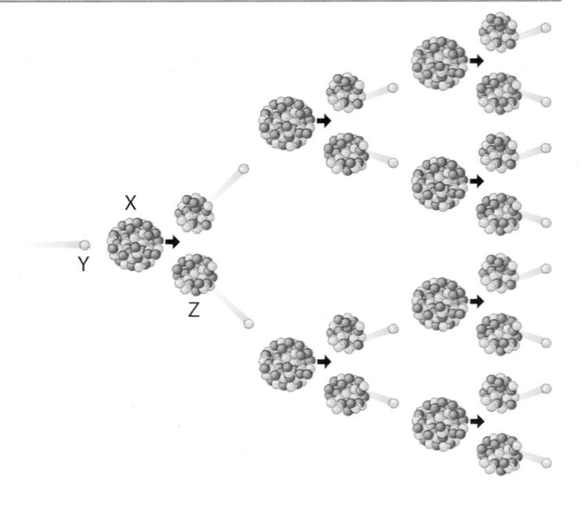

Δ Fig. 7.25 A nuclear chain reaction.

NUCLEAR FUSION

Nuclear fusion occurs when two nuclei join to form a larger nucleus. Fusion of lighter nuclei, such as hydrogen, results in a loss of mass accompanied by a release of energy.

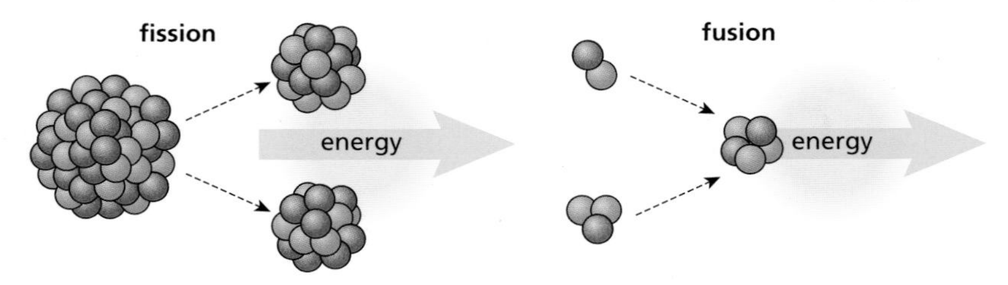

Δ Fig. 7.26 Nuclear fission and nuclear fusion.

Nuclear fusion of lighter elements releases the energy that keeps stars hot. The same fusion reaction releases a destructive amount of energy when a man-made hydrogen bomb explodes.

Nuclear fusion and the Sun

The Sun is vital to life on Earth. Without solar radiation there would be no weather, photosynthesis or rain. The energy radiated from the Sun comes from a nuclear fusion reaction. This reaction forces two small hydrogen nuclei together and this produces a larger helium nucleus, a neutron and a large amount of energy. This energy comes from the small amount of mass that is lost in the reaction.

Fusion is the energy source for all stars, and this will be covered again in Section 8.

Conditions for fusion

The nucleus of any atom is positive as it contains at least one proton. So forcing two positive nuclei together is not easy as the nuclei repel each other with electrostatic forces. The Sun has very high temperatures which means the nuclei of gases in the Sun are moving around at extremely high speeds. The pressure is also very high, which keeps the nuclei close together. These two factors can force the nuclei together, overcoming their electrostatic forces and causing nuclear fusion.

Researchers are trying to produce usable energy from fusion reactions on Earth. This could mean large amounts of energy could be cheaply produced without the hazards of the radioactive waste produced in nuclear fission. At present, they are having trouble overcoming the electrostatic repulsive forces between the nuclei. They know that high temperatures and high pressures are required (as in the Sun) but these conditions may be too impractical to create and sustain on Earth. Currently, the experimental reactors require more energy than they produce.

Attempts have also been made to create fusion reactions at low and manageable temperatures and pressures. These 'cold fusion' reactions are gathering interest. The idea that energy could be created on a large scale inexpensively and with low health and safety risk is without doubt an attractive one. Scientists, however, are unconvinced about how realistic the idea is. Nevertheless, research by NASA and well-established companies continues and may unearth a future for this developing or hopeful technology.

End of topic checklist

In **nuclear fission**, a large nucleus absorbs a neutron and then splits into two large fragments, releasing energy and further neutrons. These neutrons can then be absorbed by other nuclei, creating a chain reaction.

A **chain reaction** is a series of nuclear fission reactions where neutrons released from one reaction cause another nuclear fission reaction and so on.

Nuclear fusion is when smaller nuclei combine to create a larger nucleus. Some of the mass of the nuclei is converted to energy.

The facts and ideas that you should understand by studying this topic:

- ○ Understand that a nucleus of uranium-235 can be split following a collision with a neutron and that this process is called nuclear fission.
- ○ Understand that in nuclear fission, further neutrons are released which can then cause further fissions in a chain reaction.
- ○ Understand that a controlled chain reaction is the energy source for a nuclear power station.
- ○ Understand the functions of the shielding, control rods and moderator in a nuclear reactor.
- ○ Understand that in the Sun a fusion reaction forces two hydrogen nuclei together and this produces a helium nucleus, a neutron and a large amount of energy.
- ○ Be able to explain the difference between nuclear fission and nuclear fusion.
- ○ Be able to describe the difficulties of using nuclear fusion to produce energy for electrical generation on Earth.

End of topic questions

1. Nuclear power stations produce alpha, beta and gamma radiation and also neutron radiation. Describe how these different emissions can be shielded by different materials. **(4 marks)**

2. Describe briefly the difference between nuclear fission and nuclear fusion. **(2 marks)**

3. Describe what happens in nuclear fission. **(3 marks)**

4. Describe how splitting an atom of uranium can lead to a chain reaction. **(3 marks)**

5. Describe how a fission reaction becomes unstoppable. **(3 marks)**

6. What makes an unstoppable fission reaction so damaging? **(3 marks)**

7. How can the energy from a nuclear fission reaction be harnessed safely? **(3 marks)**

8. Large quantities of energy are released in a chain reaction. Explain why. **(6 marks)**

9. Use the words below to explain how nuclear fission is controlled in a nuclear power station. **(6 marks)**

   colliding neutrons prevent fission control absorbed nuclei

10. Many scientists are hopeful that nuclear fusion will, one day, be used for energy generation. Explain some of the difficulties in using nuclear fusion for energy generation. **(5 marks)**

11. There is a report on the local news about a proposal to build a nuclear power station near your town.

    The company spokesperson says, 'The station is designed to be safe. The uranium fuel rods are alternated with control rods to manage the rate of heat production. At the first sign of danger, these drop into the reactor and shut it down completely in seconds. The method of producing electricity is very clean. Many people think the cooling towers are giving off poisonous gases, but in fact it's just water vapour, steam, as part of the heat-flow system that drives the generators. It's much cleaner than a coal-fired power station and is the only realistic way to meet our energy needs.'

A member of an environmental lobbying group was also interviewed and said, 'It's an outrage that they want to build a nuclear power station here! The sites are expensive to build and have a huge decommissioning cost after a lifespan of only 40 or so years. Add to that the impact on the local environment, and the fact the nuclear waste will have to be transported through the area and then stored for thousands of years. We should be building a more sustainable method of generating electricity.'

Based on the information above, answer the following questions.

a) The environmental lobbying group say that nuclear power stations are bad for the environment.

i) Is there any evidence in the news article that supports this view? **(2 marks)**

ii) What evidence is there in the text against this view? **(2 marks)**

iii) Why might the environmental lobbying group make such a statement? **(2 marks)**

b) If you were a householder 5 km away from the site of a proposed nuclear plant, would you be happy for it to be built there? Explain the reasons for your answer, taking into account safety and environmental effects. **(6 marks)**

c) Radioactive waste needs to be stored for a long time. Research how this is done safely and write a paragraph or produce a labelled diagram to show this – including how we will let future generations know what is there. **(6 marks)**

Exam-style questions

Sample student answers

Question 1

Americium-241 is a radioactive isotope that decays mainly by emitting alpha particles.

It has a half-life of 432 years.

a) What are isotopes?

Atoms with the same number of protons ✓ but a different number of neutrons. ✓ (2)

(2)

b) What are alpha particles?

The nucleus of a helium atom. ✓✓ (2)

(2)

c) What does 'it has a half-life of 432 years' mean? **(2)**

The radiation loses half its power in 432 years. XX

d) Americium-241 is used in smoke detectors.

If there is no smoke present, the alpha particles ionise a path through the air and a small current passes.

i) Describe how alpha particles can ionise particles in the air.

The alpha particles hit the atoms in the air. X **(2)**

ii) Suggest what happens when particles of smoke enter the detector.

The smoke blocks the radiation so the alarm goes off. ✓XX (1) **(3)**

EXAMINER'S COMMENTS

a) Fine – standard definition. Answers in terms of atomic number or proton number and mass number or nucleon number would also be accepted.

b) Fine – answers in terms of protons and neutrons would also be accepted.

c) A very common mistake – this answer refers to the radiation given off (the alpha particles) and not the radioactive source itself. Half- life describes how the activity of the source will change over time; the radiation emitted does not decay.

d) These answers are typical of a candidate who knows about the situation but who does not give enough detail to gain the highest marks. Both answers here begin correctly but fail to develop the ideas sufficiently to gain full marks. In part i) the idea of removing electrons is required, in part ii) the consequences of blocking the radiation (no ionisiation, no current) are missing.

Exam-style questions continued

e) i) 'Cause mutations' just about scores the mark, although it does suggest that people may grow extra arms or something. 'Cause mutations in cells' would have been better.

ii) The candidate has missed the point of the question – it is in part **e)** for a reason. The examiner wanted the idea given in part **i)** to be developed further – in this case explaining why these harmful effects are NOT caused by the smoke detector. 'It's on the ceiling' gives the correct idea, but more detail is needed at this level.

e) Radioactive emissions can be harmful to humans.

i) Describe one harmful effect that radioactive emissions can have on humans.

Cause mutations ✓ (1) **(1)**

ii) Suggest why it is safe to have a radioactive source in a smoke detector.

It doesn't last very long ✗ *and it's on the ceiling.* ✗ **(2)**

(Total 14 marks)

Question 2

A nucleus of uranium-235 can absorb a neutron.

This causes the nucleus to be very unstable and it 'splits' into two main parts, releasing energy.

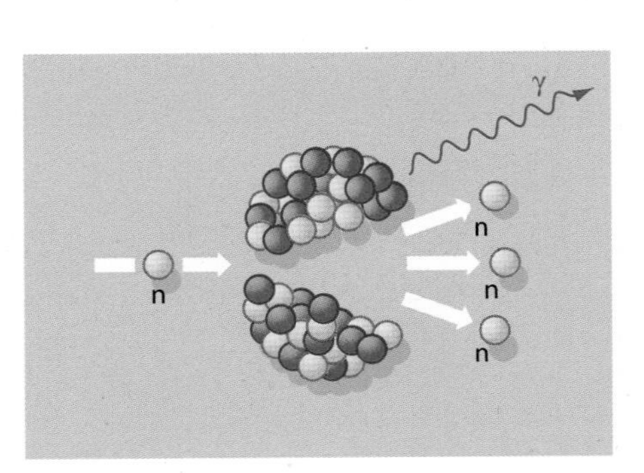

a) What is the name of this process? **(1)**

b) In what form is the energy released? **(2)**

c) Describe how this process can lead to a chain reaction.

You may draw a diagram to help your explanation. **(3)**

d) This process provides the energy source in a nuclear reactor.

In a nuclear reactor, describe the purpose of:

i) the control rods **(2)**

ii) the moderator. **(2)**

(Total 10 marks)

From simple viewing of the night sky to breathtaking images from the Hubble Space Telescope, humans have tried to make sense of what they see beyond the Earth. Many theories have been developed and many have been questioned and discarded as new evidence comes to light. This progress has been partly driven by new technologies from telescopes to space travel. But humans are very curious about our sky and that curiosity is unlikely to diminish.

You have met forces before and we will consider their effects on the movement of the planets, including Earth, and other celestial bodies. The vast amounts of energy released by nuclear fusion in stars must come to an end when temperatures and pressures are no longer great enough to sustain fusion reactions, and so a star's life cycle must come to an end. The wavelengths of light emitted by stars and the microwave radiation coming from every direction in space provide evidence for how the Universe began.

STARTING POINTS

1. What is the Solar System made of?
2. What are the differences between galaxies and stars, a red giant and a white dwarf?
3. What is the Universe and how might it have started?
4. What is meant by the life cycle of a star?

CONTENTS

8 Astrophysics

Δ More than 100 years ago in 1901, an ordinary star suddenly became one of the brightest stars in the sky. This image combines data from NASA's Chandra X-ray Observatory with radio and optical telescope data to look at the supernova it left. These pictures give insight into how scientists can study the birth, life and death of

Units

INTRODUCTION

For the topics included in astrophysics you will need to be familiar with:

Quantity	Unit	Symbol
mass	kilogram	kg
length, distance	metre	m
time	second	s
speed, velocity	metre per second	m/s
acceleration	metre per second^2	m/s^2
force	newton	N
gravitational field strength	newton per kilogram	N/kg

You have probably met all these units previously, in Section 1, Forces and motion.

LEARNING OBJECTIVES

✓ Use the following units: kilogram (kg), metre (m), metre/second (m/s), metre/second^2 (m/s^2), newton (N), second (s), newton/kilogram (N/kg)

Motion in the Universe

Δ Fig. 8.1 Edwin E Aldrin, Jnr is one of only a few humans who have stood on the surface of the Moon. Many unmanned spaceships have explored our solar system since. In the future, there is a real possibility of manned spaceships to Mars.

INTRODUCTION

Our developing understanding of our Sun, Solar System, galaxy and stars beyond the Milky Way Galaxy has been largely based on observations. Many predictions were made about the Moon – its gravity, the nature of its surface and its surface temperature. When people landed on the Moon scientists largely confirmed what they thought they already knew.

Many other space missions to planets and moons in our Solar System have taken place since, while the Hubble telescope has sent back many images of objects in space since its launch. Other space missions look for stars with planetary systems.

KNOWLEDGE CHECK

✓ Be able to name different types of object in the Solar System and in the wider Universe.
✓ Know how weight is related to mass and gravitational field strength.
✓ Know that gravity is a force of attraction between objects that have mass.

LEARNING OBJECTIVES

✓ Know that:
- the Universe is a large collection of billions of galaxies
- a galaxy is a large collection of billions of stars
- our Solar System is in the Milky Way galaxy.

✓ Understand why gravitational field strength, g, varies and know that it is different on other planets and the Moon from that on the Earth.
✓ Be able to explain that gravitational force:
- causes moons to orbit planets
- causes the planets to orbit the Sun
- causes artificial satellites to orbit the Earth
- causes comets to orbit the Sun.

✓ Describe the differences in the orbits of comets, moons and planets.
✓ Use the relationship between orbital speed, orbital radius and time period:π

$$\text{orbital speed} = \frac{2 \times \pi \times \text{orbital radius}}{\text{time period}}$$

$$v = \frac{2 \times \pi \times r}{T}$$

THE SOLAR SYSTEM

The **Solar System** is the general name for the Sun and the objects that orbit it. An orbit is a path where one object moves in a regular path around another.

Planets are large objects in orbit around a star. In the Solar System there are eight planets, which follow nearly-circular orbits. The four planets closest to the Sun are Mercury, Venus, Earth and Mars. These are balls of solid rock. The four outer planets are Jupiter, Saturn, Uranus and Neptune. These are very large balls of gases called gas giants. Pluto is no longer classified as a planet because it is too small according to the revised definition of a planet. Instead, Pluto is classed as one of 44 'dwarf planets' (more may be discovered in the future). According to the new definition, a planet is an object that orbits the Sun and is large enough to have become round due to the force of its own gravity. A planet also has to dominate the neighbourhood around its orbit.

Pluto has been demoted because it does not dominate its neighbourhood. It is only about double the size of its large 'moon' Charon, but all the true planets are far larger than their moons. Also, planets that dominate their neighbourhoods, 'sweep up' asteroids, comets, and other debris, clearing a path along their orbits. By contrast, Pluto's orbit is somewhat untidy.

More than 100 planets have been found in orbit around other stars.

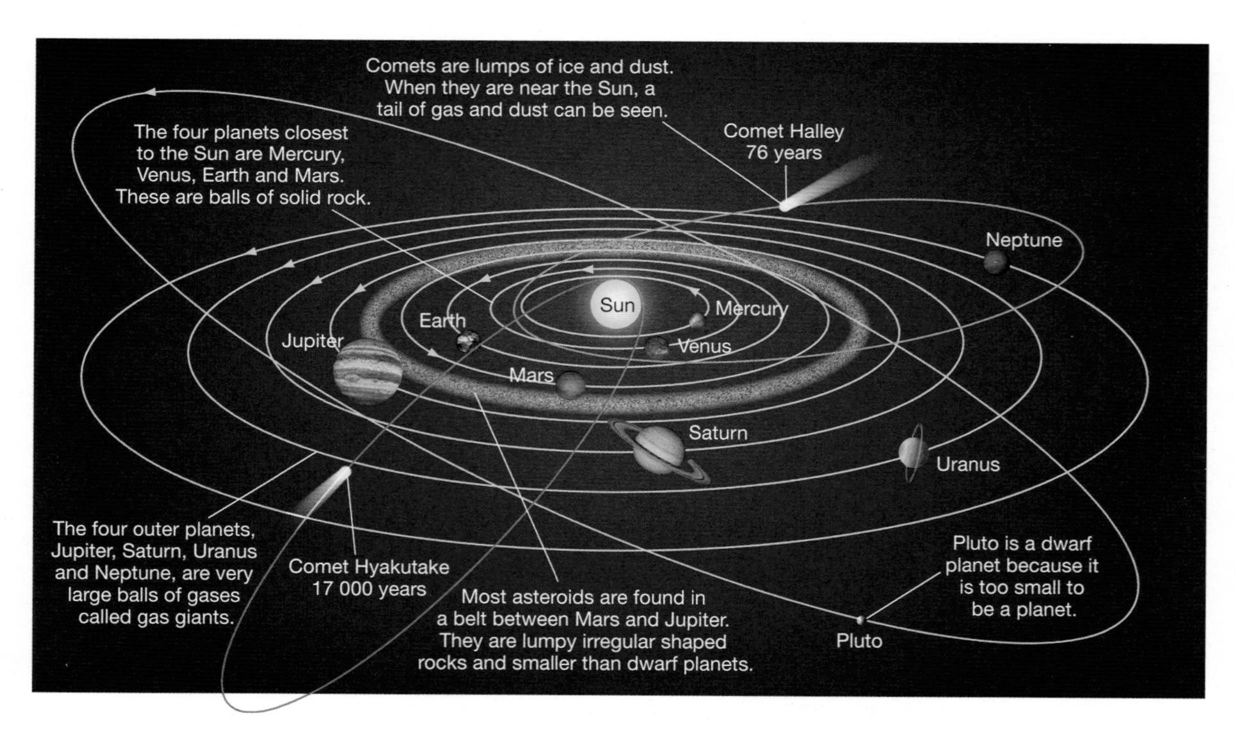

Δ Fig. 8.2 The planets of the Solar System.

QUESTIONS

1. What are the eight planets in the Solar System?
2. What is the main physical difference between the four inner planets and the four outer planets?
3. What is the definition of a planet?
4. Why has Pluto now been classed as a dwarf planet?

THE KEPLER MISSION

In March 2009, NASA launched its Kepler mission to look for stars with planetary systems. Research suggested that near-Earth-size planets might exist in habitable zones, but it was difficult to confirm this. Since 2009, at least nine potentially habitable Earth-sized planets have been found, as well as many thousands of other planets.

Ellen Stofan, chief scientist at NASA Headquarters in Washington, said 'This gives us hope that somewhere out there, around a star much like ours, we can eventually discover another Earth.'

The Kepler space telescope detects planets from other solar systems by sensing a change in brightness from a star. This is due to light from the star being blocked when a planet moves across its own sun. If the change in brightness occurs in a regular pattern this might indicate orbiting planets.'

△ Fig. 8.3 Engineers working on the Kepler spacecraft before it was launched.

COMETS AND MOONS

Comets are also in orbit around the Sun, but they are much smaller and follow a much more oval (elliptical) path (see Figure 8.2). Comets are made of ice and a mixture of other chemicals. As the comet gets closer to the Sun, some of the material evaporates and forms a tail. The tail points away from the Sun as it is pushed by the solar wind – a stream of particles emitted by the Sun. The speed of a comet changes as it follows its orbit: it moves faster when it is closer to the Sun. Comet Halley has an orbit of 76 years and Comet Hyakutake has an orbit of 17 000 years.

Moons are objects that orbit planets rather than the Sun, as planets and comets do. Our Moon is the one that orbits the Earth. Scientists think that the Moon was formed when a small planet collided with the Earth soon after it was formed 4.5 billion years ago. Most planets in the Solar System have moons in orbit around them. For example, Jupiter has 64, four of which were discovered by Galileo Galilei as long ago as January 1610.

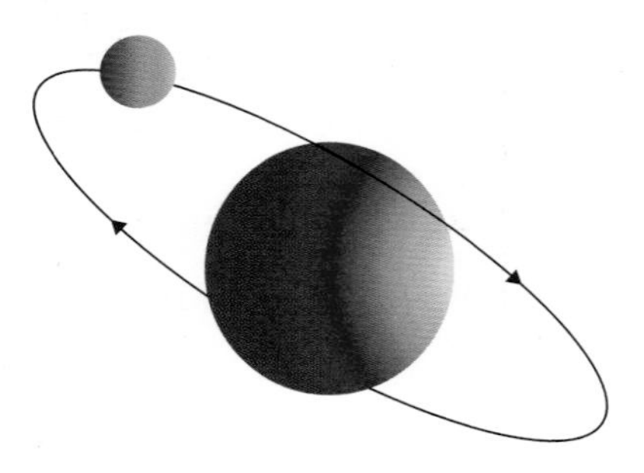

Δ Fig. 8.4 Moons orbit planets.

THE CASSINI MISSION: MOONS OF SATURN

In October 1997 the spacecraft Cassini was launched. It completed its original mission, in which it explored the Saturn System, in June 2008. Its first extended mission finished in September 2010. It is currently on its second extended mission, which is due to end in September 2017.

The moons of Saturn, Titan and Enceladus, are among the most important targets of the mission. Cassini has investigated the structure and chemistry of Titan's atmosphere. On the surface of Titan, Cassini's Huygens probe found huge lakes of methane and large stretches of hydrocarbon sand dunes, which have been sculpted by wind. Researchers have also deduced that there is an internal liquid water-ammonia ocean. On Enceladus, which is of interest because it is surprisingly active, Cassini discovered an icy plume shooting from the moon, which contains complex organic chemicals. Enceladus is being kept warm by tidal heating. Enceladus may be a place where primitive forms of life could evolve since it has heat, organic chemicals and, potentially, liquid water. Enceladus's 'astrobiological potential' is at the centre of many investigations being carried out on the extended mission.

QUESTIONS

1. Draw a diagram to explain the differences in the orbital motions of planets, moons and comets.
2. What are moons?
3. How do scientists think that Earth's Moon was formed?

SATELLITES

A **satellite** is any object that is in orbit around another. So planets are satellites around the Sun and the Moon is a satellite around the Earth. Over the past 50 years, many artificial satellites have been put into orbit around the Earth using rockets to launch them into space. These satellites follow a variety of paths and have a variety of uses, from communications to astronomical observations to monitoring the weather on Earth to and testing theories about the effects of global warming.

SATNAV AND GPS

Satellite navigation systems (SatNavs) rely on the Global Positioning System (GPS). GPS is maintained by the government of the United States of America and is freely accessible by anyone who has a GPS receiver, such as a SatNav. The whole system is based on satellites. It provides information about location anywhere on or near the Earth, where there is an unobstructed line of site to four or more GPS satellites. The system is not affected by weather.

△ Fig. 8.5 GPS (global positioning system) satellites allow people to check where they are at any time using just a mobile phone. SatNavs in cars can use this information to plot a route to any location.

The GPS system was developed in 1973 in order to overcome the limitations of previous navigation systems. It became fully operational in 1994 and was originally run using 24 satellites.

Orbits and gravitational force

Natural and artificial satellites are kept in orbit by a **gravitational force**. In the case of the Solar System it is this force of attraction between the Sun and the Earth that keeps the Earth in a very nearly circular orbit. Similarly, it is the force of gravity between the Earth and the Moon that keeps the Moon in orbit.

The orbits of artificial satellites are dependent on the gravitational force on them. The strength of the Earth's gravitational force decreases with distance above the surface. This means that the higher the satellite, the lower its orbital speed. Very high satellites take a day or more to make one orbit. In contrast, low satellites orbit close to Earth at high speed due to the larger gravitational force and can orbit Earth in about 100 minutes.

Comets orbit the Sun in a large elliptical orbit. When they are near to the Sun they move much faster because the gravitational force is greater.

Orbital speed

When launching an artificial satellite, a variety of calculations have to be carried out, including working out its **orbital speed**. For any object following a circular orbit, the speed of the object can be worked out using:

$$\text{orbital speed} = \frac{2 \times \pi \times \text{orbital radius}}{\text{time period}}$$

In symbols this is:

$$v = \frac{2 \times \pi \times r}{T}$$

WORKED EXAMPLE

The Earth orbits the Sun in one year. The average radius of the Earth's orbit is 150 million km. Find the average orbital speed of the Earth in km/h.

First, some unit conversions: One year = 365 × 24 hours = 8760 hours.

Radius of orbit = 150 million km = 150 000 000 km.

So:

$$\text{orbital speed} = \frac{2 \times \pi \times 150\,000\,000}{8760} = 108\,000 \text{ km/h}$$

QUESTIONS

1. A satellite orbits the Earth once every 90 minutes. The radius of its orbit is 6700 km.

a) Work out its orbital speed in km/h.

b) Work out its orbital speed in m/s.

2. A satellite orbits the Earth at a height of 100 km above the surface of the Earth. Work out the orbital speed of the satellite. ($r_{earth} = 6.4 \times 10^6$ m, $T = 5176$ s)

3. The period of the Moon is approximately 27.2 days (2.35×10^6 s). The radius of its orbit is 3.82×10^8 m. Work out the orbital speed of the Moon.

REMEMBER

Be careful with big numbers – make sure you are confident using your calculator.

Questions often jump back and forth between metres, kilometres, hours, days and so on.

GRAVITATIONAL FIELD STRENGTH

As you have seen earlier, the weight of an object is a force that measures how strongly gravity is pulling on the object. How big this force is depends on two things: the mass of the object, m, and the strength of the gravitational field, g, at that place.

Weight W = mass (m) × **gravitational field strength** (g).

Gravitational field strength is measured in newtons per kilogram (N/kg). It tells you what the force would be on a 1 kg mass placed at that point.

The strength of the gravitational field changes, depending on where you are in the Universe. Since gravity is generally quite a weak force, the most important factor is whether or not you are close to large masses such as moons, planets and stars.

The table shows the gravitational field strength near some different objects in the Solar System.

Object	Gravitational field strength (N/kg)
Sun	274
Mercury	3.6
Earth	9.8
Moon	1.6
Jupiter	26
Neptune	14

△ Table 8.1 Gravitational field strength in our Solar System.

REMEMBER

For International GCSE, the value of g for the Earth is taken to be 10 N/kg to make the calculations a little easier.

QUESTIONS

1. Calculate the weight of a 60 kg person on the Moon and on Jupiter.
2. Explain why astronauts on the Moon are not 'weightless'.
3. How much weight does a person of mass 65 kg 'lose' if they travel from Earth to the Moon? Use g = 10 N/kg.

MOVING OUTWARDS FROM THE SOLAR SYSTEM

Moving outwards from the Solar System, stars are not spread evenly throughout space. They collect together in large groupings called **galaxies**. The Sun and the Solar System are part of a galaxy called the Milky Way. Galaxies are collections of billions of stars along with all the planets, comets and other objects that orbit them.

Thanks to the work of astronomers such as Edwin Hubble, we know that there are billions of galaxies, separated by huge distances, and all these galaxies are in motion.

Moving out to even larger scales, there are clusters (groups) of galaxies and eventually everything there is – the entire Universe.

End of topic checklist

Gravity is a force that causes objects in the Solar System to travel in repeating cycles of motion called orbits.

Orbital speed is calculated from

orbital speed = $(2 \times \pi \times \text{orbital radius})/\text{time period}$.

The **gravitational field strength** is different on different planets.

The **Solar System** is part of the **Milky Way galaxy.**

There are billions of galaxies in the Universe.

The facts and ideas that you should understand by studying this topic:

- ❍ Know that gravity is an attractive force between masses which causes
  - moons to orbit planets
  - planets to orbit the Sun
  - artificial satellites to orbit the Earth
  - comets to orbit the Sun
- ❍ Know that gravitational field strength is different on different planets
- ❍ Use the relationship orbital speed = $(2 \times \pi \times \text{orbital radius}) / \text{time period}$
- ❍ Describe differences in the orbits of comets, moons and planets
- ❍ Understand how our Solar System is related to our local galaxy (the Milky Way) and how galaxies are related to the Universe as a whole.

End of topic questions

1. a) Describe the orbit of a planet. **(1 mark)**

   b) How does the orbit of a comet differ from that of a planet? **(1 mark)**

2. Comets have long orbital periods. Describe and explain how the speed of a comet changes throughout its orbit. **(4 marks)**

3. a) What is a satellite?

   b) Artificial satellites are used for weather reporting and TV broadcasting. Explain how satellites remain in orbit, above the Earth, without falling down. **(4 marks)**

4. a) A satellite orbits the Earth once every 24 hours. The height of the orbit above the centre of the Earth is 42 000 km. What is the orbital speed of the satellite?

   b) The period of Titan, one of Saturn's moons, is 16 days. Its orbit radius is approximately 1.2×10^6 km. What is its approximate orbital speed? **(2 marks)**

5. a) Why does an object weigh less on the Moon than it does on Earth? **(1 mark)**

   b) The gravitational field strength on the Moon is 1.6 N/kg. What is the weight of a person of mass 70 kg on the Moon? **(3 marks)**

6. Describe the differences between the orbits of moons and planets. **(2 marks)**

Δ Fig. 8.6 Cassiopeia A, a star in the constellation Cassiopeia, is a remnant of a supernova and appears as a shell of expanding material.

Stellar evolution

INTRODUCTION

When you look at the stars on a clear night you can see up to about 5000 without a telescope or binoculars. Most stars are larger than our Sun despite being visible only as tiny specks of light. Our Sun is much closer to the Earth than any other star so it seems larger. It takes about 8 minutes for light to reach us from the Sun. Our next nearest star is about 4 light years away. This is the distance that light travels in four years. Given that light travels at 300 000 000 metres every second, the nearest star is about $60 \times 60 \times 24 \times 365.25 \times 4 \times 300\,000\,000$ m away from us, or about 3×10^{16} m.

The sky and its stars stay much the same for long periods of time. But the sky does change and some stars are being made and some are ending their existence – sometimes in spectacular fashion.

KNOWLEDGE CHECK

✓ Know that the energy in stars is produced by nuclear fusion.
✓ Know that the Universe is massive and consists mainly of space.
✓ Know that the Universe changes over time.

LEARNING OBJECTIVES

✓ Understand how stars can be classified according to their colour.
✓ Describe the evolution of stars of similar mass to the Sun through the following stages:
- nebula
- star (main sequence)
- red giant
- white dwarf.

✓ Describe the evolution of stars with a mass larger than the Sun.
✓ Understand how the brightness of a star at a standard distance can be represented using absolute magnitude.
✓ Draw the main components of the Hertzsprung–Russell diagram (HR diagram).

COLOURED STARS

When we first look at stars, they all seem white. If you look for a while longer, you will see that some are slightly blue or red. When we heat a metal bar with a Bunsen burner it will glow red-hot. This indicates the frequency of the radiation that is emitted at this temperature. If we heated the bar with an acetylene torch it would become hotter and the light from it would be slightly blue. This relationship between the metal bar's colour and its temperature is also shown by stars. Cool stars appear red as most of the radiated energy is in the red to infrared range of the electromagnetic spectrum. Hot stars appear white or blue as most of their radiated energy is in the blue to ultraviolet range of the electromagnetic spectrum.

The colour of a star is related to its surface temperature and scientists use this to group stars into different types. Our Sun is a G type star.

Type of star	Colour	Approximate surface temperature (K)
O	blue	> 25 000
B	blue	11 000–25 000
A	blue	7500–11 000
F	white to blue	6000–7500
G	yellow to white	5000–6000
K	red to orange	3500–5000
M	red	< 3500

QUESTIONS

1. Stars can be classed as anything between blue → white → yellow → orange → red.

a) Explain and describe what causes the different colours of the stars.

b) Draw a table showing the type, colours, and approximate surface temperature in K for stars.

BRIGHTNESS OF STARS

The colours of stars are important as we have seen. It is also important to look at the brightness of stars.

The two main ways of describing the brightness of a star or galaxy are **apparent magnitude** and **absolute magnitude**.

The **apparent magnitude** is how bright the object is when seen by an observer on Earth. This can depend on two factors:

1. the amount of energy the star radiates per second (its **luminosity**)

2. the distance of the star from Earth.

The distance of stars from the Earth varies greatly. Therefore, very bright but distant stars can have a lower apparent magnitude than a less bright, nearby star.

The **absolute magnitude** of a star depends only on its luminosity. Absolute magnitude is the apparent magnitude of an object if it was a standard distance from the Earth. In this way, we can compare the actual brightness of stars regardless of how distant they are. The standard distance for absolute magnitude is 32.6 light years (about 3×10^{17} m). There is no need to travel to a distance of 3×10^{17} m from a star to measure its luminosity as there is a formula which relates the apparent magnitude to absolute magnitude, if its distance is known. This is based on the fact that light follows an inverse square law. This law simply means that when the light source is twice as far away, it will appear four times less bright.

EVOLUTION OF A STAR

A star is formed from a huge cloud of dust and gas called a **nebula**.

Δ Fig. 8.7 The Tarantula Nebula is a star-forming region in the Large Magellanic Cloud, the nearest galaxy to our own Milky Way. The most dense areas of gas and dust will be the birthplace of future generations of stars.

The gas clouds in a nebula contain hydrogen gas. The gas particles are gradually pulled in towards the centre of the nebula by gravitational attraction. The nebula becomes a much more dense structure called a proto-star. Compression causes the gas molecules to increase speed and the gas becomes very hot, up to about 15 million °C. Eventually, the great pressures involved also force the hydrogen nuclei together. This initiates a **nuclear fusion** reaction, fusing hydrogen nuclei to form helium. A very large amount of thermal energy is released. The thermal energy generated prevents the proto-star shrinking any more.

The star then continues fusing hydrogen into helium for many millions of years until the hydrogen starts to run out. What happens next depends on the star's mass. Our Sun is in this steady phase of fusion (**main sequence star**) and it has about 5 billion years left before the next stage.

When stars with a similar mass to our Sun start to run out of hydrogen they start fusing helium. The core of the star gets much hotter and expands. The star is no longer on the main sequence. The star appears red because the observed surface temperature of the star decreases as the outer layers spread out, so it is called a **red giant**. When fusion ends, the outer layers are pushed away and the hot, small core of the star collapses to become a **white dwarf**.

The Sun, although it appears large to us, is in fact a relatively small star. When stars with a mass larger than the Sun run low on hydrogen they start fusing helium to carbon. The star gets hotter and hotter and expands, leaving the main sequence. It becomes a **red supergiant**. Eventually this massive star explodes as a **supernova**. The explosion blows the outer layers of the star into space. The remaining core of the star contracts and may form an extremely dense **neutron star** or, if the original star was massive enough, even a **black hole**. Black holes are small, very dense massive concentrations of matter. Their gravitational pull is huge. The gravity is so great that even light cannot escape from it.

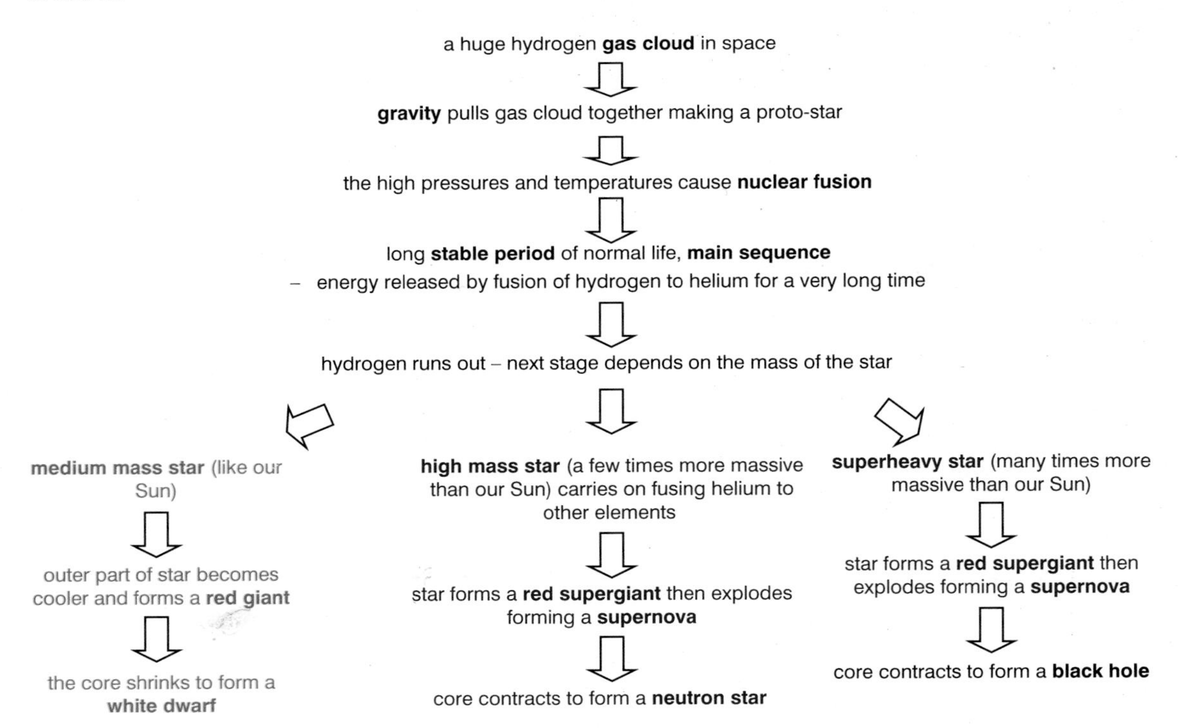

Δ Fig. 8.8 Stars are 'born', have a finite 'life' and eventually 'die'. This is the life-history of a star.

HERTZSPRUNG-RUSSELL (HR) DIAGRAM

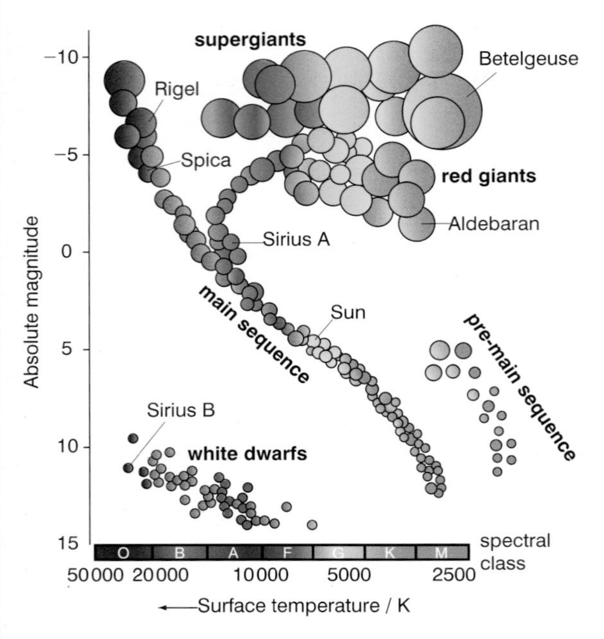

Δ Fig. 8.9 The Hertzsprung-Russell diagram, showing examples of different types of star.

The **Hertzsprung–Russell (HR) diagram** is a graph that shows the luminosity of a star plotted against its surface temperature. When the data for many stars are plotted, they are not randomly scattered but fall into distinct groups. The majority of stars are shown as a curved band. These are the main sequence stars. Stars on the main sequence towards the lower right are cooler (reddish) and dimmer and stars at the upper left are hotter (blue) and more luminous. Other groups are the red giants and white dwarfs. The HR diagram is useful to scientists as it shows what stage a star is at in its life cycle, from its position on the diagram.

QUESTIONS

1. Black holes and neutron stars are not shown on the HR diagram. Suggest a reason why.

2. a) Where are hot, bright stars found on the HR diagram?

b) What will happen to these hot, bright stars in the future?

c) Where will these hot bright stars move to on the diagram in the future?

3. Our Sun is a medium temperature and brightness star. What will happen to the Sun and its position on the diagram in the future?

End of topic checklist

A **nebula** is a huge gas cloud in space that is pulled together by gravity to form a star.

Nuclear fusion is the energy source for all stars.

In the **main sequence** stage of a star it spends a long time in a stable state undergoing fusion of hydrogen and producing a steady output of energy.

A **red giant** forms from a low mass star after the energy output from fusion of hydrogen ends and the star expands to a much greater size.

A **white dwarf** forms from a low mass star after a red giant cools and contracts.

High mass stars can explode as a **supernova** after which they can form a **neutron star** or **black hole**.

A **Hertzsprung–Russell (HR) diagram** is a diagram that shows the luminosity of a star plotted against the surface temperature.

The **absolute magnitude** is the apparent magnitude of an object if it was a standard distance from the Earth.

The facts and ideas that you should understand by studying this topic:

- ○ Understand how the temperature of stars can affect their colour.
- ○ Know how stars are formed and how they end.
- ○ Describe the differences in the life cycle of a star similar in size to our Sun and of a star much larger than our Sun.
- ○ Understand how fusion drives the energy output of stars.

End of topic questions

1. Describe the life cycle of a small star about the size of our Sun.
2. Describe the life cycle of a star with a mass much greater than the Sun.
3. Why are black holes difficult for scientists to detect?
4. Describe, using a diagram, the nuclear fusion reaction that takes place in stars.
5. Explain how a HR diagram can be useful to scientists.
6. What is meant by the term 'apparent magnitude of a star'?
7. Explain why the absolute magnitude of a star can often be more useful to scientists than its apparent magnitude.

Cosmology

INTRODUCTION

Cosmology is the study of the Universe. So you might think that it covers all topics in physics, chemistry and biology. If that was the case cosmology would be the biggest subject of them all. The Universe itself being infinite would mean that its subject matter would be infinite too! Luckily cosmology confines itself to the big picture view of the Universe: how it began, how it has evolved and how it will change in the future.

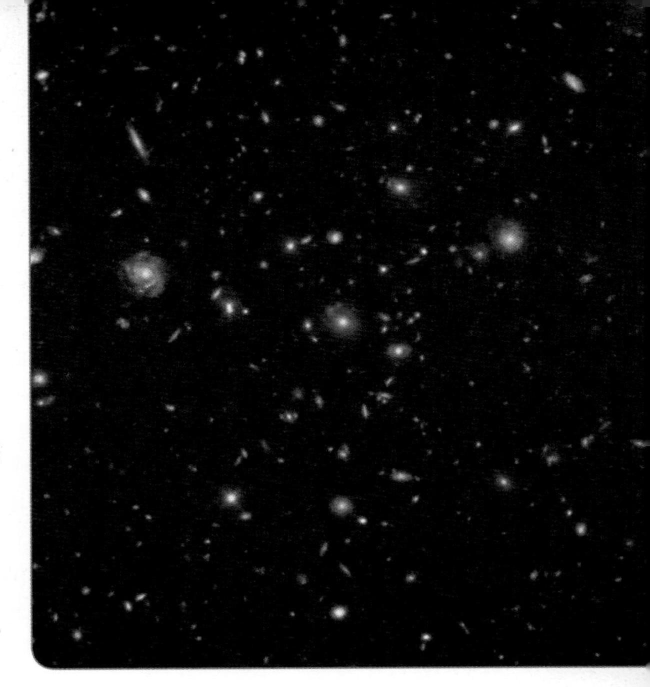

Δ Fig. 8.10 This image from NASA's Hubble space telescope shows one of the most distant galaxies we know. It dates from 750 million years after the Big Bang that created our Universe. The light took 12.9 billion years to reach us.

KNOWLEDGE CHECK

✓ Know about the Doppler effect and how movement of a light source relative to an observer can affect the wavelength and frequency of waves detected by the observer.
✓ Know that theories and models change as scientists to try to explain new evidence.

LEARNING OBJECTIVES

✓ Be able to describe the past evolution of the Universe and the main arguments in favour of the Big Bang.
✓ Be able to describe evidence that supports the Big Bang theory (red-shift and cosmic microwave background (CMB) radiation).
✓ Be able to describe that if a wave source is moving relative to an observer there will be a change in the observed frequency and wavelength.
✓ Be able to use the equation relating change in wavelength, wavelength, velocity of a galaxy and the speed of light:

$$\frac{\text{change in wavelength}}{\text{reference wavelength}} = \frac{\text{velocity of a galaxy}}{\text{speed of light}}$$

$$\frac{\lambda - \lambda_0}{\lambda_0} = \frac{\Delta\lambda}{\lambda_0} = \frac{v}{c}$$

✓ Be able to describe the red-shift in light received from galaxies at different distances away from the Earth.
✓ Be able to explain why the red-shift of galaxies provides evidence for the expansion of the Universe .

THE BIG BANG

One theory which tries to explain the origin of the Universe is known as the **Big Bang theory**. The basic idea is that all the matter in the Universe was originally concentrated in one extremely small, dense place. This matter began to expand very rapidly because of a very hot explosion.

In time the Universe continued to expand but also cooled down. The matter in the Universe then started to group together or coalesce into many large gas clouds. These were then subject to gravitational forces and then galaxies, stars and planets were eventually formed.

The Universe is older now but it is still expanding today. But what is the evidence that supports this theory? Without evidence to back it up it would just be an idea or an opinion and few would be convinced by it.

Supporting evidence: Red-shift

Scientists can observe the Universe as it is today. They look at other galaxies in space. Scientists observe light from other galaxies and they notice that it seems to be of a longer wavelength than expected. It appears to be **red-shifted**.

We know from our work on waves (section 3) that when a sound source is moving towards us, the sound waves from it are reduced in wavelength and increased in frequency. This causes the pitch of the sound to increase as the source (such as an ambulance siren) approaches. This is known as the **Doppler effect**.

The Doppler effect also occurs for moving light sources. There will be a change in the observed frequency and wavelength of light from a star that is moving towards or away from the Earth.

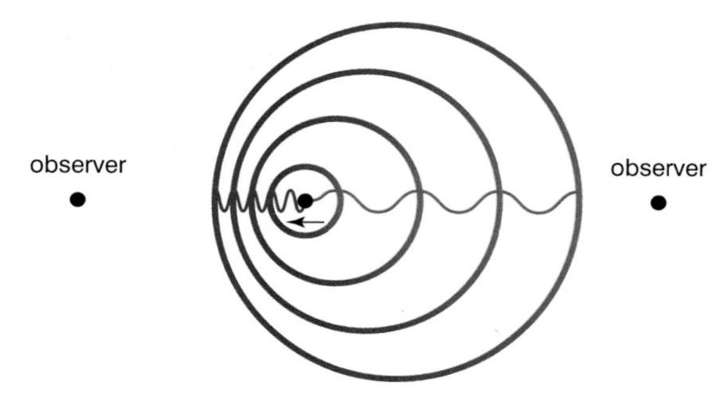

Δ Fig. 8.11 The Doppler effect for light waves.

The Sun produces helium by nuclear fusion. Atoms of helium in the atmosphere of the Sun absorb particular wavelengths of light emitted by the core of the Sun. This top spectrum in the diagram is the spectrum of the light from our Sun. The black lines show where helium has absorbed light of particular wavelengths. Less light escapes from the Sun at these wavelengths – that part of the spectrum is dark.

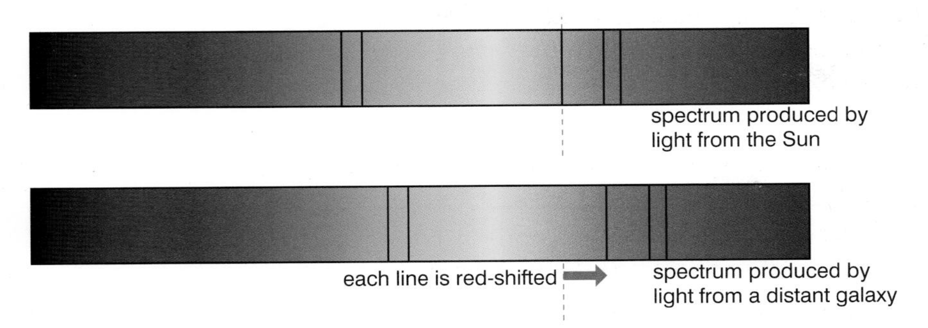

△ Fig. 8.12 The whole pattern of lines from a distant galaxy has been red-shifted compared with light from the Sun.

If we compare the Sun's spectrum with that of a star from another galaxy we can see that the absorption lines due to helium gas in the star's atmosphere are there but they have moved towards the red coloured part of the spectrum. The red light has a longer wavelength than blue and so we can say that the absorption lines have been **red-shifted**. This is like the increase in frequency of sound waves of an ambulance that moves away from us at speed. This shows that the star is moving away from us.

QUESTION

1. Most galaxies that have been observed show red-shift. A few galaxies show blue-shift.
   a) Use the Doppler effect to explain how some galaxies show blue-shift.
   b) Why is it a relief to know that the great majority of galaxies show red-shift rather than blue-shift?

Using red-shift

It is possible to work out the speed of a moving ambulance by measuring the frequency (or wavelength) of sound waves heard by a stationary observer. It is also possible to calculate the speed at which a star is moving away from Earth by looking at how much the wavelength of the light has shifted towards the red end of the spectrum.

The equation for the relationship between the change in wavelength, wavelength, velocity of a galaxy and the speed of light is:

$$\frac{\text{change in wavelength}}{\text{reference wavelength}} = \frac{\text{velocity of a galaxy}}{\text{speed of light}}$$

$$\frac{\lambda - \lambda_0}{\lambda_0} = \frac{\Delta\lambda}{\lambda_0} = \frac{v}{c}$$

Where:

λ = wavelength observed

λ_0 = reference wavelength

v = velocity of galaxy

c = velocity of light

WORKED EXAMPLE

A spectral line for hydrogen on Earth has a wavelength of 760 nm. The same spectral line for hydrogen observed from a distant quasar is 645 nm. (Quasars are very distant, extremely bright objects that drown out the light from all the other stars in the same galaxy).

The velocity of light in free space, $c = 3.0 \times 10^8$ m/s.

Calculate the velocity at which the quasar is moving away from the Earth. Give your answer in m/s.

Write down the equation:

$$\frac{\text{change in wavelength}}{\text{reference wavelength}} = \frac{\text{velocity of a galaxy}}{\text{speed of light}}$$

$$\frac{\lambda - \lambda_0}{\lambda_0} = \frac{\Delta\lambda}{\lambda_0} = \frac{v}{c}$$

change in wavelength: $\Delta\lambda = 760 - 645 = 115$ nm

rearranging the equation: $v = \frac{\Delta\lambda \times c}{\lambda_0}$

substitute the values: $v = \frac{115 \times 3 \times 10^8}{645}$

$v = 53\,488\,372$ m/s

Velocity of the quasar is 5.35×10^7 m/s (to 3 significant figures)

QUESTION

1. A spectral line for hydrogen on Earth is 760 nm. The same spectral line for hydrogen for a distant quasar is 680 nm.

   The velocity of light in free space, $c = 3.0 \times 10^8$ m/s.

   Calculate the velocity of the quasar relative to the Earth. Give your answer in m/s.

How red-shift provides evidence for the Big Bang

Scientists have measured the red-shift from many galaxies and found that:

- most galaxies show red-shift, hence most galaxies are moving away from Earth.
- distant galaxies show greater red-shift, hence distant galaxies are moving away from us more quickly that nearby ones (since $\frac{\Delta\lambda}{\lambda} = \frac{v}{c}$).

Edwin Hubble was the first astronomer to show the relationship between the recession velocity of galaxies and their distance, and to deduce that the Universe is expanding. If the Universe is expanding now, then in the past it must have been smaller. Ultimately, the idea goes back to the Universe starting at a single point in a violent 'explosion' – the Big Bang theory.

THE UNIVERSE

In the 1920s, the astronomer Edwin Hubble (1889–1953) was trying to answer the question of how big the Universe is. Was the Milky Way, our own galaxy, everything there is, or were there other objects outside it? Using the Mount Wilson telescope in California, Hubble studied small 'fuzzy' objects in the sky. Building on the work of Henrietta Leavitt (1868–1921), which described how to measure the distance to a type of star called a Cepheid Variable, Hubble discovered that these 'fuzzy' objects were complete galaxies of their own, clearly outside the Milky Way. This settled the argument and showed that the Universe (everything there is) was much bigger than anybody previously thought.

Hubble also discovered, through observations of red shift (a process where the frequency of light waves changes depending on whether the object is moving), that distant galaxies are all moving away from us – the Universe is expanding.

If the Universe is expanding now, then in the past it must have been smaller. Ultimately, the idea goes back to the Universe starting at a single point in a violent 'explosion'. This is the Big Bang theory. It is also supported by observations of the cosmic microwave background radiation – the 'echoes' of the Big Bang that can still be detected. Ever since then, gravity has been acting to slow down the expansion. It is unclear whether the Universe will expand forever or not.

However, observations in the past 20 years have suggested that the Universe has expanded faster we would expect from our current theories. There appears to be much more to the Universe than we have previously detected. Only when we add in factors called dark matter and dark energy do the observations make sense. At this time, the exact nature of these materials is not known and they have not been detected directly.

From the rate of expansion scientists have calculated that the Universe started from a single point about 13.8 billion years ago. This is the estimated age of the Universe.

Supporting evidence: Cosmic microwave background radiation

In the 1960s scientists detected a low energy, low frequency electromagnetic radiation coming from all parts of the Universe. It is detected by using radio telescopes on Earth and is not associated with any star, galaxy or other object. Radio telescopes are very large and can collect and concentrate the electromagnetic waves that come from space. Unlike optical telescopes, radio telescopes are not affected by clouds in the Earth's atmosphere. Clouds do not block radio waves as they do for visible light. So radio telescopes can be used day and night regardless of the weather conditions on Earth.

This background radiation is in the microwave range of the electromagnetic spectrum and is called the **cosmic microwave background** (CMB) radiation. The CMB radiation is very uniform – it comes from all directions in space. The COBE satellite was launched in 1989 by NASA to measure CMB radiation. The evidence from COBE also supports the uniformity of the CMB detected on Earth.

In 1948 the Big Bang theory predicted that this microwave radiation should exist as a relic of the thermal radiation from the original Big Bang explosion. The theory also predicted that the observed wavelength of the radiation today would be much greater than the wavelength of the light emitted shortly after a hot Big Bang. This is due to the expansion of the Universe, which stretches the wavelength of the light. The Big Bang theory predicted that the present-day radiation should be equivalent to a temperature slightly above absolute zero at 2.7 K. That is about 270 °C.

In the 1960s this background radiation was detected and the temperature was confirmed as 2.7 K. The COBE satellite has also confirmed these predictions. The thermal radiation from shortly after the Big Bang 13.8 billion years ago has been travelling through space ever since that time. Big Bang theorists continue to interpret this evidence.

The future of the Universe

Scientists are unsure how the Universe will progress. It depends on two things:

- how quickly the galaxies are moving apart
- how much mass there is in the Universe.

End of topic checklist

The **cosmic microwave background** (CMB) radiation is microwave radiation left over from the thermal radiation of the Big Bang.

The **Doppler effect** is how the wavelength or frequency of waves changes if the source of the waves is moving relative to the observer.

Red-shift occurs when stars are moving away from us at great speeds. This movement increases the wavelength of the light that is emitted by the star.

Observations of red-shift and CMB radiation provide evidence to support the **Big Bang** theory.

The facts and ideas that you should understand by studying this topic:

- ❍ Know that if a wave source is moving relative to an observer there will be a change in the observed frequency and wavelength.
- ❍ Use an understanding of the Doppler effect to explain red-shift.
- ❍ Know and use the relationship between change in wavelength, wavelength, velocity of a galaxy and the speed of light:

$$\frac{\text{change in wavelength}}{\text{reference wavelength}} = \frac{\text{velocity of a galaxy}}{\text{speed of light}}$$

$$\frac{\lambda - \lambda_0}{\lambda_0} = \frac{\Delta\lambda}{\lambda_0} = \frac{v}{c}$$

- ❍ Describe the past evolution of the Universe and the main arguments in favour of the Big Bang.
- ❍ Describe evidence that supports the Big Bang theory (red-shift and cosmic microwave background (CMB) radiation).
- ❍ Describe the red-shift in light received from galaxies at different distances away from the Earth.
- ❍ Explain why the red-shift of galaxies provides evidence for the expansion of the Universe.

End of topic questions

1. Explain the difference between red-shift and blue-shift. **(5 marks)**

2. The Big Bang theory predicted the existence of cosmic microwave background radiation today at an equivalent temperature of 2.7 K.

a) Describe how this radiation can be detected on Earth and in space. **(2 marks)**

b) Describe how scientists confirmed these predictions. **(5 marks)**

3. 'Distant galaxies show more red-shift than nearby ones'.

a) What does this say about the speeds and directions of distant and nearby galaxies? **(3 marks)**

b) How does this evidence support the idea that the Universe started at a point in the past? **(1 mark)**

EXAMINER'S COMMENTS

a) The student's answer is not specific enough. A better answer would have named the event and described it. E.g. It started with a massive explosion called the Big Bang.

b) This part is answered well. It is a good thing the student mentioned moving 'away'. It secured the mark for that marking point.

c) i) Well answered

ii) this is partly correct but the candidate needs also to state the equation:

$$\frac{\text{change in wavelength}}{\text{reference wavelength}} = \frac{\text{velocity of a galaxy}}{\text{speed of light}}$$

$$\frac{\lambda - \lambda_0}{\lambda_0} = \frac{\Delta\lambda}{\lambda_0} = \frac{v}{c}$$

iii) A good and full answer.

Exam-style questions

Sample student answer

Question 1

a) How do many scientists think the Universe began? **(1)**

With an explosion ✗

b) Most galaxies are red-shifted. Explain what this means. **(2)**

When a light source moves away ✓ (1) *at speed the wavelengths of light from it gets stretched out and they get larger.* ✓ (1)

c) Distant and nearby galaxies show different degrees of red-shift.

i) Describe and explain the differences in red shift for distant and nearby galaxies. **(3)**

Nearby galaxies show less red shift than distant ones. ✓ (1)

Distant ones are moving away faster ✓ (1)

so the wavelengths of the light become even longer ✓ (1)

ii) How can this information about red-shift be used to calculate the speed of a galaxy? **(3)**

If they know the wavelength of the light from the galaxy ✓ (1) *and the wavelength we see from Earth* ✓ (1) *then they can work it out.* ✗

iii) How can scientists use the results of these calculations to predict the starting point of the Universe? **(5)**

If they know the speeds of the galaxies ✓ (1) *they can predict where they have come from.* ✓ (1)

Scientists can work out if they seem to have come from the same place ✓ (1) *at the same time.* ✓ (1)

If they know the speeds they should be able to work out when the Big Bang took place. ✓ (1)

Exam-style questions continued

Question 2

The diagram shows the paths taken by some objects in the Solar System.

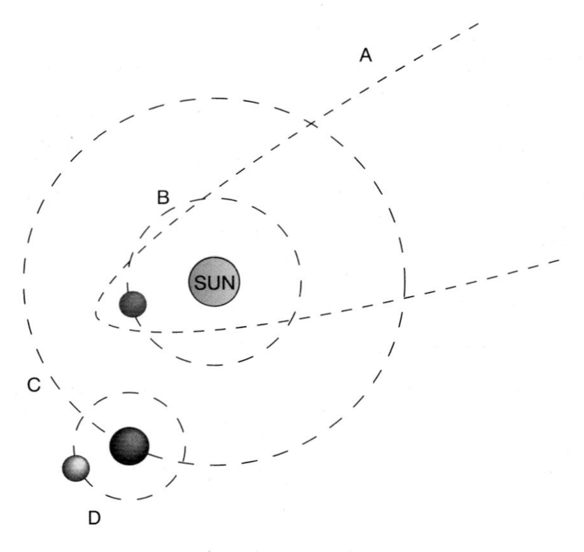

a) i) Which path represents the orbit of a comet? **(1)**

ii) Which path represents the orbit of a moon? **(1)**

b) The planet Mercury orbits the Sun once every 88 days.

Its average distance from the Sun is 58 million km.

Calculate the average orbital speed of Mercury in m/s. **(3)**

$$\text{orbital speed} = \frac{2 \times \pi \times \text{orbital radius}}{\text{time period}}$$

c) Explain the difference between a galaxy and the Universe. **(2)**

(Total 7 marks)

Question 3

Scientists find a very large type-M star that has a mass similar to our Sun.

a) Use the data from the table on coloured stars (on page 387) to describe the star and describe what stage the star is likely to be at. **(3)**

b) The star will evolve further. Suggest the ways this star could develop in the future. **(2)**

(Total 5 marks)

The International GCSE examination

INTRODUCTION

The International GCSE examination tests how good your understanding of scientific ideas is, how well you can apply your understanding to new situations and how well you can analyse and interpret information you have been given. The assessments are opportunities to show how well you can do these.

To be successful in exams you need to:

✓ have a good knowledge and understanding of science

✓ be able to apply this knowledge and understanding to familiar and new situations

✓ be able to interpret and evaluate evidence that you have just been given.

You need to be able to do these things under exam conditions.

OVERVIEW

The International GCSE course is designed to provide a basis for progression to further study in GCE Advanced Subsidiary Physics, Advanced Level Physics and the International Baccalaureate. The relationship of the assessment to the qualifications available is shown below:

Biology Paper 1 2 hours	+	Biology Paper 2 1 hour	→	International GCSE in Biology
+				
Chemistry Paper 1 2 hours	+	Chemistry Paper 2 1 hour	→	International GCSE in Chemistry
+				
Physics Paper 1 2 hours	**+**	**Physics Paper 2 1 hour**	**→**	**International GCSE in Physics**
↓				
International GCSE in Science (Double Award)				

Paper 1 is marked out of 110 and contributes 61.1% of the total International GCSE marks.

Paper 2 is marked out of 70 and contributes 38.9% of the International GCSE marks.

On the Edexcel Double Award Science course Physics Paper 1 accounts for 33.3% of the overall marks.

There is no separate assessment of investigative skills – the assessment is included in the two written papers.

There will be a range of short-answer structured questions, along with some multiple choice questions and a few questions requiring longer answers in both papers. You will be required to perform calculations, draw graphs and describe, explain and interpret physics ideas and information. In some of the questions the content may be unfamiliar to you; these questions are designed to assess data-handling skills and the ability to apply physics principles and ideas in unfamiliar situations.

ASSESSMENT OBJECTIVES AND WEIGHTINGS

The assessment objectives and weightings are as follows:

- ✓ AO1: Knowledge and understanding (38–42%)
- ✓ AO2: Application of knowledge and understanding, analysis and evaluation (38–42%)
- ✓ AO3: Experimental skills, analysis and evaluation of data and methods (19–21%).

The types of questions in your assessment fit the three assessment objectives shown in the table.

Assessment objective	**Your answer should show that you can...**
AO1 recall the science	Recall, select and communicate your knowledge and understanding of science.
AO2 apply your skills and knowledge	Apply skills, including evaluation and analysis, knowledge and understanding of scientific contexts.
AO3 use experimental skills	Use the skills of planning, observation, analysis and evaluation in practical situations.

EXAMINATION TIPS

To help you get the best results in exams, there are a few simple steps to follow.

Check your understanding of the question

- ✓ **Read the introduction to each question carefully before moving on to the questions themselves.**
- ✓ Look in detail at any **diagrams**, **graphs** or **tables**.
- ✓ Underline or circle the **key words** in the question.
- ✓ **Make sure you answer the question that is being asked** rather than the one you wish had been asked!
- ✓ Make sure that you understand the meaning of the '**command words**' in the questions.

REMEMBER

Remember that any information you are given is there to help you to answer the question.

EXAMPLE

- ✓ **'Give', 'state', 'name'** are used when recall of knowledge is required, for example you could be asked to give a definition or make a list of examples.
- ✓ '**State what is meant by**' is used when the meaning of a term is expected but there are different ways for how these can be described.
- ✓ '**Identify**' is usually used when you have to select some key information from a text or diagram in the question.
- ✓ **'Describe'** is used when you have to give the main feature(s) of, for example, a physics process or structure. You do not need to include a justification or reason.
- ✓ **'Explain'** is used when you have to give reasons, for example, for some experimental results or a physics fact or observation. You will often be asked to '**justify** or explain your answer', i.e. give reasons for it.
- ✓ **'Suggest'** is used when you have to come up with an idea to explain the information you're given – there may be more than one possible answer, no definitive answer from the information given, or it may be that you will not have learnt the answer but have to use the knowledge you do have to come up with a sensible one.
- ✓ **'Calculate'** means that you have to work out an answer in figures, showing relevant working.

- ✓ '**Determine**' means you must use data from the question, or must show how the answer can be reached quantitatively. To gain maximum marks, there must be a quantitative element to the answer.
- ✓ '**Estimate**' means find an approximate value, number or quantity from a diagram/given data or through a calculation.
- ✓ **'Plot', 'Draw a graph'** are used when you have to use the data provided to produce graphs and charts. This includes drawing a line of best fit through the points you have plotted. A suitable scale and appropriately labelled axes must be included if these are not provided in the question.
- ✓ '**Sketch**' means produce a drawing by hand. For a graph, this would need a line and labelled axes with important features indicated. The axes are not scaled.
- ✓ '**Draw**' means produce a diagram either using a ruler or by hand.
- ✓ '**Add/Label**' is used when you have to add or label something given in the question, for example, labelling a diagram or adding units to a table.
- ✓ '**Complete**' means complete a table/diagram given in the question.
- ✓ '**Comment on**' is used when you have to bring together a number of variables from data/information to form a judgement.
- ✓ '**Deduce**' means draw/reach conclusion(s) from the information provided.
- ✓ '**Discuss**' is used when you have to:
  - Identify the issue/situation/problem/argument that is being assessed within the question.
  - Explore all aspects of an issue/situation/problem/argument.
  - Investigate the issue/situation etc. by reasoning or argument.
- ✓ '**Evaluate**' means review information (e.g. data, methods) then bring it together to form a conclusion, drawing on evidence including strengths, weaknesses, alternative actions, relevant data or information. Come to a supported judgement of a subject's quality and relate it to its context.
- ✓ '**Give a reason/reasons**' is used when a statement has been made and you only have to give the reason(s) why.
- ✓ '**Justify**' means give evidence to support (either the statement given in the question or an earlier answer).
- ✓ '**Design**' means plan or invent a procedure from existing principles/ideas.
- ✓ '**Predict**' means give an expected result.
- ✓ '**Show that**' means verify the statement given in the question.

Check the number of marks for each question

- ✓ Look at the **number of marks** allocated to each question.
- ✓ Make sure you include at least as many points in your answer as there are marks.
- ✓ Look at the **space provided** to guide you as to the length of your answer. However, be aware that there may be more space than you need.

What to do if you need extra space to answer

- ✓ If you need more space to answer than provided, then either use the nearest available space, for example, at the bottom of the page, or ask for extra paper.
- ✓ However, do NOT use any blank pages in the examination paper as these will not be scanned for the Examiner.
- ✓ You MUST state clearly WITHIN the marked-out answer space where you have continued your answer, for example, 'continued at the bottom of page 12', otherwise your additional material may not be seen by the Examiner.
- ✓ You MUST make it clear IN YOUR ANSWER which question you are answering.
- ✓ If you use extra paper you MUST make sure you also include your name, candidate number and centre number.

REMEMBER

Beware of continually writing too much because it probably means you are not really answering the questions.

Use your time effectively

- ✓ Don't spend so long on some questions that you don't have time to finish the paper.
- ✓ You should spend approximately **one minute per mark**.
- ✓ If you are really stuck on a question, leave it, finish the rest of the paper and come back to it at the end.
- ✓ Even if you eventually have to guess at an answer, you stand a better chance of gaining some marks than if you leave it blank.

ANSWERING QUESTIONS

Multiple choice questions

- ✓ Select your answer by placing a cross (not a tick) in the box.

Short and long answer questions

- ✓ In short-answer questions, **don't write more than you are asked for**.
- ✓ You will not gain any marks, even if the first part of your answer is correct, if you've written down something incorrect later on or which

contradicts what you've said earlier. This just shows that you haven't really understood the question or are guessing.

- ✓ In some questions, particularly short-answer questions, answers of only one or two words may be sufficient, but in longer questions you should aim to use **good English** and **scientific language** to make your answer as clear as possible.
- ✓ Present the information in a logical sequence.
- ✓ Don't be afraid to also use **labelled diagrams** or **flow charts** if it helps you to show your answer more clearly.

Questions with calculations

- ✓ **In calculations always show your working**.
- ✓ Even if your final answer is incorrect you may still gain some marks if part of your attempt is correct.
- ✓ If you just write down the final answer and it is incorrect, you will get no marks at all.
- ✓ Write down your answers to as many **significant figures** as are used in the numbers in the question (and no more). If the question doesn't state how many significant figures then a good rule of thumb is to quote 3 significant figures.
- ✓ Don't round off too early in calculations with many steps – it's always better to give too many significant figures than too few.
- ✓ You may also lose marks if you don't use the correct **units.** In some questions the units will be mentioned, for example, calculate the mass in grams; or the units may also be given on the answer line. If numbers you are working with are very large, you may need to make a conversion, for example, convert joules into kilojoules, or kilograms into tonnes.

Finishing your exam

- ✓ When you've finished your exam, **check through** your paper to make sure you've answered all the questions.
- ✓ Check that you haven't missed any questions at the end of the paper or turned over two pages at once and missed questions.
- ✓ Cover over your answers and read through the questions again and check that your answers are as good as you can make them.

REMEMBER

In the two written papers, you will be asked questions on investigative work (Assessment objectives AO3 and AO2). It is important that you understand the methods used by scientists when carrying out investigative work.

More information on carrying out practical work and developing your investigative skills are given in the next section.

Developing experimental skills

INTRODUCTION

As part your International GCSE Physics course, you will develop practical skills and have to carry out investigative work in science.

This section provides guidance on carrying out an investigation.

Many investigations follow a common route:

1. Planning the investigation and assessing the risk
2. Carrying out the practical work safely and skilfully
3. Making and recording observations and measurements
4. Analysing the data and drawing conclusions
5. Evaluating the data and methods used

1. Planning and assessing the risk

Learning objective: to devise and plan investigations, drawing on physics knowledge and understanding in selecting appropriate techniques.

Questions to ask:

What do I already know about the area of physics I am investigating and how can I use this knowledge and understanding to help me with my plan?

✓ Think about what you have already learned and any investigations you have already done that are relevant to this investigation.

✓ List the factors that might affect the process you are investigating.

What is the best method or technique to use?

✓ Think about whether you can use or adapt a method that you have already used.

✓ A method, and the measuring instruments, must be able to produce **valid** measurements. A measurement is valid if it measures what it is supposed to be measuring.

You will make a decision as to which technique to use based on:

✓ The accuracy and precision of the results required. Investigators might require results that are as accurate and precise as possible but if you are doing a quick comparison, or a preliminary test to check a range over which results should be collected, a high level of accuracy and precision may not be required.

- ✓ The simplicity or difficulty of the techniques available, or the equipment required; is this expensive, for instance?
- ✓ The scale, for example, using standard laboratory equipment or on a micro-scale, which may give results in a shorter time period.
- ✓ The time available to do the investigation.
- ✓ Health and safety considerations.

What am I going to measure?

- ✓ The factor you are investigating is called the **independent variable**. A **dependent variable** is affected or changed by the independent variable that you select.
- ✓ You need to choose a range of measurements that will be enough to allow you to plot a graph of your results and so find out the pattern in your results.
- ✓ You might be asked to explain why you have chosen your range rather than a lower or higher range.

How am I going to control the other variables?

- ✓ These are **control variables**. Some of these may be difficult to control.
- ✓ You must decide how you are going to control any other variables in the investigation and so ensure that you are using a fair test and that any conclusions you draw are valid.

What equipment is suitable and will give me the accuracy and precision I need?

- ✓ The **accuracy** of a measurement is how close it is to its true value.
- ✓ **Precision** is related to the smallest scale division on the measuring instrument that you are using, for example, when measuring a distance, a rule marked in millimetres will give greater precision than one divided into centimetres only.
- ✓ A set of precise measurements also refers to measurements that have very little spread about the mean value.
- ✓ You need to be sensible about selecting your devices and make a judgement about the degree of precision. Think about what is the least precise variable you are measuring and choose suitable measuring devices. There is no point having instruments that are much more precise than the precision you can measure the variable to.

What are the potential hazards of the equipment and technique I will be using and how can I reduce the risks associated with these hazards?

- ✓ In the exam, be prepared to suggest safety precautions when presented with details of a physics investigation.

EXAMPLE 1

You have been asked to design and plan an investigation to explore the motion of a trolley down a ramp. In a previous investigation you have investigated such motion using ticker tape so you are familiar with what happens and the measurements you need to take.

What do I already know?

Previously you have investigated the motion of a trolley down a ramp. You know that you can use ticker tape to measure the distance the trolley travels in a given time.

What is the best method or technique to use?

The technique you used in your previous investigation can be re-used. You set up the apparatus as shown in the diagram.

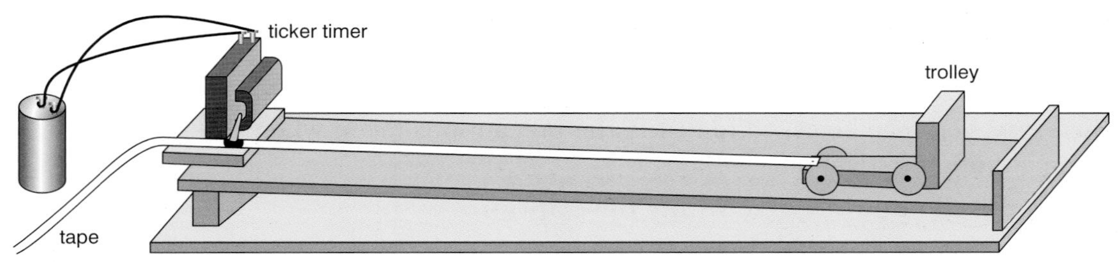

Δ Fig. 8.1 Apparatus for experiment.

What am I going to measure?

You are investigating the motion of a trolley down a ramp. You will measure the length of each 5-dot strip of ticker tape with a ruler.

How am I going to control the other variables?

It is important that you decide on the angle at which to set the ramp at the start. As you have carried out this investigation before, you can look back and see what angle you used previously and decide whether you will use the same angle, or increase or decrease it.

What equipment is suitable and will give me the accuracy and precision I need?

You now know what you will need to measure and so can decide on your measuring devices.

Measurement	Quantity	Device
Length of ticker tape	5-dot strips	ruler so can measure to nearest cm

Δ Table 8.1 Suitable equipment for experiment.

Choosing a ruler which can measure to the nearest mm would not be appropriate as the width of the ticker tape dots is of the order of mm.

What are the potential hazards and how can I reduce the risks?

The hazards are as follows: trolley and ramp. These indicate that there are no specific hazards you need to be aware of.

In terms of the equipment and technique, the major hazard will be the trolley rolling off the end of the ramp. You can limit this hazard by putting a buffer at the end of the ramp as shown in Fig. 8.1.

2. Carrying out the practical work safely and skilfully

Learning objective: To demonstrate and describe appropriate experimental and investigative methods, including safe and skilful practical techniques.

Questions to ask:

How shall I use the equipment safely to minimise the risks – what are my safety precautions?

✓ When writing a Risk Assessment, investigators need to be careful to check that they've matched the hazard with the technique used.

✓ In the exam, you may be asked to justify the precautions taken when carrying out an investigation.

How much detail should I give in my description?

✓ You need to give enough detail so that someone else who has not done the experiment would be able to carry it out to reproduce your results.

How should I use the equipment to give me the precision I need?

✓ You should know how to read the scales on the measuring equipment you are using.

✓ You need to show that you are aware of the precision needed.

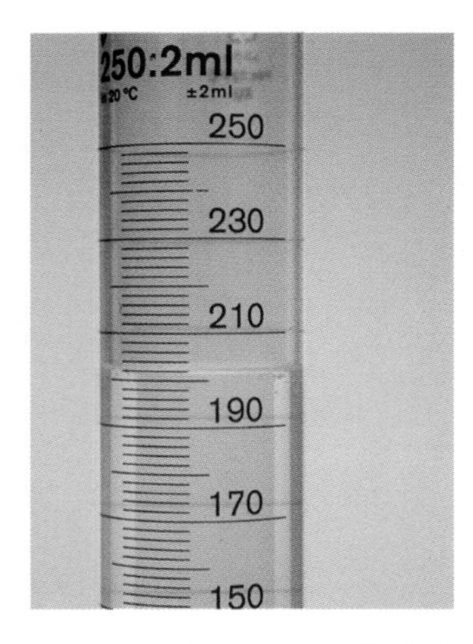

Δ Fig. 8.2 The volume of liquid in a measuring cylinder must be read to the bottom of the meniscus. The volume in this measuring cylinder is 202 cm^3 (ml), not 204 cm^3.

EXAMPLE 2

This is an extract from a student's notebook. It describes how she investigated terminal velocity.

Safety precautions

a) *Equipment.*

I will be using a measuring cylinder made from glass so I will need to handle it carefully

COMMENT

The student has realised the possible risks with using glass equipment.

I will need to handle the oil carefully, have a damp cloth ready to wipe up any spills.

COMMENT

The student has suggested some sensible precautions.

The student's method is given below

1. *The tube was marked every 10 cm using tape.*
2. *The ball was released carefully from the surface of the oil.*
3. *At the same time, a stopclock was started.*
4. *As the ball passed each mark, the time was noted.*
5. *Since the marks are 10 cm apart, the speed of the ball in each section of the tube can be calculated.*

COMMENT

The method is well written and detailed. Point 1 could have been improved if the student had noted the width of the tape used.

Precision and accuracy. An example from the notebook is:

The speed measured to the nearest 0.1 cm/s

The student has appreciated the accuracy that can be achieved using this method.

3. Making and recording observations and measurements

Learning objective: to make observations and measurements with appropriate precision, record these methodically, and present them in a suitable form.

Questions to ask:

How many different measurements or observations do I need to take?

✓ Sufficient readings have been taken to ensure that the data are consistent.

✓ It is usual to repeat an experiment to get more than one measurement. If an investigator takes just one measurement, this may not be typical of what would normally happen when the experiment was carried out.

✓ When repeat readings are consistent they are said to be **repeatable**.

Do I need to repeat any measurements or observations that are anomalous?

An **anomalous result** or **outlier** is a result that is not consistent with other results.

✓ You want to be sure a single result is accurate (as in the example below). So you will need to repeat the experiment until you get close agreement in the results you obtain.

✓ If an investigator has made repeat measurements, they would normally use these to calculate the arithmetical mean (or just mean or average) of these data to give a more accurate result. You calculate the mean by adding together all the measurements, and dividing by the number of measurements. Be careful though, anomalous results should not be included when taking averages.

✓ Anomalous results might be the consequence of an error made in measurement. But sometimes outliers are genuine results. If you think an outlier has been introduced by careless practical work, you should omit it when calculating the mean. But you should examine possible reasons carefully before just leaving it out.

✓ You are taking a number of readings in order to see a changing pattern. For example, measuring the speed every 10 cm for 60 cm (so 6 different readings). It is likely that you will plot your results onto a graph and then draw a **line of best fit**.

✓ You can often pick an anomalous reading out from a results table (or a graph if all the data points have been plotted, as well as the mean, to show the range of data). It may be a good idea to repeat this part of the practical again, but it's not necessary if the results show good consistency.

✓ If you are confident that you can draw a line of best fit through most of the points, it is not necessary to repeat any measurements that are obviously inaccurate. If, however, the pattern is not clear enough to draw a graph then readings will need to be repeated.

How should I record my measurements or observations – is a table the best way? What headings and units should I use?

✓ A table is often the best way to record results.

✓ Headings should be clear.

✓ If a table contains numerical data, do not forget to include units; data are meaningless without them.

✓ The units should be the same as those that are on the measuring equipment you are using.

✓ Sometimes you are recording observations that are not quantities. Putting observations in a table with headings is a good way of presenting this information.

EXAMPLE 3

The student from Example 2 has recorded the results in a table as shown below.

Distance fallen through oil (cm)	Speed 1st experiment	Speed 2nd experiment	Speed 3rd experiment
0	0.0	0.1	0.1
10	2.4	2.4	2.3
20	4.4	4.3	4.4
30	5.6	5.6	5.7
40	6.0	5.9	5.9
50	6.4	6.4	6.3
60	6.4	6.3	6.4

Δ Table 8.2 Readings from investigation.

EXAMPLE 4

In an experiment to measure the efficiency of a small motor the student has sensibly recorded her results in a table. Notice each column has a heading *and* units.

Mass lifted (g)	Distance lifted (m)	Useful work done (J)	Voltage of motor (V)	Current in motor (A)	Time to lift the mass (s)	Electrical energy supplied (J)
0.01	1.0		2.4	0.20	22.0	
0.03	1.0		2.4	0.22	24.4	
0.05	1.0		2.4	0.25	26.5	
0.07	1.0		2.3	0.28	27.6	
0.09	1.0		2.3	0.29	28.7	

Δ Table 8.3 Table of results

EXAMPLE 5

In another experiment the student has recorded his results obtained in an experiment to investigate the strength of an electromagnet as the current in the coil varies.

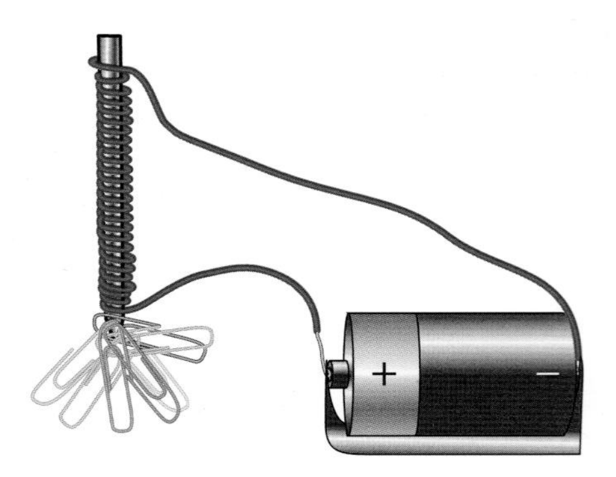

Δ Fig. 8.3 Apparatus for experiment.

Current (A)	Number of paper clips held
0	0
0.3	2
0.5	5
0.7	6
0.9	9
1.0	9

Δ Table 8.4 Results of experiment.

COMMENT

In this table of results:

The description of each measurement is clear.

The units are given in each case.

4. Analysing the data and drawing conclusions

Learning objectives: to analyse and interpret data to draw conclusions from experimental activities which are consistent with the evidence, using physics knowledge and understanding, and to communicate these findings using appropriate specialist vocabulary, relevant calculations and graphs.

Questions to ask:

What is the best way to show the pattern in my results? Should I use a bar chart, line graph or scatter graph?

✓ Graphs are usually the best way of demonstrating trends in data.

✓ A bar chart or bar graph is used when one of the variables is a **categoric variable**, for example when the melting points of the oxides of the group 2 elements are shown for each oxide the names are categoric and not continuous variables.

✓ A line graph is used when both variables are continuous, for example, time and temperature, time and volume.

✓ Scatter graphs can be used to show the intensity of a relationship, or degree of *correlation*, between two variables.

✓ Sometimes a line of best fit is added to a scatter graph, but usually the points are left without a line.

When drawing bar charts or line graphs:

✓ Choose scales that take up most of the graph paper

✓ Make sure the axes are linear and allow points to be plotted accurately. Each square on an axis should represent the same quantity. For example, one big square = 5 or 10 units; not 3 units

✓ Label the axes with the variables (ideally with the independent variable on the *x*-axis)

✓ Make sure the axes have units

✓ If more than one set of data is plotted use a key to distinguish the different data sets.

If I use a line graph should I join the points with a straight line or a smooth curve?

✓ When you draw a line, do not just join the dots!

✓ Remember there may be some points that don't fall on the curve – these may be incorrect or anomalous results.

✓ A graph will often make it obvious which results are anomalous and so it would not be necessary to repeat the experiment (see Example 6).

Do I have to calculate anything from my results?

✓ It will be usual to calculate means from the data.

✓ Sometimes it is helpful make other calculations, before plotting a graph, for example you might calculate 1/time for a rate of reaction experiment.

✓ Sometimes you will have to make some calculations before you can draw any conclusions.

Can I draw a conclusion from my analysis of the results, and what physics knowledge and understanding can be used to explain the conclusion?

✓ You need to use your physics knowledge and understanding to explain your conclusion.

✓ It is important to be able to add some explanation which refers to relevant scientific ideas in order to justify your conclusion.

What is the best way to show the pattern in my results?

If the experiment involves **continuous variables** and so a line graph is needed.

Straight line or a smooth curve?

The results obtained will either require a smooth curve or a straight line of best fit. The shape of the results should show you which is needed. In physics experiments you will often need to draw a line of best fit.

Do I have to calculate anything from my results?

If you have to calculate a quantity you will often be able to do this by looking at the change in steepness/ gradient of the curve.

Can I draw a conclusion from my analysis of the results?
You need to write a sentence summarising what you have learnt from your investigation.

Make sure that you write a clear statement. You might refer, for example, to 'the gradient of the line' at points 1, 2 and 3 to make your conclusion even more precise.

What physics knowledge and understanding can be used to explain the conclusion? You need to be able to explain your results using your knowledge of the physics of the situation.

COMMENT

A good conclusion will make direct links to scientific knowledge in relation to the topic.

5. Evaluating the data and methods used

Learning objective: to evaluate data and methods.

Questions to ask:

Do any of my results stand out as being inaccurate or anomalous?

✓ You need to look for any anomalous results or outliers that do not fit the pattern.

✓ You can often pick this out from a results table (or a graph if all the data points have been plotted, as well as the mean, to show the range of data).

What reasons can I give for any inaccurate results?

✓ When answering questions like this it is important to be specific. Answers such as 'experimental error' will not score any marks.

✓ It is often possible to look at the practical technique and suggest explanations for anomalous results.

✓ When you carry out the experiment you will have a better idea of which possible sources of error are more likely.

✓ Try to give a specific source of error and avoid statements such as 'the measurements must have been wrong'

Your conclusion will be based on your findings, but must take into consideration any uncertainty in these introduced by any possible sources of error. You should discuss where these have come from in your evaluation.

Error is a difference between a measurement you make, and its true value.

The two types of errors are:

✓ random error

✓ systematic error.

With **random error**, measurements vary in an unpredictable way. This can occur when the instrument you're using to measure lacks sufficient precision to indicate differences in readings. It can also occur when it's difficult to make a measurement.

With **systematic error**, readings vary in a controlled way. They're either consistently too high or too low. One reason could be down to the way you are making a reading, for example, taking a burette reading at the wrong point on the meniscus, or not being directly in front of an instrument when reading from it.

What an investigator *should not* discuss in an evaluation are problems introduced by using faulty equipment, or by using the equipment inappropriately. These errors can, or could have been, eliminated, by:

✓ checking equipment

✓ practising techniques before the investigation, and taking care and patience when carrying out the practical.

Overall was the method or technique I used precise enough?

✓ If your results were good enough to provide a confident answer to the problem you were investigating the method probably was good enough.

✓ If you realise your results are not precise when you compare your conclusion with the actual answer it may be you have a **systematic error** (an error that has been made in obtaining all the results.) A systematic error would indicate an overall problem with the experimental method.

If your results do not show a convincing pattern then it is fair to assume that your method or technique was not precise enough and there may have been a **random error** (i.e. measurements vary in an unpredictable way).

If I were to do the investigation again what would I change or improve upon?

✓ Having identified possible errors it is important to say how these could be overcome. Again you should try and be absolutely precise.

✓ When suggesting improvements, do not just say 'do it more accurately next time' or 'measure the volumes more accurately next time'.

✓ For example, if you were measuring small lengths, you could improve the method by using a vernier scale to measure the lengths rather than a ruler.

EXAMPLE 6

A student was measuring how current varies with voltage. He used the circuit shown in Fig. 8.4.

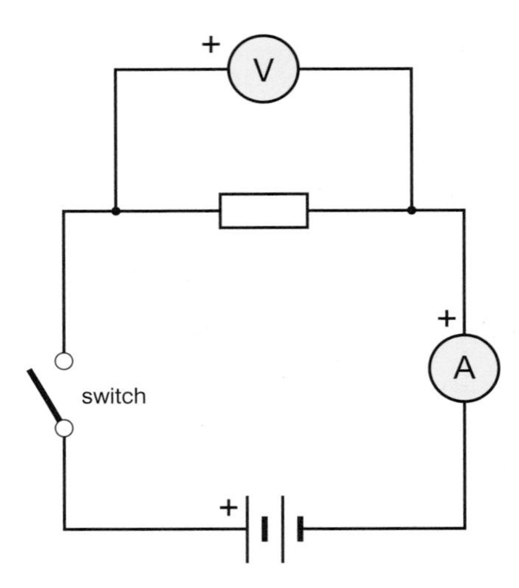

Δ Fig. 8.4 Circuit used to measure how current varies with voltage.

Do any of my results stand out as being inaccurate or anomalous?

The student plotted his results on a graph, as shown in Fig. 8.5. An inaccurate result stands out from the rest, as shown by the circle on the graph. Given the pattern obtained with the other results there is no real need to repeat the result – you could be very confident that the result should have followed the pattern set by the others. A result like this is referred to as an anomalous result. It was an error but not a systematic error.

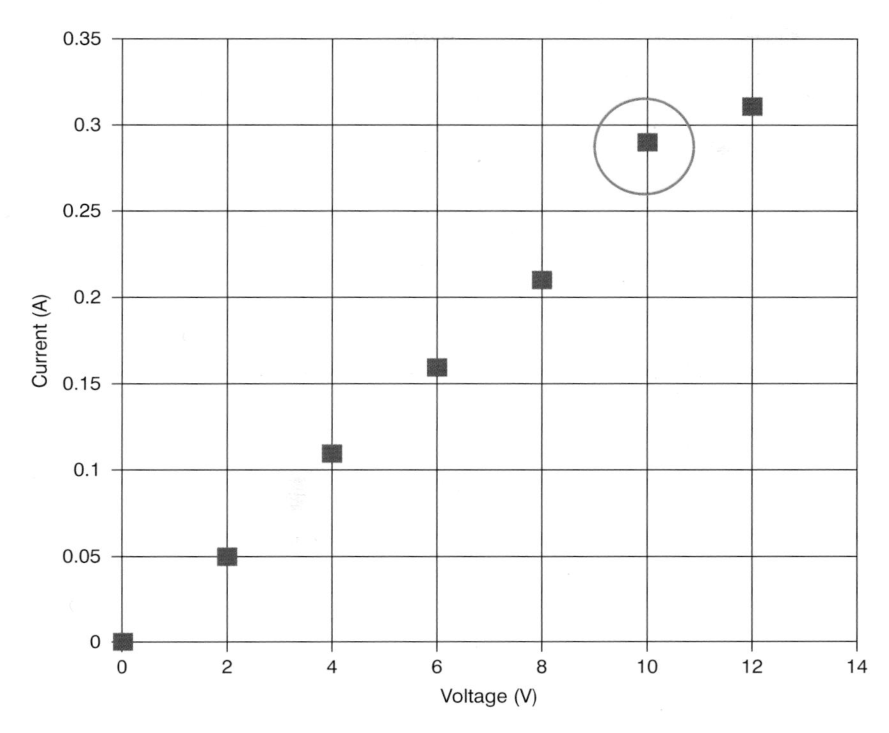

Δ Fig. 8.5 Graph of results.

What reasons can I give for any inaccurate results?

There are two main possible sources of error: either one of the variables was noted incorrectly.

Was the method or technique I used precise enough?

You can be reasonably confident that using digital meters for the current and voltage readings will give you precise measurements.

How can I improve the investigation?

For example, you could say, take readings after the component has heated up so that the steady state resistance is noted.

Mathematical skills

The table below shows the mathematical skills that you will need to use during your International GCSE course. You will also need to be able to use these skills during your examinations.

		P
1	**Arithmetic and numerical computation**	
A	Recognise and use numbers in decimal form	✓
B	Recognise and use numbers in standard form	✓
C	Use ratios, fractions, percentages, powers and roots	✓
D	Make estimates of the results of simple calculations, without using a calculator	✓
E	Use calculators to handle $\sin x$ and $\sin^{-1} x$, where x is expressed in degrees	✓
2	**Handling data**	
A	Use an appropriate number of significant figures	✓
B	Understand and find the arithmetic mean (average)	✓
C	Construct and interpret bar charts	✓
D	Construct and interpret frequency tables, diagrams and histograms	✓
E	Understand simple probability	✓
F	Use a scatter diagram to identify a pattern or trend between two variables	✓
G	Make order of magnitude calculations	✓
3	**Algebra**	
A	Understand and use the symbols $<$, $>$, $\propto$, $\sim$	✓
B	Change the subject of an equation	✓
C	Substitute numerical values into algebraic equations using appropriate units for physical quantities	✓
D	Solve simple algebraic equations	✓

4	**Graphs**	
A	Translate information between graphical and numerical form	✓
B	Understand that $y = mx + c$ represents a linear relationship	✓
C	Plot two variables (discrete and continuous) from experimental or other data	✓
D	Determine the slope and intercept of a linear graph	✓
E	Understand, draw and use the slope of a tangent to a curve as a measure of rate of change	✓
F	Understand the physical significance of area between a curve and the x-axis, and measure it by counting squares as appropriate	✓
5	**Geometry and trigonometry**	
A	Use angular measures in degrees	✓
B	Visualise and represent 2D and 3D objects, including two dimensional representations of 3D objects	✓
C	Calculate areas of triangles and rectangles, surface areas and volumes of cubes	✓

Glossary

absolute magnitude The magnitude of a star as it would appear at a standard distance from the Earth, taken as 32.6 light-years.

absolute zero The lowest temperature possible, –273 °C. All molecules stop moving at this temperature.

acceleration A change in speed divided by the time taken to change.

air resistance The drag caused by something moving through air.

alloy A mixture of a metal and one or more other elements.

alpha particle A particle emitted from the nuclei of radioactive atoms, consisting of two protons and two neutrons.

alternating current (a.c.) Electrical current that repeatedly reverses its direction, like mains electricity.

ammeter An instrument that measures electrical current in amperes.

ampere A unit of current measuring the electric charge that flows during one second.

amplitude The maximum change of the medium from normal in a wave. For example, the height of a water wave above the level of calm water.

angle of incidence The angle between the incident light ray and the normal to the surface.

angle of refraction The angle between the refracted light ray and the normal to the surface inside the material.

angle of reflection The angle between the reflected light ray and the normal to the surface of the material.

anode A positive electrode.

apparent magnitude The brightness of a star as seen by an observer on Earth.

atomic number The number of protons found in the nucleus of an atom.

average speed The distance an object has moved, divided by the time taken.

background radiation The level of radiation found due to natural processes in the environment.

balanced forces The forces acting on an object are balanced when the resultant force is zero. For example, when two forces acting on an object are equal in size but act in opposite directions.

becquerel The units for the activity of a substance. One becquerel is one decay per second.

beta particle A type of nuclear radiation emitted as an electron by a radioactive nucleus.

Big Bang theory A theory that tries to explain the origin of the Universe. It concludes from observations of evidence that the Universe started with a massive explosion.

black hole A very compact object that has a high mass concentrated in a very small volume. This creates a very dense object with a strong gravitational field. Its gravity is so great that light (and other radiation) cannot escape its pull forces.

blue-shift The wavelength of the radiation that is emitted from a light source moving towards us decreases because of the Doppler effect. These objects appear more blue because shorter wavelengths of light are at the violet end of the visible light spectrum.

boiling point The temperature at which a substance changes state from a liquid to a gas.

Boyle's law If the temperature of the gas stays constant, then the volume of the gas is inversely proportional to the pressure.

centre of gravity The point in a body from which the force of gravity appears to be acting on the body.

centre of mass Another name for the centre of gravity.

chain reaction A series of nuclear fission reactions where neutrons released from one reaction cause another nuclear fission reaction and so on.

charge A fundamental property of matter that produces all electrical effects. It is equal to current × time.

chemical energy Energy stored in molecules. Batteries and fuels contain stored chemical energy.

circuit breaker A device that breaks a circuit when there is an increase in current.

comet An object with an oval orbit that orbits the Sun.

commutator The part of an electric motor that allows the coils to be connected to the opposite terminals each time the motor rotates through 180°.

compression The squashing together of particles in a particular region.

condensation The change of state from a gas to a liquid.

conduction (electrical) The process by which electrical charge passes through a material.

conduction (thermal) The process by which thermal energy is transferred in a substance without the substance itself moving.

conductors, electricity Substances that conduct electricity well.

conductors, thermal Substances that conduct thermal energy very well.

control rods These are rods or tubes that contain a neutron absorber, such as boron, that can be put into, or taken out of, the core of a nuclear reactor. Their presence can control the rate of a fission reaction.

convection Heat transfer in a liquid or gas – when particles in a warmer region gain energy and move into cooler regions carrying this energy with them.

conventional current Movement of positive charge that is imagined to move from the positive terminal to the negative terminal of a battery. Equivalent in effect to the real flow of negative charge in the opposite direction.

cosmic microwave background Electromagnetic radiation at microwave wavelengths that comes from every direction in the Universe. It is thought this radiation is the result of the Big Bang.

coulomb The unit of electric charge.

count rate The number of decays recorded each second by a detector (e.g. Geiger-Muller tube).

crest The highest part of a wave.

critical angle The angle of incidence where the angle of refraction becomes greater than 90°.

current, electric A flowing electric charge caused by the flow of electrons.

count rate The number of decays recorded each second by a detector (e.g. Geiger-Muller tube).

daughter nucleus The isotope formed during a radioactive decay process.

density The mass, in kilograms, of a one metre cube of a substance: mass divided by volume.

diffusion Molecules moving from an area of high concentration to an area of low concentration.

diode A device that lets electricity flow through it one way only.

direct current (d.c.) Current that always flows in the same direction.

dispersion Splitting white light into the colours that it is made of.

displacement The distance and direction of one point from another.

distance–time graph A visual representation of how distance travelled varies with time.

double-insulated When an device has a casing that is made of an insulator and does not need an earth wire.

dynamo A simple current generator.

efficiency Useful energy output divided by total energy output.

earth wire A wire connecting the case of an electrical appliance, through the earth pin on a three-pin plug, to earth.

elastic Describes materials that go back to their original shape and size after you stretch them.

elastic behaviour When a material returns to its original shape when the forces have been removed.

elastic energy A form of stored energy from stretching or compressing an object like a spring.

electric current Flowing electric charge.

electromagnetic induction A changing magnetic field can induce a potential difference across and an electric current in a wire.

electromagnetic spectrum The 'family' of electromagnetic radiations (from longest to shortest wavelength): radio, microwave, infrared, visible light, ultraviolet, X-rays, gamma rays. In order of frequency, the order is reversed. They all travel at the same speed in a vacuum.

electromagnetic wave A wave that transfers energy – it can travel through a vacuum and travel at the speed of light.

electromagnets Magnets made from a coil of wire. The magnetic force is made when electric current flows in the coil. The magnetic force is stronger when the coil is wrapped around a piece of iron.

electron Negatively charged particles with almost no mass that orbit the nucleus in all atoms.

electrostatic effects Effects caused by having extra electrons or a shortage of electrons. They happen when insulators have gained or lost electrons.

electrostatic forces Forces between charged particles.

element A substance that contains only one kind of atom.

equilibrium When there is no resultant force and no turning effect on an object.

evaporation The change of state from a liquid to a gas.

extension The increase in length when something is stretched.

Fleming's left-hand rule Predicts the direction of the force on a wire that is carrying a current when the thumb, and first and second fingers are held at right angles. The first finger points in the direction of the field, the second finger in the direction of the current and the thumb gives the direction of the force (movement).

fluid Any liquid or gas.

force A push or a pull, measured in newtons (N). Also a change in momentum divided by time taken.

fossil fuel Non-renewable energy resource such as coal, oil or natural gas.

free fall Movement under the effect of the force of gravity alone.

frequency The number of vibrations per second or number of peaks or troughs that pass a point each second, measured in hertz (Hz). It is equal to 1/time period.

friction The force that resists when you try to move something. It can cause insulators to become charged.

fuse A special wire that protects an electric circuit. If the current gets too large, the fuse melts and stops the current.

galaxy A large group of stars.

gamma ray Ionising electromagnetic radiation – radioactive and dangerous to human health.

Geiger–Müller (G–M) tube A device that detects radioactivity.

geothermal power Generating electricity using the heat in underground rocks to heat water.

gradient The slope of a curve.

gravitational field strength The force of gravity on a mass of one kilogram. The unit is the newton per kilogram, and it is different on different planets.

gravitational potential energy A form of stored energy given by mass $\times g \times$ height.

gravity A force that causes objects in the Solar System to travel in repeating cycles of motion called orbits.

half-life The time it takes for half of the radioactive nuclei in a sample to decay.

heavy water Water that contains an isotope of hydrogen called deuterium.

hertz Unit of frequency, symbol Hz.

Hertzsprung–Russell diagram The graph that shows the luminosity plotted against the surface temperature for a group of stars.

Hooke's law The extension of a spring is in direct proportion to the force applied to it, as long as the force is smaller than the material's elastic limit.

hydroelectric power Generating electricity from the transfer of energy in water flowing downhill.

ideal gas A theoretical gas that has particles that have perfectly elastic collisions, negligible volume and zero intermolecular forces. An ideal gas would exactly obey the ideal gas law. An ideal gas, of course, does *not* exist.

induction When something is affected without touching it. An electric force can induce charge in a conductor. A changing magnetic force can induce electric current in a wire.

infrared The part of the electromagnetic spectrum that has a slightly longer wavelength than the visible spectrum.

insulators (thermal) Substances that do not conduct thermal energy very well.

inversely proportional The relationship between two quantities if one doubles when the other halves.

ion An atom (or group of atoms) with a positive or negative charge, caused by losing or gaining electrons.

ionising radiation Charged particles or high-energy electromagnetic waves that ionise the material they travel through.

isotope Atoms of the same element that contain different numbers of neutrons. Isotopes have the same atomic number but different mass numbers.

joule Unit of energy. One joule is the energy needed to push an object through one metre with a one newton force.

Kelvin scale A temperature scale that starts at absolute zero.

kinetic energy The energy of moving objects, equal to $\frac{1}{2} \times$ mass $\times$ (speed)^2.

kinetic molecular model The theory describing the movement of particles in solids, liquid and gases.

latent heat (or energy) The energy required for a substance to change state.

law of conservation of energy Energy cannot be created or destroyed.

light dependent resistor A resistor with a resistance that decreases when light is shone on it.

limit of proportionality When a material is stretched, the point at which the extension is no longer directly proportional to the load applied.

longitudinal wave A wave in which the change of the medium is parallel to the direction of the wave. Sound is an example.

luminosity The amount of energy radiated by a star every second.

main sequence star A phase in the life cycle of a star. In this phase, a star fuses hydrogen nuclei to form helium nuclei in its centre. This phase is stable and lasts for a very long time.

magnetic field The region in which magnetic materials feel a force.

magnetic field lines The lines that show the path a free North pole would follow.

magnetic materials Materials that are attracted to magnets and can be made into magnets. Iron, cobalt, and nickel are magnetic materials.

magnetic hardness When materials retain their magnetism.

magnetic softness When materials lose their magnetism.

mass The amount of material in an object, measured in kilograms.

mass number The number of protons and neutrons in the nucleus of an atom.

melting The change of state from a solid to a liquid.

microwave High frequency radio wave. Part of the electromagnetic spectrum.

moderator A substance, such as water or graphite, that is used in a nuclear reactor to slow down fast neutrons.

moment Force × perpendicular distance from the pivot.

momentum Mass × velocity.

monochromatic (light) Light that has one wavelength.

moon An object that orbits a planet.

mutation A mutation is a change in the genetic material in the nucleus of a cell.

neutron A particle in the nucleus of atoms that has mass but no charge.

neutron radiation A type of radiation emitted during nuclear fission; fast-moving neutron.

neutron star A phase in the life cycle of a high-mass star. A neutron star is a very dense, compact star that forms after a supernova. It is made mostly of neutrons. It is a dense object but not as dense as a black hole.

newton Unit of force, symbol N.

Newton's first law of motion For a body to change the way it is moving, a resultant force needs to act on it.

Newton's second law of motion When a force acts on a body, the body is accelerated in the direction of the force. It is given by the equation $F = ma$.

Newton's third law of motion Forces always come in pairs that are equal in size and opposite in direction.

non-renewable (resources) An energy resource that will run out, such as oil or natural gas.

normal Imaginary line at 90° to the surface of a material; used to draw ray diagrams.

nuclear equation Equation that shows what happens in a change in the nucleus such as radioactive decay.

nuclear fission The process where a large nucleus absorbs a neutron and then splits into two large fragments, releasing energy and further neutrons.

nuclear fusion When the nuclei of small atoms join to form larger atoms, releasing energy. Very high temperatures and pressures are required.

nucleon Particles found in the nucleus, that is protons and neutrons.

nucleus, atomic The tiny centre of an atom, made up of protons and neutrons.

nuclide A nucleus that contains a particular number of protons and neutrons.

ohm Unit of resistance, symbol Ω; a component with a resistance of 1 Ω allows a current of 1 A when a potential difference of 1 V is applied.

Ohm's law The current flowing through a component is proportional to the potential difference between its ends, providing temperature is constant.

orbital speed How fast one object orbits another. It is calculated from orbital speed = (2 × π × orbital radius)/time.

parallel Describes a circuit in which the current splits up into more than one path.

pascal SI unit of pressure, symbol Pa. 1 Pa of pressure arises when 1 N of force is applied over an area of 1 m^2.

permanent magnet Object or material that produces its own magnetic field even if it is not within the magnetic field of another object.

photons Particles of light and other electromagnetic radiations. Sometimes radiation behaves like waves, sometimes like particles.

pitch Whether a note sounds high or low to your ear.

plane mirror A flat mirror.

planet An object that orbits a star.

plastic flow When a material stretches permanently.

polarity The polarity of an object is where it is a north pole or a south pole or positively or negatively charged.

potential difference (PD) The energy transferred from one coulomb of charge between two points. Measured in volts. Often called the 'voltage'.

potential energy A form of stored energy.

power The amount of energy transferred every second, equal to work done/ time taken. Power can be transferred from somewhere (such as a power station) or to somewhere (such as an electric kettle).

power rating Measure of how fast an electrical appliance transfers energy supplied as an electrical current.

pressure The effect of a force spread out over an area. Pressure is equal to force/area. Pressure difference (in a fluid) = density Ü g Ü height difference.

pressure law If the volume of the gas stays constant, then the pressure of the gas is proportional to its absolute temperature.

primary coil The input coil of a transformer. You connect it to the voltage you want to change.

principle of conservation of momentum In any collision, the total momentum before the collision is the same as the total momentum after the collision.

principle of moments When an object is not turning, the sum of the clockwise moments about any pivot equals the sum of the anticlockwise moments about the same pivot.

prism A solid object that has two flat, parallel ends, and the cross-section remains the same all along its length. E.g. a triangular prism, a rectangular prism.

proportional *see* directly proportional, inversely proportional.

proton Positively charged, massive particles found in the nucleus of an atom.

radiation Energy, such as electromagnetic rays, that travels in straight lines.

radio wave The part of the electromagnetic spectrum that has a long wavelength and is used for communications.

radioactive carbon dating Using the half-life of carbon-14 to calculate the age of plant or animal material.

radioactive decay Natural and random change of a nucleus.

radioactive Describes a substance that has nuclei that are not stable.

radioactivity The emission of particles or energy from an unstable nucleus.

radiotherapy Using radioactivity to kill cancer cells.

rarefaction Region where particles are stretched further apart than normal.

ray diagram Line diagram showing how rays of light travel.

red giant A large star that has a low surface temperature for its size. Red giants are a phase in the life cycle of stars when there is no hydrogen left to undergo nuclear fusion.

red-shift The wavelength of the radiation that is emitted from a light source moving away from an observer increases because of the Doppler effect. These objects appear more red because longer wavelengths of light are at the red end of the visible light spectrum.

reflection When waves bounce off a surface. The angle of incidence is the same size as the angle of reflection.

refraction When waves change direction because they have gone into a different medium. They change direction because their speed changes.

refractive index Indicates how strongly a particular material changes the direction of light, where $n = \sin i / \sin r$ and $n = 1/\sin c$.

renewable (resource) An energy resource that is constantly available or can be replaced as it is used, such as solar power or wind power.

resistance The property of an electrical conductor that limits how easily an electric current flows through it. Measured in ohms.

resistor An electrical component that has a particular resistance.

resultant force A single imaginary force that is equivalent to all the forces acting on an object, equal to mass × acceleration.

rheostat A resistor whose resistance can be varied.

Sankey diagram A diagram that shows the different types of energy a device transfers.

satellite An object that orbits a larger object.

scalar A quantity that only has magnitude.

semiconductor A material that does not conduct electricity as well as metal, for example, but conducts electricity better than an insulator, such as plastic.

series Describes a circuit in which the current travels along one path through every component.

shielding A material around a nuclear reactor that will absorb alpha particles, beta particles, gamma rays and neutrons.

short circuit The unwanted branch of an electrical circuit that bypasses other parts of the circuit and causes a large current to flow.

Solar System The Sun and all the objects that orbit it.

solenoid A coil of current-carrying wire that generates a magnetic field.

solidification The change of state from a liquid to a solid.

speed A measure of how far something moves every second.
Average speed = distance travelled ÷ time taken.

specific heat capacity The energy needed to raise the temperature of 1 kg of a substance by 1 °C. Its units are J/kg °C.

spectrum The 'rainbow' of colours that make up white light: red, orange, yellow, green, blue, indigo and violet.

state of matter Form that particles of a substance take depending on temperature; different states include solid, liquid and gas.

supernova When a high-mass star ends fusion, gravitational collapse can cause the star to explode. Its luminosity can increase by up to twenty times and most of the star's matter is ejected at very high speeds.

terminal velocity Maximum velocity an object can travel at – at terminal velocity, forward and backward (or upward and downward) forces are the same.

thermal conductor Material that allows thermal energy to transfer through it quickly.

thermal insulator Material that does not allow thermal energy to transfer through it quickly.

thermistor A resistor made from semiconductor material: its resistance decreases as temperature increases.

time period The time taken for each complete cycle of the wave motion.

total internal reflection When rays of light are entirely reflected back inside a medium.

transformer A machine that changes the voltage of AC electricity. The ratio of the number of turns in the coils is the same as the ratio of the voltages produced. A step-up transformer increases the voltage. A step-down transformer decreases the voltage.

transverse wave A wave in which the change of the medium is at 90° to the direction of the wave. Light is an example.

trough The lowest part of a wave.

turbine A machine that rotates. It is pushed by the movement of a substance such as air or water.

ultrasound Sound with frequencies that are above the range of human hearing.

ultraviolet The part of the electromagnetic spectrum that has a slightly longer wavelength than the visible spectrum.

unbalanced forces Forces are unbalanced when the resultant force is not zero.

vacuum A space with no particles in it.

vapour Another term for gas.

variable resistor A component with a resistance that can be manually altered.

vector A quantity that has both magnitude and direction.

velocity The speed and direction of an object.

velocity–time graph A graph of how velocity varies with time.

visible spectrum The range of light that can be seen by human eyes.

volt A unit of voltage. The energy carried by one coulomb of electric charge.

voltage A measure of the energy carried by an electric current.

voltmeter Instrument for measuring voltage or potential difference.

watt A unit of power, symbol W. One watt is one joule transferred every second.

wave equation Wave speed = frequency × wavelength.

wavefront The moving line that joins all the points on the crest of a wave.

wavelength The distance between the same points of successive waves, for example, the distance from one crest to the next.

wave speed Equal to frequency × wavelength.

weight The force of gravity on a mass, equal to mass × gravitational field strength. The unit of weight is the newton.

white dwarf A star about half the mass of our Sun will eventually turn into a white dwarf at the end of its 'life'. They consist of helium.

work The energy transferred when a job is done, equal to force × distance moved in the direction of the force.

X-ray Part of the electromagnetic spectrum that has a high energy.

Answers

SECTION 1 PRINCIPLES OF PHYSICS

Units

Page 11

1. Length, time, mass.
2. So that the unit remains constant and the science remains comparable.
3. **a)** Base: kilogram, second, metre.
   b) Derived: metre/second, newton.

Movement and position

Page 15

1. Stationary – the distance does not change as time increases.
2. A straight line – positive gradient for moving away, negative gradient for moving towards.
3. Student's own graph – likely to be a curve, as shown in Fig. 1.2.

Page 16

1. 10 000/(15 × 60) = 10 000/900 = 11. 1 m/s
2. 22.5 m
3. 3000 s = 50 minutes

Page 17

1. Speed describes how far an object travels in a particular time. Velocity is numerically identical to speed, but it also includes a statement of the direction of the travel.
2. The sign of a velocity indicates a direction. 'Positive' and 'negative' velocities are in opposite directions to each other. We still need to explain which direction we are taking to be positive – this will change depending on the motion we are studying. For example, if we are looking at the motion of objects acting under the force of gravity, we might say upwards is the positive direction, so that downwards becomes the negative direction.
3. **a)** Diagram showing tennis ball moving upwards with a speed of +5 m/s.
   b) Diagram showing tennis ball moving downwards with a speed of –5 m/s.

Page 20 (top)

1. Acceleration = (change in velocity)/ (time taken) = rate of change of velocity.
2. Deceleration.
3. 10 m/s^2
4. –15 m/s^2

Page 20 (bottom)

1. $v = ?$ $a = 1.6$ $s = 1.5$ $u = 0$
   $v^2 = u^2 = 2as$
   $v^2 = 0^2 + 2 \times 1.6 \times 1.5$
   $v^2 = \sqrt{4.8}$
   $v = 4.8 = 2.19$ or 2.2 m/s

Page 22

1. Constant velocity or speed in a straight line.
2. The line will be a straight line (constant gradient) – positive gradient for speeding up, negative gradient for slowing down.
3. Acceleration.

Page 24–25

1. The acceleration is the gradient of the graph.
2. The distance travelled is the area under the graph.
3. **a)** Athlete A: 8 m/s, athlete B: 6.25 m/s.
   b) Athlete A: horizontal line at 8 m/s, starting at 0 s and finishing at 50 s. Athlete B: horizontal line at 6.25 m/s, starting at 0 s and finishing at 64 s.
   c) Area under speed–time graph is the same, i.e. 8 m/s × 50 s = 6.25 m/s × 64 s
4. 45 m
5. **a)** 10 m/s
   b) The object is travelling at a steady speed.
   c) C and D
   d) It is decelerating.
   e) 120 m
   f) 10 m/s^2

Forces, movement, shape and momentum

Page 31

1. Effects – forces can change the speed of an object, the shape of an object, the direction the object is moving in. Types of force – gravitational, electric, magnetic, (electromagnetic), strong nuclear force.
2. Gravity.
3. The mass of objects and the distance between their centres.
4. Electromagnetic force.

Page 34 (top)

1. Vector has size and direction, scalar just has size.
2. **a)** resultant force = 1000 – 100 = 900 N forward
   b) resultant force = 1000 – 500 = 500 N forward

c) resultant force = 1000 – 1000 = 0 N

d) resultant force = 1000 – 1200 = –200 N forwards (or 200 N backwards)

3. 2560 N

Page 34 (bottom)

1. Walking, driving – if there is no friction, you skid.
2. Where energy is transferred to thermal energy which is lost to the surroundings.

Page 36

1. The object will be stationary or moving in a straight line at constant speed.
2. The object will be changing speed and/or direction.
3. His weight.

Page 38

1. Force = mass × acceleration
2. The force must be the resultant force, the mass must remain constant.
3. 600 N
4. 800 m/s^2

Page 39

1. Mass amount of matter, weight force
2. 600 N
3. 50 kg

Page 41

1. The different planets have different masses and so they have different gravitational field strengths. Since your weight = mass × gravitational field strength you will have a different weight (but the same mass) on different planets.
2. 96 N
3. 50 kg
4. 20 N on left hand side.

Page 42 (top)

1. The velocity that an object, such as a skydiver, has when the forces are balanced so that the object travels at a maximum constant speed.
2. 10 m/s
3. 15 N upwards
4. 585 N downwards
5. 9.75 m/s^2

Page 42 (bottom)

1. a) velocity
   b) force
   c) acceleration
2. The velocity graph decreases from initial zero until it reaches a constant value, which is the terminal velocity. The force graph includes the constant weight (the drag), which increases with time and then becomes constant; and the resultant, which is found by adding the weight and the drag. The acceleration graph increases initially and then reaches zero because the velocity has reached terminal velocity, which is not changing, so there is no acceleration.

Page 43

1. Thinking distance: Any one of: Speed, fatigue, use of medication/drugs/alcohol, level of distraction. Stopping distance: Speed, weather condition of road, amount of tread tyres, friction from brakes, mass of car.
2. 36 m
3. 18
4. 7.5 m
5. It would increase.

Page 43 and 44

1. Thinking and braking distances are both less.
2. Yes, reaction time is faster.
3. Velocity at 30 km/h = 8.3 m/s, at 45 km/h = 12.5 m/s. In both cases acceleration = $6.5 \times 10^3/1000$ = 6.5 m/s^2. At 30 km/h, time = 8.3/6.5 = 1.28 s, so distance = 5.3 m. At 45 km/h, time = 12.5/6.5 = 1.92 s, so distance = 12 m.

Page 45

1. The size of the force and the distance between the line of the force and the turning point, which is called the pivot.
2. 2 N m
3. 1.25 N m
4. 0.4 m

Page 47

1. The centre of gravity is the point where we can assume all the mass of an object is concentrated.
2. Student's own description based on method described in the text.
3. No, it will fall over.

Page 49

1. 1.5 m
2. 20 N each
3. F_4 = 16 N, F_3 = 24 N

Page 54

1. 0.01 N
2. 20 N/m
3. 20 cm

Page 56

1. 2.32 kg m/s
2. 50 000 kg m/s
3. 1000 kg
4. 25 m/s

Page 60

1. Momentum = mass × velocity. Momentum is a vector quantity (it includes a direction) because velocity is a vector quantity.
2. From Newton's second law, for a particular change in momentum, the longer the change takes the smaller the force will be. So if a parachutist bends their knees when they land, the landing takes a longer time, the momentum change is fixed and so the force is reduced.
3. 2.5 m/s
4. 4 m/s
5. **a)** 100 000 kg m/s

   b) 100 000 kg m/s

   c) 10 m/s
6. When the ball hits the goalkeeper the padding changes shape and absorbs the kinetic energy of the ball. The ball will take a longer distance and time to stop. The change in momentum of the ball takes place over a longer time.

   $F = m(v - u)/t$

   If the time increases, then the force reduces, resulting in less injury.

Page 61

1. Newton's third law states that the forces on each skater will be equal and opposite, so both experience a force of 200 N. Newton's second law states that $F = ma$. Rearranging the formula: $a = F \div m$. So for the same force, the skater with more mass will have a lower acceleration. The accelerations are also in opposite directions.

SECTION 2 ELECTRICITY

Mains electricity

Page 82

1. Brown wire: live, connected to fuse. Blue wire: neutral wire. Green and yellow wire: earth wire.
2. It melts when the current gets too high.
3. The student should not choose the 'nearest' fuse, but the 'nearest above'. If the appliance requires 6 A, the 5 A fuse will break whenever the appliance is used.
4. The earth connection needs to be a *low resistance* path (see later in the chapter). This means that a *high current* will pass through the wire and this will melt the fuse (or trip the circuit breaker).
5. The casing cannot become live because it is not a conductor.

Page 83

1. A resistor is a device that opposes the flow of current.
2. It increases.
3. Electrical energy to heat energy.

Page 85

1. 220 W
2. 12 V
3. 0.11 A
4. **a)** 1.6 A, 3 A fuse

   b) 3.47 A, 5 A fuse

   c) 0.4 A, 3 A fuse

   d) 9.1 A, 13 A fuse

   e) 3.9 A, 5 A fuse

Page 86–87

1. The charger will still work, although the battery may not charge as effectively. The transformer in the charger (see section 6) will reduce the mains voltage by the same factor, so the battery may not be able to charge fully.
2. The charger will reduce the mains voltage by the same factor as usual, but starting with 230 V, the resulting voltage may be high enough to damage the circuitry in the laptop.
3. 2160 J
4. 3.3 A
5. 218 s
6. 12 V

Energy and voltage in circuits

Page 95

1. There is probably a break in the circuit.
2. In a parallel circuit, more energy is transferred each second – the bulbs are brighter. The energy available from the battery is limited by the chemicals used – running two bulbs at once at full power will transfer the chemical energy more quickly and the battery will become flat sooner.
3. So that they can be switched independently.
4. **a)** In a series, dimmer because battery energy shared between two bulbs.

   b) In parallel, full energy of battery given to each bulb.

Page 97

1. Bulb B has the higher resistance. The brighter bulb must have the higher current (since, in parallel, both bulbs are connected to the same voltage), so it must also have the lower resistance.
2. In series, bulb B will be brighter. It has a higher resistance, so in series it will have a bigger share of the voltage and hence the energy.
3. Use a circuit like the one shown in Fig. 2.22. Vary the voltage and record the current flowing in the circuit.
4. The ammeter should be connected in series and the voltmeter in parallel with the component.

Page 100

1. In brighter light, the resistance of the LDR decreases, so the current in the circuit will increase. The ammeter reading will therefore increase in brighter light.
2. A diode has a low resistance if the current moves in one direction and a very high resistance in the other. A waterfall allows water to move easily in one direction, but the water cannot go 'backwards' up the waterfall. In this way the analogy is reasonable.
3. Increasing the temperature reduces resistance.

Page 103

1. 10 V
2. 6.0 V
3. The current would rise ever more slowly as the potential difference increased. This is because increasing the current makes the wire hotter, increasing the resistance.

Page 105

1. **a)** 15 C
   b) 20 C
   c) 92 C
   d) 45 C
2. 9000 C
3. 120 s

Page 106

1. Opposite directions.

Page 110 (top)

1. The battery or power source.
2. They transfer the energy to the components in the circuit.
3. The difference in voltage between the two points.
4. Yes, potential difference is the difference in energy of a coulomb of charge between two points in a circuit.
5. The electrons at the point with the higher voltage will have more energy than the electrons at the point with the lower voltage.

Page 110 (bottom)

1. Students' own answers. One possible answer is that the current is the amount of water flowing, the potential difference is height the water is pumped up. The ammeter measures the amount of water flowing past the water wheel, the voltmeter measures the height difference. A series circuit would have two or more water wheels connected one after the other, a parallel circuit would have two or more water wheels connected side by side.
2. Students' own answers. One possible answer is that it is a good model, but does not model the way current is affected by resistance in series and parallel circuits. The amount of water flowing relies on how much water the pump is pumping.

Electric charge

Page 120

1. You are the insulator. You will have rubbed electrons either onto or off yourself, perhaps by sliding your feet over a carpet. The metal handrail is a conductor – when you touch it, it allows the electrons to move to restore the balance and this is what you feel as a shock.

Page 121

1. Electrons are extremely small and negatively charged. Electrons are also around the outside of atoms. These key ideas make it much easier for electrons to be moved about (than the positively-charged parts of the atom) to account for all the electric effects we know.
2. The electron clouds in your feet repel the electron clouds in the floor. The force of repulsion is easily enough to stop you falling through.
3. **a)** Polythene rod should be negatively charged and cloth positively charged. Perspex rod should be positively charged and cloth negatively charged.
   b) They will be equal and opposite.

Page 125

1. Any suitable example such as combing your hair, touching a door knob or a car.
2.

Situation	Danger	Prevention
Refuelling a car	Sparks could form as charge builds up causing an explosion.	Tyres made of graphite – a conductor that transfers electrons to earth.
Making sensitive electronic equipment	High voltages can damage (low voltage) electronic equipment.	Workers use earth straps to transfer electrons to the ground and prevent them becoming charged.
Refuelling an aeroplane full of passengers	High voltages can cause sparks which can ignite fuel and cause an explosion.	Fixing an earth strap between the aeroplane and ground will transfer electrons to the ground preventing the aeroplane becoming charged.

3. Any suitable example such as electrostatic scrubbers in power station chimneys, vacuum cleaners, inkjets and photocopiers.
4. They can both attract or repel objects.

SECTION 3 WAVES

Properties of waves

Page 139

1. **a)** The waves travel by vibrations in the direction of travel of the wave.
   b) The waves travel by vibrations at a right angle to the direction of travel of the wave.
2. **a)** The distance between consecutive peaks or troughs of the wave.
   b) The number of vibrations per second or number of peaks or troughs that pass a point each second.
   c) The size of the vibrations.
3. 15 m
4. Energy.

Page 140

1. 1.2 m
2. 3×10^8 m/s
3. 1500×10^6 Hz

Page 141

1. 0.00227 s
2. 8.33×10^{-7} s
3. 4×10^{-8} s

Page 143

1. An echo is a single reflection from a hard surface.
2. $v = f\lambda$, f constant, so if v reduced so must λ.
3. It depends on the change in speed.
4. **a)** D
   b) Natural frequency of sound from the train is 440 Hz.
   Frequency is increased when train approaches. Frequency is reduced when train moves away. Steady speed means the increase in frequency must be the same as the decrease in frequency (i.e. 30 Hz is gained on approach and 30 Hz is reduced when moving away).

The electromagnetic spectrum

Page 150

1. Shine a ray of white light at an angle to one of the faces of the prism. A prism splits light by using the fact that different wavelengths of light slow down inside the glass by different amounts so they are refracted through different angles.
2. 3×10^8 m/s
3. X-rays.
4. Gamma, UV, visible, infrared, microwaves.

Page 153

1. Gamma rays are produced by radioactive nuclei. X-rays are produced by firing high energy electrons at a metal target.
2. An image of infrared radiation.
3. Water particles in food absorb the energy carried by microwaves. They vibrate more making the food much hotter.

Page 154

Student's own answers.

Light and sound

Page 162

1. 48°
2. The 'normal' is the imaginary line at 90° to a mirror.

Page 165 (top)

1. The light from a submerged object is refracted away from the normal as it leaves the pond, so it approaches your eye at a shallower angle. Your brain is fooled into thinking this shallower angle is the true angle to the object.
2. Being more dense than air, the carbon dioxide in the balloon will slow down the sound wave. This will tend to focus the sound together, like a lens. (A balloon filled with hydrogen will do the opposite.)
3. 30.7°
4. 36.3°
5. 1.43

Page 165 (bottom)

1. Slower.
2. Diamond, because its refractive index is greatest.
3. Air: 2.99×10^8 m/s; water: 2.25×10^8 m/s; glass: 1.97×10^8 m/s; diamond: 1.24×10^8 m/s

Page 170 (top)

1. The angle of incidence at which the angle of refraction becomes equal to 90°.
2. 45.6°
3. By repeated internal reflection.
4. It always hits the surface at an angle greater than the critical angle.
5. Infrared rays.
6. In endoscopes.

Page 170 (bottom)

1. Longitudinal waves.
2. Small differences in air pressure.

Page 174

1. 20 – 20 000 Hz
2. Ultrasound is sound above range for human hearing.
3. Method should have the features of the Developing Investigative Skills box on measuring the speed of sound.

SECTION 4 ENERGY SOURCES

Energy

Page 192

1. Gravitational – kinetic – gravitational.
2. In springs.
3. a) Nuclear, b) chemical, c) kinetic, d) elastic, e) thermal, f) gravitational, g) electrostatic.
4. a) Sound/sound wave, b) radiation of light/ light wave, c) heating.

Page 193

1. Energy cannot be created or destroyed.
2. Ball's gravitational store decreases, ball's kinetic store increases, thermal energy store of surroundings increases; total amount of energy remains the same.
3. **a)** Chemical energy of fuel.
   b) Kinetic energy store of the car.
   c) Increased thermal energy store of the surroundings (and a small amount of energy transferred by sound).

Page 198 (top)

1. How much energy is transferred each way.
2. Sankey diagram should have 1000 J as input. There should be two output arrows, one labelled 400 J useful heat energy, the other 600 J wasted heat energy. The sizes of the arrows should be in proportion.
3. 40%
4. **a)** 39 200 J
   b) 2%

Page 198 (bottom)

Student's own Sankey diagrams.

Page 199

1. They contain electrons that can move freely and transfer energy.
2. There are no particles to transfer energy by colliding with each other.
3. Vibrations of the particles are passed on through the bonds between the particles.
4. Some wax is put on one end of a metal rod. The other end of the metal rod is heated until the wax on the other end melts.

Page 200 (top)

Use petroleum jelly to attach drawing pins at regular distances along the copper strip. Heat one end of the strip and measure the time it takes for each drawing pin to fall off. Plot a graph of time until the drawing pin falls off against distance from the point that is being heated.

Page 200 (bottom)

1. The particles are free to move.
2. Warm air expands, which makes it less dense. Less dense air floats up above more dense (cooler) air.
3. Heat some potassium manganate(VII) crystals in water.
4. Fibres in the insulation create air pockets. This restricts the movement of the air and so convection currents cannot form.

Page 202

1. It can travel through a vacuum.
2. A hot object.
3. Temperature and type of surface.
4. The dull, black side.

Page 205

1. Convection.
2. The top of the room.
3. a) Conduction – reduced by having a vacuum (no particles to connect).
   b) Convection – vacuum space has no particles to form convection currents.
   c) Radiation – silver surfaces reflect the energy and do not absorb or emit it.
4. The flask reduces energy transfer in both directions, so hot drinks do not lose their energy and cold drinks are not warmed by energy entering from outside.

Page 208

1. More than half the energy wasted is through these two features.
2. Answer should refer to trapped air in the walls, roof, windows, etc.
3. Air is a bad conductor (good insulator) so keeping a layer near the body reduces heat loss by conduction. Keeping the layer trapped reduces heat loss by convection.
4. In hotter climates, you may want to lose heat, so loose clothing allows the air to move and transfer heat away.

Work and power

Page 214 (top)

1. 250 J
2. 500 N
3. 80 J
4. 50 m
5. 400 N

Page 214 (bottom)

1. $8.45 \times 10^{12}/8 \times 10^{3} = 1.06 \times 10^{9}$ N
2. Series of manoeuvres used before landing to get rid of excess energy.
3. 1650 °C. The orbiter is covered with ceramic insulating materials designed to protect it from this heat. The materials include:
   Reinforced carbon-carbon (RCC) on the wing surfaces and underside.
   High-temperature black surface insulation tiles on the upper forward fuselage and around the windows
   White Nomex blankets on the upper payload bay doors, portions of the upper wing and mid/aft fuselage.
   Low-temperature white surface tiles on the remaining areas.

Page 216

1. 100 J
2. 4 J
3. 812.5 J
4. 8 kg
5. 3 m/s

Page 218 (top)

1. 100% efficiency.
2. Heat energy from friction.
3. 8.9 m/s

Page 218 (bottom)

Student's own answers, but should include key points from the text.

Page 223

1. 240 W
2. The man lifts twice the mass up the stairs, so he does twice the work. They take the same time, so the man must be providing twice the power.
3. It can transfer a lot of energy each second.
4. The watt.
5. a) 3000 J
   b) 50 W

Human influences on the environment

Page 230

Student's own answers.

Page 234

1. Light energy to electrical energy and heat energy.
2. Kinetic energy in wind to kinetic energy in turbine to electrical energy and heat energy.
3. A fuel is used to turn water into steam. The steam has kinetic and thermal energy and drives a turbine. The turbine drives a generator which transfers the kinetic energy to electrical energy.
4. Using heat energy from the Earth.

SECTION 5 SOLIDS, LIQUIDS AND GASES

Density and pressure

Page 249

1. a) $40\ cm^3$
   b) $7.8\ g/cm^3$
2. The bread contains more air spaces, making the overall density less.
3. It is lower than the density of sea water.

Page 252–253

1. When the object floats, absorbs the liquid (like a sponge) or is damaged by it (such as dissolving).
2. Narrow as possible to give greater movement of level, fine scale to help with precision.
3. $1.25\ g/cm^3$
4. 78 g
5. $7.14\ cm^3$

Page 255

1. For the pin, the force is concentrated over a smaller area – there is a greater pressure.
2. 500 Pa
3. 80 N
4. $0.78\ m^2$

Page 258

1. Particles colliding with other particles and the walls of the container.
2. The same air pressure is in our lungs and presses outwards.
3. The pressure outside the bottle is greater than the pressure inside the bottle.

Page 261

1. Depth, density of the fluid, gravitational field strength.
2. 80 000 Pa (80 kPa)
3. 1 030 000 Pa (1030 kPa)
4. 400 kPa
5. 20 400 Pa (20.4 kPa)

Change of state

Page 267

1. Increases – either as a larger vibration in solids or as faster translational motion in liquids and gases.
2. Compressing a gas pushes the particles closer together, in a liquid they are already close to each other and will repel if pushed closer.
3. a) Regular pattern, closely packed together, particles held in place.
   b) Irregular, closely packed together, particles able to move past each other.
   c) Irregular, widely spaced, particles able to move freely.
4. The size of the container it is put in.

Page 269

1. Energy is needed to break the bonds between the particles in the ice.
2. All the particles have the energy to escape. In evaporation, only a fraction has.

Page 271 (top)

1. Higher temperature, flow of air across the surface.
2. Larger area for the water molecules to evaporate from.

Page 271 (bottom)

Student's own answers, but should be along the following lines:

1. I changed from liquid state to gaseous state.
2. In the liquid state I was reasonably close to my neighbours, but not so close that I was right beside them – they were there if I needed them but we didn't live in each other's pockets! I was given some extra energy. The extra energy meant that I moved more quickly and was able to break completely away from my neighbours.
3. I escaped to the surface of the liquid and became a gas particle.

Ideal gas molecules

Page 279 (top)

1. The particles are colliding with the walls of the container.
2. Faster molecules hit the walls harder, so create a bigger force.

Page 279 (bottom)

1. The molecules are always moving about and spread out throughout the container.
2. The pressure of the gas increases as temperature increases.
3. A higher temperature means the molecules move more quickly, so the force on the walls will be higher.

Page 280

1. a) 323 K
   b) 2273 K
2. a) −173 °C
   b) 1227 °C
3. They stop moving/vibrating.

Page 283

1. The temperature is proportional to the average kinetic energy, as long as the temperature is measured in Kelvin.
2. Any relevant units can be used for pressure, as long as the same units are used throughout.
3. a) 289 K
   b) 318 kPa

Page 285

1. The pressure is inversely proportional to volume.
2. The mass of the gas must remain constant (that is, no particles move in or out of the system). The temperature must be measured using the Kelvin scale. The gas must be ideal (not liquefy or solidify).
3. 519.8 cm^3

SECTION 6 MAGNETISM AND ELECTROMAGNETISM

Magnetism

Page 296

1. They attract each other.
2. They repel each other.
3. It will line up approximately north–south.

Page 297

1. Magnetically hard materials retain their magnetism; magnetically soft materials lose their magnetism if the outside magnetic influence is removed.
2. A permanent magnet that is made of magnetically hard materials.
3. In electromagnets and relays.

Page 300 (top)

1. A region of space where their magnetism affects other objects
2. The path that a free north pole would take from a north pole to a south pole
3. Using iron filings on a thin sheet of plastic or a plotting compass

Page 300 (bottom)

1. It has magnetism induced in it and is attracted to the magnet.
2. A south pole.
3. Attraction only tells you the material is a magnetic material. Only two magnets can repel, so that is the test.
4. Two north poles together or two south poles together.

Page 302

1. The magnetic field lines radiate out from the north pole and go round to the south pole.
2. They attract each other and the field lines go from a north pole to a south pole.
3. The field lines go from a north pole to a south pole, not between like poles.

Page 303

1. Magnetic field lines show the direction of a force (on a free north pole); if field lines crossed it would indicate a force in more than one direction, which makes no sense.
2. Even, the same strength at all places, etc.
3. Between the poles of a U-shaped magnet, or when opposite poles of two magnets are placed close to each other.

Electromagnetism

Page 307

1. Make the wire into a coil. Connect one terminal of the battery to the switch and one end of the coil of wire. Connect the other end of the coil to the switch. Connect the switch to the other terminal of the battery.
2. Increase the current, increase the number of turns on the coil.
3. Diagram should be similar to Fig. 6.16.

Page 310

1. Sketch should match Fig. 6.20. Similar shape to bar magnet pattern – 'fanning out' at the poles; 'loops' along the sides, etc.
2. The field at the centre of the coil is most nearly uniform.
3. a) Sketch should match right-hand part of Fig. 6.19.
   b) Sketch should match left-hand part of Fig. 6.19.

Page 313

1. The current in the coil creates a magnetic field around it. This interacts with the magnetic field from the permanent magnet, producing forces in opposite directions on either side of the coil. Putting the coil on an axle allows these forces to spin the coil.
2. Connect a length of wire across a power supply and coil around one pole of a permanent magnet. Switch the current on and the wire jumps off the magnet.
3. The changing current in the coil produces a changing magnetic field which interacts with the field from the permanent magnet. This creates a 'backwards and forwards' motion of the coil and paper cone. This makes the air vibrate – a sound wave.

Page 316

1. The thumb, and first and second fingers are held at right angles. The first finger points in the direction of the field, the second finger in the direction of the current and the thumb gives the direction of the force (movement).
2. The commutator reverses the direction of the current in the coil each half- turn. This allows the coil to keep turning in the same direction.
3. Increase the current, increase the strength of the permanent magnets, increase the number of turns on the motor coil. (Reducing the friction on the axle would make the motor spin faster, but it doesn't increase the force, so it doesn't answer the question.)
4. Reverse the direction of the current (by turning the battery round, etc), reverse the polarity of the magnets.

Electromagnetic induction

Page 321 (top)

1. Relative motion between the magnetic field and the wire – a current will pass if there is a complete loop of wire; if there isn't, an emf (voltage) will be generated across the ends of the wire.
2. No, because the wire is not moving.
3. By changing the strength of the magnetic field and/or the speed the wire is moving at.

Page 321 (bottom)

1. A larger current is created.
2. The area of the coil, the number of turns in the coil, the strength of the magnetic field and the speed of rotation.
3. A coil of area 2 m^2

Page 322

1. The magnet is rotated inside the coil.
2. The speed of rotation.
3. To maximize the effect, increase the rate of rotation, increase the strength of the magnetic field, increase the number of turns on the coil. The output is an alternating current because the relative motion between coil and field reverses each half-turn.

Page 324

Student's own answer that should cover the following:

Equipment needed: magnet, coil of wire and sensitive meter.
Vary the current by altering the number of coils and the speed the magnet is moved in the coil.

Page 325

Student's own answer. A shaver socket contains an isolating transformer, which means that there is no direct link between the current that flows through the shaver and the mains supply.

Page 326

1. To increase the strength of the magnetic field.
2. In a step-up transformer, the output voltage is higher than the input voltage (because the secondary coil has more turns than the primary coil). In a step-down transformer, the output voltage is lower than the input voltage (because the secondary coil has less turns than the primary coil).
3. 20 V
4. 6.0 V
5. **a)** 24 V
   b) 2 A
   c) 24 A

Page 329

1. Higher voltages mean less energy is wasted in heating the transmission cables, meaning more useful power is delivered at the far end.
2. Transformers change voltages. To send energy at high voltage, firstly the voltage must be increased at the power station end (using a step-up transformer) and then the voltage must be reduced (for safety) at the far end (using a step-down transformer).
3. 40 000 W
4. 1%

SECTION 7 RADIOACTIVITY, FISSION AND FUSION

Radioactivity

Page 341

1. An atom is what all elements are made of.
2. Electrons and protons have equal and opposite charges and there are equal numbers of them in an atom.
3. Atomic number = number of protons in a nucleus (and number of electrons in a neutral atom). Mass number = number of protons + number of neutrons in a nucleus. Isotopes are nuclei with the same atomic numbers, but different mass numbers.
4. Isotopes of an element have the same number of electrons and it is these that determine the chemical behaviour.

Page 345 (top)

1. The emission of particles and/or energy from an unstable nucleus.
2. Unstable.
3. Alpha particles are relatively large, so as they travel they collide often, reducing their energy quickly, so they cannot travel very far. Beta particles are smaller and travel more quickly, having fewer collisions and losing less energy each time. This means they will travel further before losing all their energy. Gamma radiation, being electromagnetic waves, only interacts weakly with matter so it has a much larger range.

Page 345 (bottom)

Student's own answer that should include the following:

Amount of beta radiation picked up by detector is very sensitive to thickness of the paper and so accurate.

Beta radiation is absorbed by a thin sheet of aluminium, so the source can be shielded easily.

Page 347

1. Two protons and two neutrons are emitted. Atomic number reduced by 2 and mass number by 4.
2. A neutron changes into a proton. Atomic number increases by 1. Mass number not affected.
3. No change to the number or types of particles. There is a reduction in energy however.
4. Emitting alpha or beta radiation changes the number of protons in the nucleus. This means the number of outer electrons will also change, changing the chemical behavior – it is a different element.

Page 348

1. Cosmic rays, radiation from rocks, radon gas, radioactive isotopes in the body.
2. Medical sources, consumer products and others.
3. Medical sources.
4. X-rays, gamma rays (from radiotherapy).
5. Nuclear power stations, atomic bombs.

Page 352

1. Radioactivity in soil, rocks and materials like concrete, radioactive gases in the atmosphere and cosmic rays from outer space.
2. Half-life is the time taken for half of the unstable nuclei in a sample to decay: that is, the time for the activity to reduce to half its current level.
3. 2 hours.
4. 4 hours.
5. Activity is 800 Bq now, so in 8 hours it will be 400 Bq. 8 hours later it will be 200 Bq and a further 8 hours later the activity will be 100 Bq. This gives a total of 24 hours, three half lives.

Page 355

1. Sterilising medical equipment and in preserving food.
2. A radioactive substance with a half-life long enough for it to be spread out and be detected. Medical tracers are used to detect blockages in vital organs. Agricultural tracers monitor the flow of nutrients through a plant. Industrial tracers can measure the flow of liquid and gases through pipes to identify leakages.
3. For example: medical uses – gamma tracers, radioactive iodine to target the thyroid gland; non-medical – dating of rocks, smoke detectors.
4. Radioactive emissions are ionising radiations – they can ionise cells in the body which may destroy the cells or damage them (particularly hazardous is the damage involving mutations to the cell which can lead to cancerous changes).
5. Irradiation is when a material is exposed to alpha, beta or gamma radiation but it does not become radioactive itself. Contamination occurs when an organism ingests a radioactive material. This can sometimes stay in the body for many years. Other materials can also be contaminated when they have absorbed or are coated with radioactive dust for instance.

Fission and fusion

Page 364

1. The process where a large nucleus absorbs a neutron and then splits into two large fragments with the release of energy and further neutrons.
2. A lot of energy is given out, the nucleus splits into two parts and some neutrons are emitted.

3. The neutrons released during a fission can collide with further nuclei, causing further fissions which release further neutrons for further fissions, and so on.

Page 366 (top)

1. Fuel rods – where the fission reaction happens releasing energy as thermal energy.

   Control rods – act as 'sponges' absorbing neutrons to control the rate of fission.

   Moderator – slows the neutrons down so they are at the correct speed to be absorbed for further fission reactions.
2. By circulating pressurised water or another fluid through the reactor.
3. The waste is radioactive, some of it highly radioactive, so it needs to be handled, transported and stored carefully to reduce the risks to people.

Page 366 (bottom)

1. The control rods absorb some of the neutrons, which prevents them going on to split further nuclei. The number of nuclei that undergo fission is thus controlled and so the energy released is also controlled.
2. Take the control rods out of the core/further out of the core.
3. Move the control rods further into the core.

SECTION 8 Astrophysics

Motion in the Universe

Page 379

1. Mercury, Venus, Earth, Mars, Jupiter, Saturn, Uranus, Neptune.
2. Inner four = solid, outer four = gaseous.
3. A planet is an object that orbits the sun and is large enough to have become round due to the force of its own gravity. A planet also has to dominate the neighbourhood around its orbit.
4. Pluto does not dominate its neighbourhood. It is only about double the size of its large 'moon' Charon, but all the true planets are far larger than their moons. Also, planets that dominate their neighbourhoods 'sweep up' asteroids, comets, and other debris, clearing a path along their orbits. Pluto does not.

Page 380

1. Diagram of Solar System with central Sun, a planet in circular orbit around it, a moon in circular orbit around the planet and a comet in highly elliptical orbit around the Sun.
2. Moons are objects that orbit planets rather than the Sun, as planets and comets do.
3. Scientists think that a small planet collided with the Earth soon after it was formed 4.5 billion years ago, forming our Moon.

Page 382

1. **a)** 28 000 km/h

   b) 7795 m/s
2. 7900 m/s
3. 1021 m/s

Page 383

1. 96 N on the Moon, 1560 N on Jupiter.
2. They weigh less on the Moon, but there is gravity there (or else they would float off into space if truly weightless).
3. 546 N

Stellar evolution

Page 387

1. **a)** The colour of light that is emitted depends on the surface temperature of the star. Red stars are much cooler than blue stars. These stars emit lower frequency red light.

   b)

Type of star	Colour	Approximate surface temperature (K)
O	blue	> 25 000
B	blue	11 000–25 000
A	blue	7500–11 000
F	white to blue	6000–7500
G	yellow to white	5000–6000
K	red to orange	3500–5000
M	red	< 3500

Page 390

1. Radiation cannot escape from a black hole because of the extremely strong gravitational field. Neutron stars are non-luminous.
2. **a)** Top left of the HR diagram.

   b) These stars will become hotter and brighter and probably explode as a supernova.

   c) They will move towards the top left of the HR diagram.
3. It is in its main sequence phase now. After this it will cool and become a red giant. It will move from the middle lower part of the HR diagram upwards to the right.

Cosmology

Page 395

1. a) The Doppler effect explains that wavelength or frequency of waves changes if the source of the waves is moving relative to the observer. The wavelength of light emitted by a galaxy that moves towards us at speed will decrease. This means it shifts towards the blue end of the visible spectrum, i.e. has a higher frequency.

   b) Red-shifted galaxies are moving away from us. Blue-shifted galaxies are moving towards us and could collide with our galaxy at some time in the future.

Page 396

1. Change in wavelength: $\Delta\lambda = 760 - 680 = 80$

   $v = (80 \times 3 \times 10^8)/680$

   $\frac{\Delta\lambda}{\lambda_0} = \frac{v}{c}$

   Velocity of the quasar is 3.5×10^7 m/s (to 2 significant figures)

   The light is red-shifted so the direction of movement is away from the Earth.

INDEX

I

J

K

Q

R

S

T

Acknowledgements

The publishers wish to thank the following for permission to reproduce photographs. Every effort has been made to trace copyright holders and to obtain their permission for the use of copyright materials. The publishers will gladly receive any information enabling them to rectify any error or omission at the first opportunity:

(t = top, c = centre, b = bottom, r = right, l = left)

Cover & p1 DrHitch/Shutterstock, pp8-9 Neo Edmund/Shutterstock, p12 wavebreakmedia ltd/Shuttertsock, p15t Kinetic Imagery/Shutterstock, p15b silver-john/Shutterstock, p29 joyfull/Shutterstock, p32 National Geographic Image Collection/Alamy, p40 Alexander Chaikin/Shutterstock, p41 Marcel Jancovic/Shutterstock, p46 Andrew Barker/Shutterstock, p52 Martyn F. Chillmaid/Science Photo Library, pp76-77 Joshua Haviv/Shutterstock, p79 Alex Kuzovlev/Shutterstock, p80 Photoseeker/Shutterstock, p88 Slaven/Shutterstock, p92 Bart Coenders/iStockphoto, p94 littlesam/Shutterstock, p98 Art Directors & TRIP/Alamy, p107 Richard Wareham Fotografie/Alamy, p117 Tomasz Szymanski/Shuttertock, p121 Nir Levy/ Shutterstock, p122 Andrew Howe/iStockphoto, p131 BortN66/Shutterstock, pp134-135 Willyam Bradberry/Shutterstock, p137 Yuri Arcurs/Shutterstock, p142 Gustavo Miguel Fernandes/Shutterstock, p148 Fedorov Oleksiy/Shutterstock, p150t Darren Pullman/ Shutterstock, p150b nikkytok/Shutterstock, p152 Dario Sabljak/Shutterstock, p158 Aija Lehtonen/Shutterstock, p159 bluecrayola/Shutterstock, p160 Payless Images/Shutterstock, p161l Steve Heap/Shutterstock, p161r Morgan Rauscher/Shutterstock, p162 Falk Kienas/ Shutterstock, p171t Losevsky Pavel/Shutterstock, p171b Kick the beat/WikiMedia Commons, pp186-187 r.nagy/Shutterstock, p189t Monkey Business Images/Shutterstock, p189b Adrian Hughes/Shutterstock, p190r sizov/Shutterstock, p190l chatchai/Shutterstock, p193 clearlens/ Shutterstock, p201 Ulrich Mueller/Shutterstock, p206 Vixit/Shutterstock, p212 Abramova Kseniya/Shutterstock, p214 NASA/JSC, p227 ssuaphotos/Shutterstock, p229 zoia Kostina/ Shutterstock, p230 Ocean Power Delivery/Look at Sciences/Science Photo Library, p245 Rido/Shutterstock, pp244-245 Katrina Leigh/Shutterstock, p247 Andrea Danti/Shutterstock, p248 Michal Vitek/Shutterstock, p250l Roxana Bashyrova/Shutterstock, p250r JIANG HONGYAN/Shutterstock, p251 Jerritt Clark/Stringer/Getty Images, p252 Charles D. Winters/ Science Photo Library, p254t Marcel Jancovic/Shutterstock, p254c DenisNata/Shutterstock, p257 Charles D. Winters/Science Photo Library, p258 Mike Heywood/Shutterstock, p259 JonMilnes/Shutterstock, p265 Geoffrey Kuchera/Shutterstock, p266t Alan Freed/ Shutterstock, p266b Geoffrey Kuchera/Shutterstock, p271 Tomas Pavelka/Shutterstock, p277 jele/Shutterstock, p279 goldenangel/Shutterstock, p283 adam36/Shutterstock, pp292-293 Chris Alleaume/Shutterstock, p295 Nadezhda Bolotina/Shutterstock, p297t REDAV/Shutterstock, p297b Matthew Cole/iStockphoto, p299 Pi-Lens/Shutterstock, p302 epa european pressphoto agency b.v./Alamy, p306 oksana2010/Shutterstock, p307l Giphotostock/Science Photo Library, p307r Giphotostock/Science Photo Library, p310 xpixel/Shutterstock, p311l Trevor Clifford Photography/Science Photo Library, p311r Trevor Clifford Photography/Science Photo Library, p312 Giphotostock/Science Photo Library, p315 Ahellwig/WikiMedia Commons, p320 M. Niebuhr/Shutterstock, p325 MaverickLEE/Shutterstock, p326 Paul Andrew Lawrence/Alamy, p327 SuriyaPhoto/ Shutterstock, pp336-337 InnaFelker/Shutterstock, p339 Toxicotravail/WikiMedia Commons, p341 Gail Johnson/Shutterstock, p342 CERN/Science Photo Library, p344 Andrew Lambert Photography/Science Photo Library, p353c Robin Weaver/Alamy, p353b Belmonte/Science Photo Library, p356 lhervas/Shutterstock, p362 Troy GB images/Alamy, p364 Dmitry Naumov/Shutterstock, p366 Thorsten Schier/Shutterstock, pp374-375 NASA, p377 World History Archive/Alamy Stock Photo, p379 NASA Jet Propulsion Laboratory (NASA-JPL), p381 Mechanik/Shutterstock, p386 NASA, p388 NASA, p393 NASA.

NOTES

NOTES

NOTES

NOTES

NOTES

NOTES